SCIAMACHY - Exploring the Changing Earth's Atmosphere

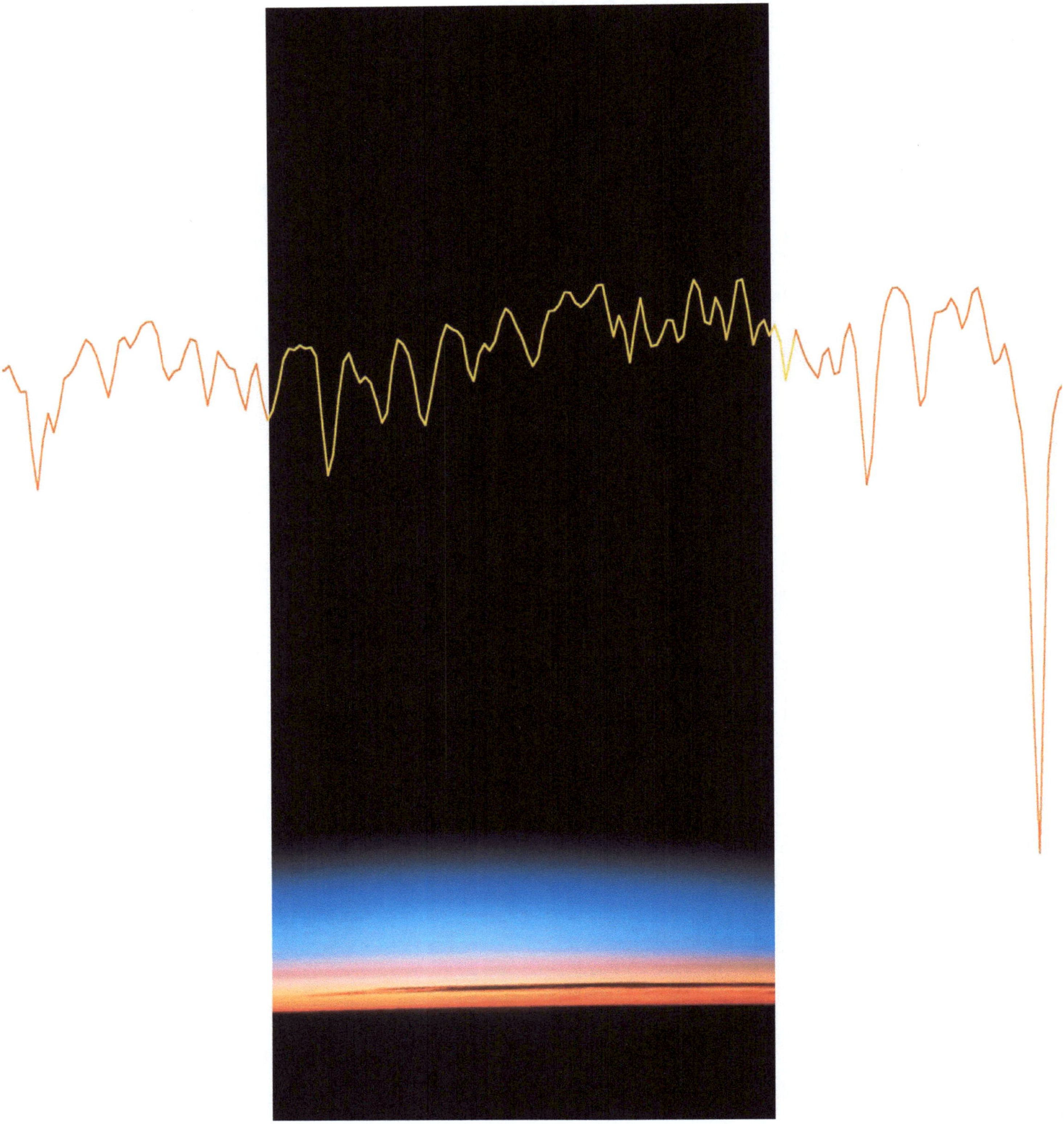

Manfred Gottwald · Heinrich Bovensmann
Editors

SCIAMACHY - Exploring the Changing Earth's Atmosphere

Springer

Editors
Manfred Gottwald
Research Scientist
Remote Sensing Technology Institute (IMF)
German Aerospace Center (DLR)
Oberpfaffenhofen
D-82234 Wessling
Germany

Heinrich Bovensmann
Research Scientist
Institute of Environmental Physics (IUP)
Institute of Remote Sensing (IFE)
University of Bremen
Otto-Hahn-Allee 1
D-28359 Bremen
Germany

Cover: Earth's limb at sunrise as seen from the ISS (photo: NASA), global NO_2 (Courtesy: A. Richter, IUP-IFE, University of Bremen) and global CO (Courtesy: Gloudemans et al. 2009) column densities.
Frontispiece: The Earth's stratosphere as seen from the Space Shuttle. Overlaid is a part of an earthshine spectrum as recorded by SCIAMACHY (photo: NASA).

ISBN 978-94-007-8989-0 ISBN 978-90-481-9896-2 (eBook)
DOI 10.1007/978-90-481-9896-2
Springer Dordrecht Heidelberg London New York

Softcover re-print of the Hardcover 1st edition 2011

Printed on acid-free paper

Springer is part of Springer Science+Business Media (www.springer.com)

We dedicate this book to our beloved SCIAMACHY team member Dr. Annemieke Gloudemans, who recently passed away at the age of 35 years. She will always be remembered for her kind personality and through her invaluable contributions to the challenging analysis and interpretation of SCIAMACHY carbon monoxide measurements.

Editorial

In March 2002 the atmospheric science instrument SCIAMACHY was launched on-board ESA's ENVISAT mission into low-Earth orbit. It was, and is still, one of the major current Earth Observation undertakings of Germany, The Netherlands and Belgium, accomplished in cooperation with the European Space Agency (ESA). Having meanwhile successfully monitored and explored the Earth's atmosphere for more than 8 years with SCIAMACHY, the involved scientific institutes and space agencies feel that it is worth to issue a comprehensive publication describing the SCIAMACHY mission – from the very first ideas to the current results. The targeted readership is not only the existing and potentially new SCIAMACHY data users from undergraduate student level up to researchers new in the fields of atmospheric chemistry and remote sensing but anyone who is keen to learn about SCIAMACHY's efforts to study the atmosphere and its responses to both, natural phenomena and anthropogenic effects.

The first chapter explains briefly why it is necessary to make measurements from low-Earth orbit to study the atmosphere. The global views from an altitude of 800 km open new windows to observe large-scale phenomena which are of prime importance to understand today's changing atmosphere and climate. This introductory chapter sets the stage for the rest of the publication. Having summarised why SCIAMACHY was selected to be launched into space, Chapter 2 takes a closer look at the ENVISAT mission, which hosts SCIAMACHY. The purpose of Chapter 2 is to describe those aspects of ENVISAT being of relevance to the SCIAMACHY mission. The location on and the environment of the platform, together with the ENVISAT orbit, determines many aspects of the instrument's design and operation. Similarly the concepts for data downlink and handling in the ground segment specify how measurement data is received, processed and disseminated in general.

A detailed description of the instrument concept is the subject of Chapter 3. It permits insight into optical, thermal and electronic subsystems. Main emphasis is given to the optical paths since they collect and generate the spectral signals containing the information on geophysical parameters thus being prime responsible for the achieved data quality. In order to provide the reader with an idea about the challenges of instrument development, chapter 3 also outlines the history of how SCIAMACHY was built. Without a flexible operations concept however, all the sophisticated instrument functionalities would have been useless. How SCIAMACHY is operated in-orbit can be found in Chapter 4. The chosen operations approach allows, despite the instrument complexity, full exploitation of its capabilities in a well structured operations environment thus supporting the need for long, stable measurements as required in atmospheric and climate research. In Chapter 5, the various steps necessary to calibrate the instrument, on-ground and in-orbit, are presented. Calibration is required to fully characterise the optical paths. Additionally, Chapter 5 also addresses optical performance monitoring which permits quantification of the degradation of optical components. Calibration and monitoring together ensure that the recorded signals are transformed into well calibrated spectra – a prerequisite for retrieving geophysical parameters with high accuracy over the full mission lifetime.

Chapter 6 brings the more technical part of the publication to a close. It describes SCIAMACHY's in-orbit mission lifetime, starting with the launch and the Commissioning Phase and illustrates now more than 8 years of routine measurements. Various instrument characteristics, derived from the monitoring activities, are presented and show the excellent in-orbit behaviour. We also outline how in-orbit degradation, a phenomenon common to each long duration space mission, impacts SCIAMACHY's measurements and demonstrate how it can be corrected or compensated.

With Chapter 7 the science related information is introduced by summarising the principles and methods for the derivation of geophysical parameters from the measured spectra. While in the early years of the mission the retrieval methods were mostly relying on standard algorithms, the scientific ingenuity has meanwhile generated a wealth of novel techniques permitting the retrieval of geophysical parameters beyond the original ideas. This chapter can be regarded as the basis for most of the SCIAMACHY data processing and scientific results being described in the next sections.

Operational and scientific data products are the subject of Chapter 8. The reader learns which products are generated under ESA responsibility and which are provided by research institutions involved in SCIAMACHY. For the ESA generated products, the strict requirements and implementations of the operational processing environment are outlined. As every geophysical parameter retrieval requires well calibrated measurements, we also report on how calibration and monitoring information is used to derive earthshine, extraterrestrial radiance and irradiance data products from the raw signals.

Retrieved geophysical parameters do not necessarily immediately translate into atmospheric science results. It has to be proven first that the data products are of sufficient quality. This process of product validation, subject of Chapter 9, was an enormous effort in the first years of the mission and is required, at an adequate level, throughout the mission and even beyond to create long-term datasets of known quality relevant for environmental, atmospheric, and climate change research. Chapter 9 explains the selected validation procedures, associated teams and summarises results acquired so far.

Chapter 10 concludes the publication by presenting SCIAMACHY's unique view of the changing Earth's environment. The capabilities of the instrument permit studying phenomena ranging from the the atmospheric layer where we are living in, i.e. the lower troposphere, up to the mesosphere and lower thermosphere where solar-terrestrial interactions begin to prevail. However SCIAMACHY does no longer stop at the bottom or top of the atmosphere. Even Earth surface parameters like vegetation or phytoplankton properties are now within reach as well as monitoring the solar activity. Finally SCIAMACHY has proven surprisingly successful in acquiring spectral signals from the atmosphere of our neighbour Venus. The content of Chapter 10 nicely illustrates the success achieved up to now and justifies continuing investments in the SCIAMACHY mission and data usage. Although we intended to be as complete as possible in the framework of this book, Chapter 10 is still a summary only. Therefore the interested reader is referred to the references given in this book and the websites listed below allowing a deeper look into SCIAMACHY results:

http://envisat.esa.int/
http://www.sciamachy.de/
http://www.sciamachy.org/
http://www.iup.uni-bremen.de/
http://joseba.mpch-mainz.mpg.de/
http://www.temis.nl/
http://wdc.dlr.de/

The SCIAMACHY project is and has been an important milestone in Earth Observation as it has already yielded unique scientific insights into the working of the Earth system. Many people have contributed to it. With the exception of the Principal Investigator John P. Burrows, who pushed the project since over 25 years, they cannot all be named here but we intended to make clear that the mission is a team effort combining expertise from agencies, industry and science.

The editors would like to thank all involved in the preparation and careful reviewing of this publication. We hope that the result is well received by the readers.

Manfred Gottwald
Heinrich Bovensmann
November 2010

Contents

Contributors

I. Aben
SRON, Netherlands Institute for Space Research, Sorbonnelaan 2,
3584 CA Utrecht, The Netherlands

B. Aberle
Remote Sensing Technology Institute, German Aerospace Center (DLR-IMF),
Oberpfaffenhofen, 82234 Wessling, Germany

S. Beirle
Max Planck Institute for Chemistry, Johann-Joachim-Becher-Weg 27,
55128 Mainz, Germany

H. Bovensmann
Institute of Environmental Physics/Institute of Remote Sensing (IUP-IFE),
University of Bremen, Otto-Hahn-Allee 1, 28359 Bremen, Germany

A. Bracher
Alfred Wegener Institute for Polar and Marine Research, Bussestraße 24,
27570 Bremerhaven, Germany
and
Institute of Environmental Physics/Institute of Remote Sensing (IUP-IFE),
University of Bremen, Otto-Hahn-Allee 1, 28359 Bremen, Germany

K. Bramstedt
Institute of Environmental Physics/Institute of Remote Sensing (IUP-IFE),
University of Bremen, Otto-Hahn-Allee 1, 28359 Bremen, Germany

M. Buchwitz
Institute of Environmental Physics/Institute of Remote Sensing (IUP-IFE),
University of Bremen, Otto-Hahn-Allee 1, 28359 Bremen, Germany

J.P. Burrows
Institute of Environmental Physics/Institute of Remote Sensing (IUP-IFE),
University of Bremen, Otto-Hahn-Allee 1, 28359 Bremen, Germany

J. Carpay
Netherlands Space Office (NSO), Juliana van Stolberglaan 3, 2595 CA The Hague,
The Netherlands

K. Chance
Harvard-Smithsonian Center for Astrophysics (SAO), 60 Garden Street,
02138 Cambridge, MA, USA

C. Chlebek
German Aerospace Centre, Space Agency, Königswinterer Straße 522-524,
53227 Bonn, Germany

M. de Graaf
Royal Netherlands Meteorological Institute (KNMI), Wilhelminalaan 10,
3732 GK De Bilt, The Netherlands

F.-J. Diekmann
European Space Operations Centre (ESOC), European Space Agency (ESA),
Robert-Bosch-Str. 5, 64293 Darmstadt, Germany

A. Doicu
Remote Sensing Technology Institute, German Aerospace Center (DLR-IMF),
Oberpfaffenhofen, 82234 Wessling, Germany

K.-U. Eichmann
Institute of Environmental Physics /Institute of Remote Sensing (IUP-IFE),
University of Bremen, Otto-Hahn-Allee 1, 28359 Bremen, Germany

T. Fehr
European Space Research Institute (ESRIN), European Space Agency (ESA),
Via Galileo Galilei, 00044 Frascati (RM), Italy

C. Frankenberg
Jet Propulsion Laboratory, California Institute of Technology, 4800 Oak Grove Drive,
91109 Pasadena, CA, USA

A. Friker
German Aerospace Center, Space Agency, Königswinterer Str. 522-524,
53227 Bonn, Germany

A. Gloudemans
SRON, Netherlands Institute for Space Research, Sorbonnelaan 2,
3584 CA Utrecht, The Netherlands

A.P.H. Goede
FOM Institute for Plasma Physics Rijnhuizen, PO Box 1207, 3430 BE Nieuwegein,
The Netherlands

M. Gottwald
Remote Sensing Technology Institute, German Aerospace Center (DLR-IMF),
Oberpfaffenhofen, 82234 Wessling, Germany

R. Hoogeveen
SRON, Netherlands Institute for Space Research, Sorbonnelaan 2,
3584 CA Utrecht, The Netherlands

A. Kokhanovsky
Institute of Environmental Physics /Institute of Remote Sensing (IUP-IFE),
University of Bremen, Otto-Hahn-Allee 1, 28359 Bremen, Germany

K. Kretschel
Remote Sensing Technology Institute, German Aerospace Center (DLR-IMF),
Oberpfaffenhofen, 82234 Wessling, Germany

E. Krieg
Remote Sensing Technology Institute, German Aerospace Center (DLR-IMF),
Oberpfaffenhofen, 82234 Wessling, Germany

M. Krijger
SRON, Netherlands Institute for Space Research, Sorbonnelaan 2,
3584 CA Utrecht, The Netherlands

H. Kröger
Bichlmairstraße 17, 83703 Gmund am Tegernsee, Germany

S. Kühl
Max Planck Institute for Chemistry, Johann-Joachim-Becher-Weg 27,
55128 Mainz, Germany

J.-C. Lambert
Belgian Institute for Space Aeronomy (BIRA-IASB), 3 Avenue Circulaire,
1180 Brussels, Belgium

C. Lerot
Belgian Institute for Space Aeronomy (BIRA-IASB), 3 Avenue Circulaire,
1180 Brussels, Belgium

G. Lichtenberg
Remote Sensing Technology Institute, German Aerospace Center (DLR-IMF),
Oberpfaffenhofen, 82234 Wessling, Germany

P. Lützow-Wentzky
EADS Astrium GmbH, Claude Dornier Straße, 88090 Immenstaad, Germany

R. Mager
EADS Astrium GmbH, Claude Dornier Straße, 88090 Immenstaad, Germany

M. Meringer
Remote Sensing Technology Institute, German Aerospace Center (DLR-IMF),
Oberpfaffenhofen, 82234 Wessling, Germany

S. Mieruch
Institute of Environmental Physics/Institute of Remote Sensing (IUP-IFE),
University of Bremen, Otto-Hahn-Allee 1, 28359 Bremen, Germany

A. Moore
Vega IT GmbH, c/o ESA/ESOC, Robert-Bosch-Str. 5, 64293 Darmstadt, Germany

C. Muller
Belgian Institute for Space Aeronomie (BIRA-IASB), 3 Avenue Circulaire,
1180 Brussels, Belgium

S. Noël
Institute of Environmental Physics/Institute of Remote Sensing (IUP-IFE),
University of Bremen, Otto-Hahn-Allee 1, 28359 Bremen, Germany

M. Penning de Vries
Max Planck Institute for Chemistry, Johann-Joachim-Becher-Weg 27,
55128 Mainz, Germany

A. Piters
Royal Netherlands Meteorological Institute (KNMI), Wilhelminalaan 10,
3732 GK De Bilt, The Netherlands

A. Richter
Institute of Environmental Physics/Institute of Remote Sensing (IUP-IFE),
University of Bremen, Otto-Hahn-Allee 1, 28359 Bremen, Germany

A.V. Rozanov
Institute of Environmental Physics/Institute of Remote Sensing (IUP-IFE),
University of Bremen, Otto-Hahn-Allee 1, 28359 Bremen, Germany

V.V. Rozanov
Institute of Environmental Physics/Institute of Remote Sensing (IUP-IFE),
University of Bremen, Otto-Hahn-Allee 1, 28359 Bremen, Germany

D. Scherbakov
Remote Sensing Technology Institute, German Aerospace Center (DLR-IMF),
Oberpfaffenhofen, 82234 Wessling, Germany

H. Schrijver
SRON, Netherlands Institute for Space Research, Sorbonnelaan 2,
3584 CA Utrecht, The Netherlands

A. Schönhardt
Institute of Environmental Physics/Institute of Remote Sensing (IUP-IFE),
University of Bremen, Otto-Hahn-Allee 1, 28359 Bremen, Germany

B.M. Sinnhuber
Institute of Environmental Physics/Institute of Remote Sensing (IUP-IFE),
University of Bremen, Otto-Hahn-Allee 1, 28359 Bremen, Germany

M. Sinnhuber
Institute of Environmental Physics/Institute of Remote Sensing (IUP-IFE),
University of Bremen, Otto-Hahn-Allee 1, 28359 Bremen, Germany

S. Slijkhuis
Remote Sensing Technology Institute, German Aerospace Center (DLR-IMF),
Oberpfaffenhofen, 82234 Wessling, Germany

R. Snel
SRON, Netherlands Institute for Space Research, Sorbonnelaan 2,
3584 CA Utrecht, The Netherlands

P. Stammes
Royal Netherlands Meteorological Institute (KNMI), Wilhelminalaan 10,
3732 GK De Bilt, The Netherlands

R. van Hees
SRON, Netherlands Institute for Space Research, Sorbonnelaan 2,
3584 CA Utrecht, The Netherlands

M. Van Roozendael
Belgian Institute for Space Aeronomie (BIRA-IASB), 3 Avenue Circulaire,
1180 Brussels, Belgium

A. von Bargen
German Aerospace Center, Space Agency, Königswinterer Str. 522-524,
53227 Bonn, Germany

C. von Savigny
Institute of Environmental Physics/Institute of Remote Sensing (IUP-IFE),
University of Bremen, Otto-Hahn-Allee 1, 28359 Bremen, Germany

T. Wagner
Max Planck Institute for Chemistry, Johann-Joachim-Becher-Weg 27,
55128 Mainz, Germany

T. Watts
Dutch Space B.V., Newtonweg 1, 2333 CP Leiden, The Netherlands

H. Weber
Remote Sensing Technology Institute, German Aerospace Center (DLR-IMF),
Oberpfaffenhofen, 82234 Wessling, Germany

M. Weber
Institute of Environmental Physics/Institute of Remote Sensing (IUP-IFE),
University of Bremen, Otto-Hahn-Allee 1, 28359 Bremen, Germany

F. Wittrock
Institute of Environmental Physics/Institute of Remote Sensing (IUP-IFE),
University of Bremen, Otto-Hahn-Allee 1, 28359 Bremen, Germany

Chapter 1
SCIAMACHY – The Need for Atmospheric Research from Space

J.P. Burrows, A.P.H. Goede, C. Muller, and H. Bovensmann

Abstract The Earth's atmosphere is part of a complex system comprising the Sun and the Earth. Throughout the Earth's history this system responded to natural and anthropogenic phenomena. To assess the significance of changes induced by natural and anthropogenic activities, a detailed understanding of the physical and chemical processes controlling the composition of the global atmosphere is required. Space-based remote sensing instruments can contribute many of the required geophysical parameters on various spatial and temporal scales. The scientific goals of the imaging absorption spectrometer SCIAMACHY, flying on the European ENVISAT platform, are wide ranging and encompassing the need to improve our understanding of biogeochemical cycling and assess the impact of man on the Earth system. SCIAMACHY is an absorption spectrometer sensing simultaneously electromagnetic radiation in the range from the UV to SWIR. It observes the scattered and reflected spectral radiance in nadir and limb geometry, the spectral radiance transmitted through the atmosphere in solar and lunar occultation geometry and both the extraterrestrial solar irradiance and lunar radiance. Research using SCIAMACHY data includes the following topics: tropospheric ozone, air quality, the transport and transformation of pollution, global climate change, the exchange between the stratosphere and troposphere, stratospheric chemistry and dynamics, mesopheric and thermospheric composition, and solar variability. SCIAMACHY was proposed in the 1980s and is now part of the current fleet of atmospheric remote sensing missions. Due to its success, several missions with SCIAMACHY heritage will extend the important data sets far into the twenty-first century.

Keywords Earth's atmosphere • Spaceborne remote sensing • Anthropogenic changes • Pollution • Greenhouse effect • Ozone depletion • Atmospheric chemistry

It is widely recognised that the Earth's atmosphere is part of a complex system comprising the Sun and the Earth, and conditions within it are maintained by a biogeochemical equilibrium. The state of the system has responded throughout the Earth's history to natural and anthropogenic phenomena. Changes in the composition of the Earth's atmosphere and at the Earth's surface are occurring on both short and long timescales and thereby modifying the environment, its ecosystem, the biodiversity of the planet and ultimately the climate. Since the industrial revolution and driven by the supply of energy from fossil fuel combustion, the Earth's population has increased nearly exponentially. This has been accompanied by an increasing standard of living. The recognition that the resultant pollution and land use change has reached levels such that it impacts the Earth system from the local to the global scale, has led to public concern, political debate and the evolving development of national and international environmental policy, designed to minimise negative impact on the environment, biodiversity and climate. Established examples where significant change to the Earth system resulting from anthropogenic activity has been identified by the scientific community are:

- The *ozone hole*, i.e. the rapid loss of stratospheric ozone (O_3) above Antarctica and the Arctic in spring, resulting from the tropospheric release of chlorofluorocarbon compounds, (CFC, halones and hydrogenated chlorofluorcarbon compounds, HCFC – WMO 2003)

J.P. Burrows and H. Bovensmann (✉)
Institute of Environmental Physics/Institute of Remote Sensing (IUP-IFE), University of Bremen, Otto-Hahn-Allee 1, 28359 Bremen, Germany
e-mail: heinrich.bovensmann@iup.physik.uni-bremen.de

A.P.H. Goede
FOM Institute for Plasma Physics Rijnhuizen, PO Box 1207, 3430 BE Nieuwegein, The Netherlands

C. Muller
Belgian Institute for Space Aeronomie (BIRA-IASB), 3 Avenue Circulaire, 1180 Brussels, Belgium

M. Gottwald and H. Bovensmann (eds.), *SCIAMACHY - Exploring the Changing Earth's Atmosphere*,
DOI 10.1007/978-90-481-9896-2_1,

- The decline of ozone over extra polar regions, 60°S–60°N, which has slowed down as a result of the implementation of the Montreal Protocol and its amendments (WMO 2007)
- The global increase of tropospheric ozone (WMO 1995) and its adverse impact on air quality
- The trans-boundary transport and transformation of pollution resulting for example in acidic deposition and affecting air quality far from pollution sources (WMO 2004; UNECE 2007)
- Global dimming attributed to increases in aerosol and clouds (Wild et al. 2005)
- The observed increase of tropospheric greenhouse gases CO_2, CH_4, N_2O and O_3 and their impact on climate (IPCC 2007)
- The coupling between increased greenhouse gas concentrations and stratospheric ozone loss (Shindell et al. 1998; WMO 2007)

The overarching objective is to achieve sustainable development, and protect the Earth's environment and its biodiversity. This requires an adequate understanding of the complex system comprising the Sun, the atmosphere and the Earth's surface, and the physical chemical and biological processes, which determine and maintain the conditions within the biosphere. Improving our understanding of the Earth system calls on the one hand for theoretical models, describing the atmospheric dynamics, chemistry and biogeochemical cycling in the atmosphere. On the other hand there is a clear need for accurate strategies to provide the observational evidence base by means of measurements of the atmospheric composition in the troposphere, stratosphere and mesosphere ranging from local to global scale. Models are validated by such measurements, which also independently provide evidence of the changes in the atmosphere observed. In order to test our understanding and improve the predictive capability of models of environmental and climate change, global measurements of key parameters are needed.

Remote sounding instrumentation on-board space-based platforms have the potential to deliver a unique part of the required geophysical parameters on global scales. During the last decades substantial progress has been made in sounding even the lowest layer of the atmosphere, the troposphere (Burrows et al. 2011) and SCIAMACHY acted as pathfinder in that field. The SCIAMACHY (SCanning Imaging Absorption spectroMeter for Atmospheric CHartographY) project (Burrows et al. 1988a; Bovensmann et al. 1999), being Greek for *chasing or hunting shadows*, has delivered an instrument which flies on-board ENVISAT (Environmental Satellite) of the European Space Agency ESA. It is designed to yield quantitative data about of the temporal and spatial properties of atmospheric trace constituents including gases, aerosol and clouds. In addition it provides information on surface optical properties for ocean colour and land cover studies complementary to other instrumentation. The SCIAMACHY project was designed to address the scientific curiosity and to demonstrate that the measurements needed to regulate sustainable development are feasible from space.

1.1 How to Study the Earth's Atmosphere from Space?

Prior to the advent of space flight, measurements of atmospheric composition were limited in geographical coverage and the sampling was skewed to the developed world. In particular, the regions over the oceans and the southern hemisphere were undersampled. During the last 3 decades, the development of satellite platforms in low-Earth and geostationary orbit has provided the opportunity to observe the Earth and its atmosphere by utilising the novel viewing geometries afforded from space. The potential to make near-simultaneous observations at the global scale for the first time has facilitated the emergence of *Earth System Science*. This approach treats the Earth as an interdependent system, comprising the atmosphere, oceans and land, whose energy is provided by the Sun.

In particular the atmospheric and meteorological sciences have exploited satellite observations. Remote regions of the atmosphere over the land and oceans, where ground-based stations or shipborne measurements are usually rare, can now be probed regularly from space and variations of geophysical parameters on small and large scales, both spatially and temporally, can be studied.

Passive atmospheric sounding from space measures the radiation leaving the top of the atmosphere. It may be subdivided into the categories measuring *absorption* or *emission* of atmospheric molecules and atoms. The absorption and emission processes in the atmosphere produce spectra, which are characteristic for the emitting or absorbing atom or molecule. *Emission spectra* consist mainly of the thermal emission from the Earth and the atmospheric constituents, which radiate mainly in the infrared and microwave spectral range according to their characteristic rotational and vibrational excitation. They can be regarded as the thermal 'fingerprint' of the atmosphere. From the emission spectra trace gas concentrations are derived.

Measuring *absorption spectra* of molecules in the UV, visible (VIS), near (NIR) and short-wave infrared (SWIR) part of the atmospheric spectrum yields information on the electronic properties of atoms and molecules. Concentration is derived from the absorption, i.e. the reduction in the intensity of electromagnetic radiation, which is integrated along the path travelled by solar photons through the atmosphere. This is the approach utilised by SCIAMACHY. Atoms, molecules and particles absorb, emit and scatter the incoming solar electromagnetic radiation. The incoming solar radiation is described to a good approximation by the emission from a black body having a temperature of about 5800 K, modulated by atomic absorption in the solar photosphere, chromospheres and corona, known as the solar Fraunhofer lines. In the Earth system the incoming solar radiation is further modified by scattering, absorption and emission processes in the atmosphere and by reflection and scattering at the Earth's surface. When leaving the Earth again at the top of the atmosphere, the upwelling radiation carries with it the signature of the atmospheric constituents. The air mass traversed by the solar photons needs to be modelled precisely, in order to infer concentration distributions from the integrated air column sampled (see Chapter 7). This task requires a quantitative knowledge and understanding of the scattering of electromagnetic radiation in the atmosphere as well as highly accurate absorption spectroscopy data.

1.2 The Road to SCIAMACHY

The scientific community has long recognised the need for global observations of the Earth system and has proposed global observing systems for research to meet the needs of international legislation monitoring (IGACO 2004). Such systems aim to establish integrated networks of having complementary instrumentation from ground-based, airborne or shipborne platforms, as well as satellite sensors. The overall objectives are

- To improve our understanding of the physical and chemical processes determining the behaviour of the atmosphere
- To demonstrate and assess the capability and applicability of remote sensing from space for Earth System and Atmospheric Science
- To move towards a sustained global observing system capable of meeting the needs of Earth System Science and the demands for global international treaty verification, monitoring and forecast

The first measurements of atmospheric ozone and aerosols from space were made by the Soviet space programme in the middle of the 1960s from manned spacecraft. In the early 1970s NASA initiated its efforts to make global measurements of atmospheric ozone with the Backscattered Ultra Violet (BUV) instrument aboard the NASA Nimbus 4 satellite. This instrument was significantly enhanced and extended to build the Solar Backscatter Ultraviolet (SBUV) and Total Ozone Mapping Spectrometer (TOMS), which flew on NASA's Nimbus 7 satellite. Subsequently NOAA was responsible for a series of SBUV-2 instruments on its operational meteorological platforms while NASA operated several TOMS instruments on a variety of satellites (see Fig. 1.1). Useful solar occultation measurements began in 1978 with the launch of SAM-II (Stratospheric Aerosol Measurement) which was followed by the Stratospheric Aerosol and Gas Experiment (SAGE-I, SAGE-II) and the Russian occultation spectrophotometer SFM-2. The research mission SME (Solar Mesospheric Explorer), launched in 1983, performed limb scanning measurements of solar backscattered radiation to determine ozone and nitrogen dioxide. Another milestone in the exploration of the atmosphere from space is the Upper Atmosphere Research Satellite (Reber et al. 1993), which carried instruments to sound the upper atmosphere like the Microwave Limb Sounder (MLS) and the Halogen Occultation Experiment (HALOE). HALOE measurements were continued until 2005.

The European participation in remote sounding of atmospheric constituents and parameters by autonomous satellites was focused initially on the development of the geostationary METEOSAT programme being established in the late 1970s. Initiated by ESA and finally transferred to EUMETSAT, it provided measurements of meteorological parameters since the early 1980s. The first ESA payload to carry significant instruments for atmospheric trace gas sounding was SPACELAB-1 which was among the shuttle payload in 1983. One of these instruments, the GRILLE spectrometer, performed observations in solar occultation of most of the nitrogen and carbon families, HF and HCl. It flew once again in 1992 in the NASA SPACELAB ATLAS-1 mission with the JPL ATMOS solar occultation interferometer. This convinced ESA of the possibility of an atmospheric observatory in orbit.

While the first polar orbiting European Research Satellite (ERS-1), primarily a platform for microwave and radar sensors, did not address the needs of the atmospheric chemistry community, the second European Remote Sensing Satellite (ERS-2), carrying the Global Ozone Monitoring Experiment (GOME), took Europe a large step forward towards ozone and atmospheric composition measurements. GOME on ERS-2 is a smaller scale version of SCIAMACHY derived from the original SCIAMACHY concept, measuring in nadir viewing geometry the upwelling radiation at the top of the atmosphere between 240 and 793 nm. ERS-2 was launched on 20 April 1995 into a sun-synchronous orbit with an equator crossing time in descending node of 10:30 a.m. The feasibility of the SCIAMACHY instrument and retrieval concepts could be successfully demonstrated for nadir observations with GOME. The absorptions of the trace gases O_3, NO_2, BrO, OClO, H_2O, SO_2, and HCHO were observed as predicted and the retrieval of total and tropospheric column

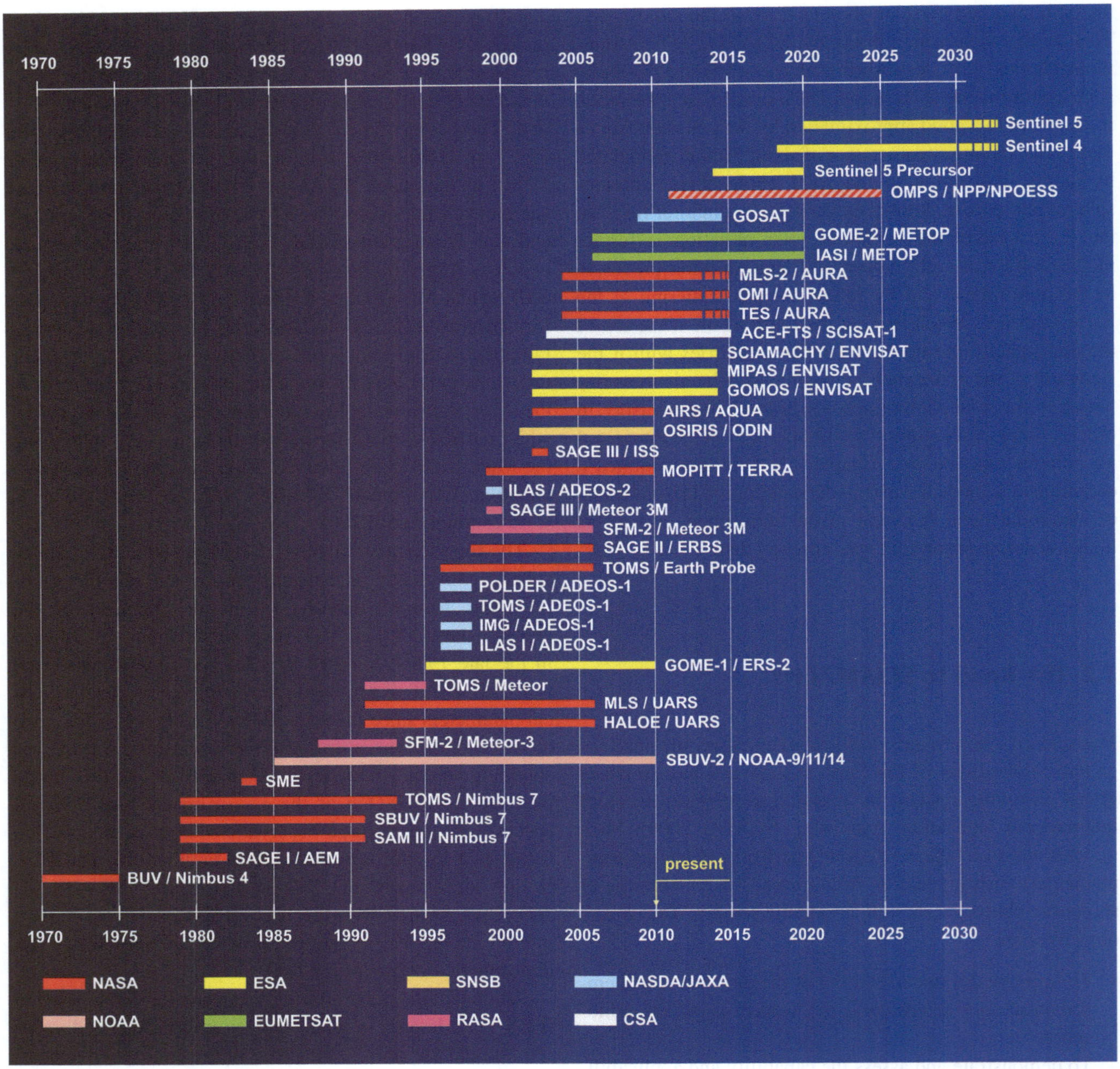

Fig. 1.1 Atmospheric science spaceborne instruments and missions since 1970 with relevance for SCIAMACHY. The list of missions is not intended to be complete but to illustrate the progress in spaceborne instrumentation for atmospheric composition monitoring (Courtesy: DLR-IMF).

information from GOME measurements was achieved (Burrows et al. 1999 and references therein). In addition, similarly to SBUV, O_3 profiles including some information on tropospheric ozone were retrieved from GOME observations (Munro et al. 1998).

The interest and the need for global observations of atmospheric constituents is reflected in the number of new missions which have been launched since the start of ERS-2 and contribute significantly to Earth system research in the fields of atmospheric chemistry and physics: NASA's Earth Observing System (EOS) satellites TERRA, AQUA, and AURA, the Japanese Advanced Earth Observing System (ADEOS1/2) and GOSAT, the Canadian/Swedish/French ODIN mission and the Canadian SCISAT mission. ESA's ENVISAT, which was launched in 2002, includes SCIAMACHY as part of its atmospheric payload. Together with EUMETSAT's series of METOP platforms, which began in late 2006, it extends the European data record started with ERS-2 to 2020 and beyond. The past 3 decades can be regarded as the pioneering golden age of Earth Observation with their research missions generally having been very much more successful than originally expected (Table 1.1).

Table 1.1 Passive satellite instruments designed to determine trace gas distributions in the atmosphere, coverage of their measurements, species measured and the satellite platform. The list of sensors refers to Fig. 1.1

Instrument	Name	Measurement altitude[a]			Target species	Observation geometry[b]				
		TR	ST	ME		N	L	SO	LO	STO
ACE-FTS	Atmospheric Chemistry Experiment		×		O_3, H_2O, CO, CFCs, HNO_3, NO, and more			×		
AIRS	Atmospheric Infrared Sounder	×[c]			O_3, H_2O, CO_2, CH_4	×				
BUV	Backscatter Ultraviolet Ozone Experiment		×		O_3	×				
GOME	Global Ozone Monitoring Experiment	×	×		O_3, NO_2, H_2O, BrO, OClO, SO_2, HCHO, CHOCHO, IO, clouds and aerosols	×				
GOME-2	Global Ozone Monitoring Experiment	×	×		O_3, NO_2, H_2O BrO, OClO, SO_2, HCHO, CHOCHO, IO, clouds and aerosols	×				
GOMOS	Global Ozone Monitoring by Occultation of Stars	×[c]	×	×	O_3, NO_2, H_2O, NO_3, aerosols, T					
GOSAT	Greenhouse Gas Observing Satellite	×			CO_2, H_2O, CH_4	×				
HALOE	Halogen Occultation Experiment		×		CO_2, H_2O, O_3, NO_2, HF, HCl, CH_4, NO			×		
IASI	Infrared Atmospheric Sounding Interferometer	×[c]	×		O_3, H_2O, CO, CH_4, N_2O, T	×				
ILAS I, II	Improved Limb Atmospheric Spectrometer		×		O_3, NO_2, N_2O, H_2O, CFC11, CH_4, aerosols			×		
IMG	Interferometric Monitor for Greenhouse Gases	×[c]	×		O_3, N_2O, H_2O, CH_4, CO, CO_2	×				
MIPAS	Michelson Inferometer for Passive Atmospheric Sounding		×	×	O_3, NO_x, N_2O_5 $ClONO_2$, CH_4, CFCs, HNO_3, and more, T and P		×			
MLS	Microwave Limb Sounder		×	×	ClO, O_3, H_2O, HNO_3, T and P		×			
MLS-2	Microwave Limb Sounder		×	×	CO, HCL, ClO, O_3, H_2O, BrO, N_2O, SO_2, HCN, CH_3CN		×			
MOPITT	Measurement of Pollution in the Troposphere	×			CO, CH_4	×				
MTG IRS	Infrared Sounder	×[c]	×		O_3, H_2O, CO, T and P	×[d]				
OMI	Ozone Monitoring Instrument	×	×		O_3, SO_2, NO_2, BrO, CHOCHO, HCHO, aerosols	×				
OMPS	Ozone Monitoring and Profiling Suite	×	×	×	O_3, SO_2, aerosols	×	×			
OSIRIS	Optical Spectrograph and Infrared Imaging System		×	×	NO, OClO, O_3, NO_2, aerosols		×			
POLDER	Polarization and Directionality of the Earth's Radiance	×			polarisation, aerosols, clouds	×				
SAGE I	Stratospheric Aerosol and Gas Experiment I		×		O_3, NO_2 aerosols			×		
SAGE II	Stratospheric Aerosol and Gas Experiment II		×		O_3, NO_2, H_2O, aerosols			×		
SAGE III	Stratospheric Aerosol and Gas Experiment III		×		O_3, OClO, H_2O, BrO, NO_2, NO_3, aerosols			×	×	
SAM II	Stratospheric Aerosol Measurement II		×		Aerosols			×		
SBUV	Solar Backscatter Ultraviolet Ozone Experiment	×	×		O_3, SO_2	×				
SBUV-2	Solar Backscatter Ultraviolet Ozone Experiment 2	×	×		O_3, SO_2	×				
SCIAMACHY	Scanning Imaging Absorption Spectrometer for Atmospheric Chartography	×	×	×	O_3, O_2, $O_2(^1\Delta)$, O_4, NO, NO_2, N_2O, BrO, OClO, H_2O, HDO/H_2O, SO_2, HCHO, CHOCHO, IO, CO, CO_2, CH_4, cloud, aerosols	×	×	×	×	
SENTINEL 4 UVN	Ultraviolet-Visible-Near Infrared	×	×	×	O_3, NO_2, SO_2, HCHO, CHOCHO, cloud, aerosols	×[d]				
SENTINEL 5 Precursor UVNS	Ultraviolet-Visible-Near Infrared-Shortwave Infrared	×	×	×	O_3, NO_2, BrO, SO_2, HCHO, CHOCHO, H_2O, HDO/H_2O, CO, CH_4, cloud, aerosols	×				

(continued)

Table 1.1 (continued)

Instrument	Name	Measurement altitude[a] TR	ST	ME	Target species	Observation geometry[b] N	L	SO	LO	STO
SENTINEL 5 UVNS	Ultraviolet-Visible-Near Infrared-Shortwave Infrared	×	×	×	O_3, NO_2, BrO, SO_2, HCHO, CHOCHO, H_2O, HDO/H_2O, CO, CO_2, CH_4, cloud, aerosols	×				
SFM-2	Spectrophotometer		×		O_3, aerosols				×	
SME	Solar Mesospheric Experiment		×	×	O_3, $O_2(^1\Delta)$, NO_2		×			
TES	Tropospheric Emission Spectrometer	×	×		HNO_3, O_3, NO, H_2O, CH_4, CO, SO_2	×	×			
TOMS	Total Ozone Monitoring Spectrometer	×	×		O_3, SO_2, aerosols	×				

[a] *TR* troposphere, *ST* stratosphere, *ME* mesosphere
[b] *N* nadir, *L* limb, *SO* solar occultation, *LO* lunar occultation, *STO* stellar occultation
[c] Mid/Upper troposphere
[d] Geostationary orbit

The heritage of the SCIAMACHY instrument derives from the ground-based measurements using Differential Optical Absorption Spectroscopy (DOAS) and previous satellite atmospheric remote sensing missions like SBUV, TOMS, SME, and SAGE. SCIAMACHY combines and extends the measurement principles and observational modes of the nadir scattered sunlight recording instruments SBUV and TOMS, the solar occultation instrument SAGE and the limb scattered sunlight measuring instrument SME within one instrument. SCIAMACHY detects in the UV-VIS-NIR-SWIR wavelength range from 214 to 2386 nm:

- The scattered and reflected spectral radiance in nadir and limb geometry
- The spectral radiance transmitted through the atmosphere in solar and lunar occultation geometry
- The extraterrestrial solar irradiance and the lunar radiance

Limb, nadir and occultation measurements are made during every orbit. Trace gases, aerosols, clouds and the surface of the Earth modify the light observed by SCIAMACHY via absorption, emission and scattering processes. Inversion of the radiance and irradiance measurements allows retrieval of the amounts and distributions of a significant number of constituents from their spectral signatures.

1.3 Determining Impacts on the Earth's Atmosphere

SCIAMACHY's Goals

The main objectives of the SCIAMACHY mission are to improve our knowledge of global atmospheric composition, its change in response to both natural and anthropogenic activity and the processes associated to it as well as the related global issues of importance to the chemistry and physics of our atmosphere such as:

- The impact of anthropogenic activity and natural processes on tropospheric ozone, air quality, and global warming
- The exchange processes between the stratosphere and troposphere
- The interaction of stratospheric chemistry and dynamics
- The natural modulations of atmospheric composition resulting from volcanic eruptions, lightning, solar output variations, e.g. solar cycle, or solar proton events

To achieve these goals, SCIAMACHY was proposed to deliver a multitude of parameters characterising the system *Earth–Atmosphere–Sun*, especially key trace gases and other parameters in the troposphere and stratosphere. The following gases were targeted for measurement: molecular oxygen (O_2), ozone $(O)_3$, the oxygen dimer (O_4), nitrogen monoxide (NO), nitrogen dioxide (NO_2), the nitrate radical (NO_3), carbon monoxide (CO), carbon dioxide (CO_2), formaldehyde (HCHO), methane (CH_4), water vapour (H_2O), nitrous oxide (N_2O), sulphur dioxide (SO_2), bromine monoxide (BrO), and chlorine dioxide (OClO). The combined use of nadir and limb observations yields tropospheric amounts of the relevant constituents down to the ground or cloud top depending on cloud cover. In addition to the trace gases, information on clouds (cloud top height, cloud optical thickness, ice-water cloud discrimination) and aerosol can be deduced from SCIAMACHY measurements. Particularly interesting, among these, are Polar Stratospheric Clouds (PSC) and Noctilucent Clouds (NLC), also referred to as Polar Mesospheric Clouds (PMC). It was recognised at the outset that whilst the spatial resolution of SCIAMACHY was low, its measurements would also provide some unique information about ocean colour and land cover parameters.

Table 1.2 Summary of parameters derived from SCIAMACHY and the relevant application areas

Subject	Parameter from SCIAMACHY data	Application areas
Surface	Spectral surface albedo, UV A/B, chlorophyll content of plants and phytoplankton	– Earth radiative budget – Ocean biology – Plant biology – Spectral surface characteristics
Troposphere (incl. boundary layer[a])	Columns of O_3, NO_2, BrO, SO_2, HCHO, CHOCHO, IO, H_2O, HDO/H_2O, CO, CO_2, CH_4, clouds, aerosol, actinic flux	– Transport and transformation of pollutants including air quality – Tropospheric ozone and oxidation potential – Carbon budget – Quantification of emissions – Climate–chemistry interaction – Hydrological cycle
Tropopause Region	Concentrations of O_3, NO_2, H_2O, CO, CH_4, clouds, thermodynamic phase of clouds	– Transport processes in the tropopause region – Water budget, including ice clouds – Impact of aviation on climate – Climate–chemistry interaction
Stratosphere	Profiles of O_3, NO_2, BrO, OClO, H_2O, aerosol, PSC	– Development of the ozone layer – Climate–chemistry interactions – Solar–terrestrial interactions
Mesosphere	Profiles of O_3, NO, OH, metal ions, temperature, NLC	– Climate–chemistry interactions – Solar–terrestrial interactions
Top of the Atmosphere	Earth spectral reflectance	– Earth radiative budget
Sun	Spectral solar irradiance, Mg-Index (solar activity)	– Earth radiative budget – Solar–terrestrial interactions – Solar physics

[a] Sensitivity to boundary layer dependent on surface albedo, cloudiness and aerosol loading

In addition to these originally targeted parameters, it became evident from the analysis of the high quality data SCIAMACHY delivered over the last years, that several new parameters could be derived, including tropospheric glyoxal (CHOCHO), tropospheric iodine oxide (IO), mesospheric metal ions, phytoplankton information, and the ratio of HDO to H_2O. Recently SCIAMACHY was also used to observe the solar backscattered radiation coming from Venus.

Table 1.2 summarises the parameters derived from SCIAMACHY measurements and their application areas while the successive chapters provide descriptions of the associated scientific problems. The data sets obtained by SCIAMACHY significantly contribute to the quantification of the complex interactions between natural and human activities, climate, atmospheric composition, and the relevant chemical and physical processes. For an illustrative summary of SCIAMACHY's achievements the reader is referred to Chapter 10.

The Atmospheric Layers

The Earth's atmosphere is a complex system comprising a set of layers, characterised by their temperature as a function of altitude. Figure 1.2 shows typical temperature and pressure profiles at mid-latitudes. The rate of temperature change in the atmosphere as a function of height can be used to

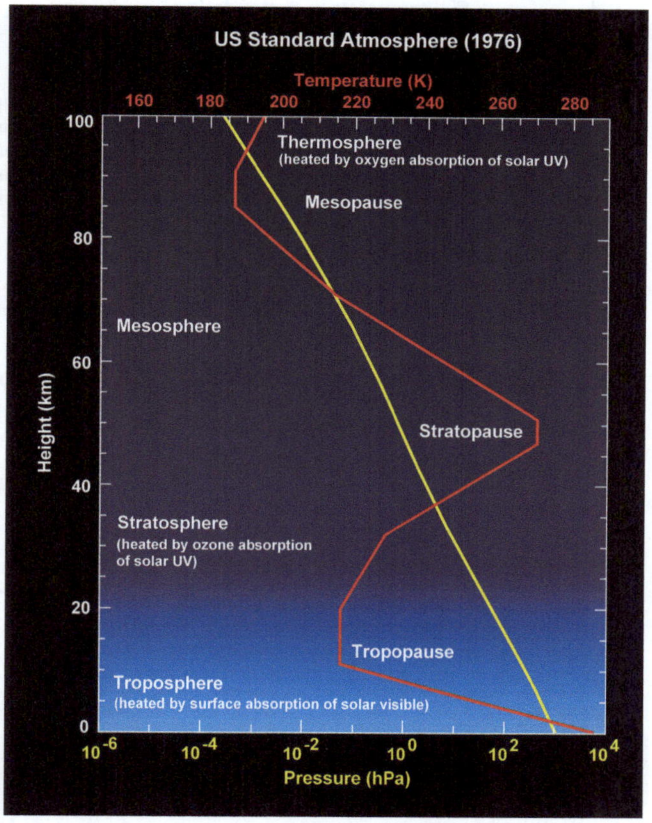

Fig. 1.2 Atmospheric pressure and temperature profiles for mid latitudes (US Standard Atmosphere).

define regions of positive and negative gradient or lapse rate. Starting at the Earth's surface, the temperature decreases up to the region known as the tropopause. The latter separates the troposphere, which is vertically well mixed, from the stratosphere, which is characterised by slow vertical mixing. In the stratosphere the temperature increases from the tropopause to the stratopause, which separates the stratosphere from the mesosphere. In the thermosphere, above the mesosphere, the temperature increases again.

The increase in temperature in the stratosphere results mainly from the absorption of solar radiation between 200 snd 300 nm by the stratospheric ozone layer. In the thermosphere a different but related mechanism results in a temperature increase, caused by the absorption of short wavelength solar radiation typically below 200 nm by molecules, atoms and ions. The temperature of the thermosphere is modulated significantly by the solar cycle.

The pressure of the atmosphere is highest at the Earth's surface and decreases exponentially with height according to the barometric formula. The height of the tropopause varies between about 8 km at the poles and 16 km in the tropics dictated by the ideal gas law and solar irradiation. The stratopause is located typically at 45–50 km and the mesopause typically at 85–90 km. Between 80% and 90% of the atmospheric mass are contained within the troposphere.

The behaviour and composition of troposphere, stratosphere and mesosphere are coupled through dynamic, radiative and chemical processes as indicated in Fig. 1.3.

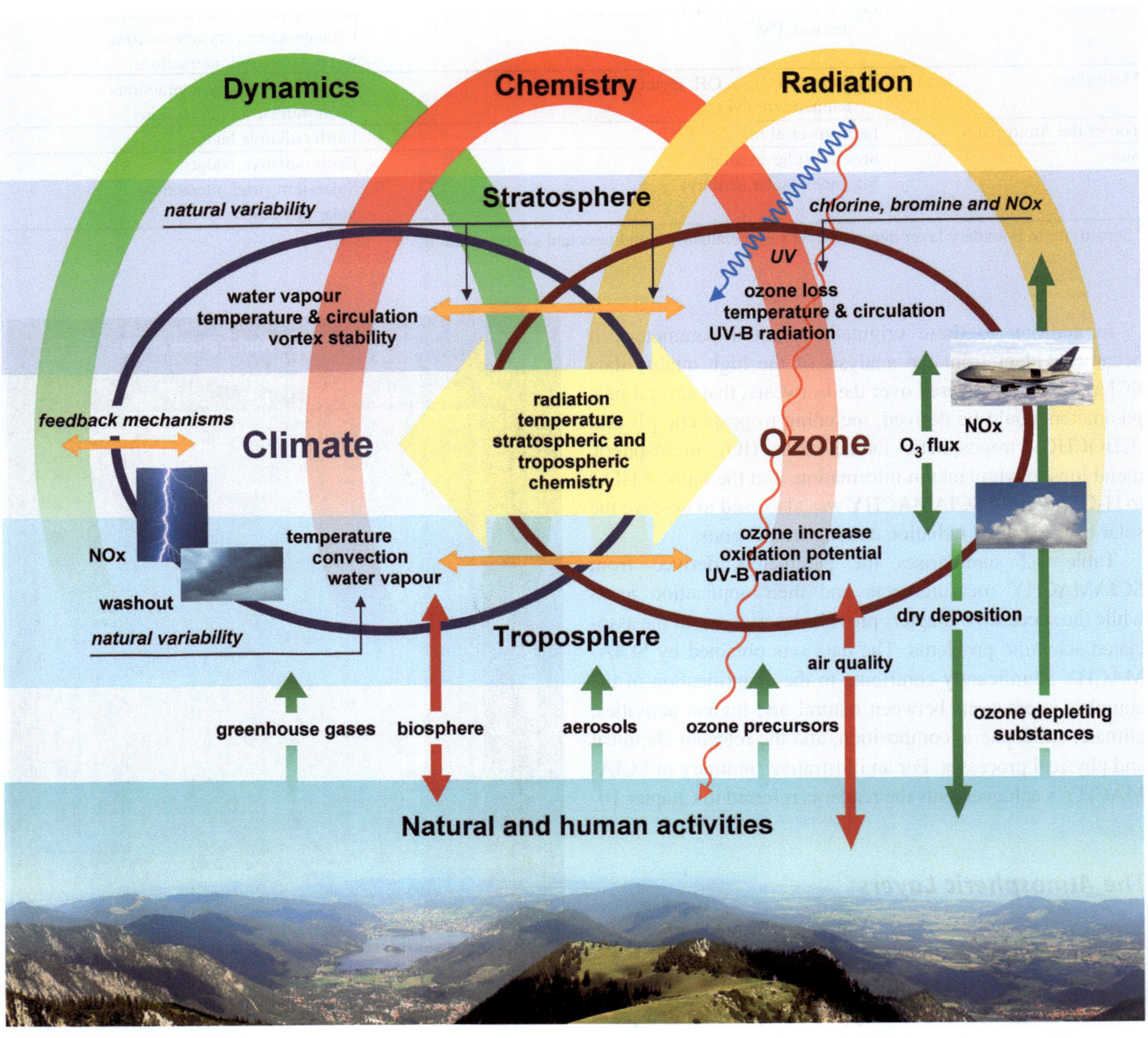

Fig. 1.3 Interactions between human activity, atmospheric composition, chemical and physical processes and climate (Courtesy: DLR-IMF, after IGACO 2004).

The conditions experienced in the biosphere, the layer around the Earth's surface where life resides, are determined in a complex manner by the physical and chemical processes occurring in all these atmospheric layers.

Anthropogenic Impact on the Earth-Atmosphere System

The composition of the Earth's atmosphere is different from that of neighbouring planets such as Mars and Venus, which are apparently lifeless. Fossil records indicate that the atmosphere evolved to its present composition as a result of life. The atmospheric increase of the concentration of molecular oxygen for several billion years indicates that it resulted from photosynthesis after the appearance of life (e.g. Wayne 1992; Holloway and Wayne 2010). With the formation of a sufficiently thick ozone layer harmful short wave radiation became attenuated so that life could spread over the Earth's surface, initially in the oceans still requiring the protecting environment of water, later also on land. According to the *Gaia hypothesis* the biosphere has played an important role in determining the composition of the atmosphere since life on Earth began (Lovelock 1979). This hypothesis also suggested that the biosphere maintains favourable conditions for life on Earth. On geological timescales the impact of anthropogenic activities on the atmosphere has been of minor significance. However, since the onset of the industrial revolution at the end of the eighteenth century accommodating a constantly growing human population has dramatically increased the need for land use, energy and food supplies with the consequence of severe impacts on the Earth environment, especially to the Earth's atmosphere. Examples relating to significant modifications of the Earth's atmosphere include (after Crutzen and Stoermer 2000):

- Fossil fuels, generated over several hundred million years, burnt and exhausted into the atmosphere within a few generations.
- The release of SO_2, globally about 160 Tg/year, by coal and oil burning, over twice the sum of all natural emissions, mainly marine dimethyl-sulfide from the oceans.
- Human activity transformed 30–50% of the land surface. More nitrogen is now fixed synthetically and applied as fertilizer in agriculture than is fixed naturally in all the terrestrial ecosystems.
- The discharge of CO and NO_x (NO and NO_2) into the atmosphere by fossil fuel and biomass combustion, much larger than natural inputs, giving rise to photochemical ozone (smog) formation and degraded air quality in many regions of the world.
- The substantially increased concentration of atmospheric greenhouse gases: CO_2 by more than 30% and CH_4 by more than 100%, enhancing radiative forcing and causing changing climate. CO_2 and CH_4 are currently regulated by the Kyoto Protocol until 2012.
- New and not naturally existing substances released into the environment. Some of these substances, the chlorofluorocarbons, have caused the Antarctic *ozone hole* to open up and would have destroyed much of the ozone layer if no international regulatory measures, the Montreal Protocol, had been enforced to end their production.

Considering the major and still growing impacts of human activities on the Earth and the atmosphere on global scales, Crutzen and Stoermer (2000) proposed to use the term *Anthropocene* for the current geological epoch.

Tropospheric Chemistry

Most atmospheric gases, such as greenhouse gases (CO_2, CH_4, etc.) or the pollutants (NO_2, CO, SO_2, etc.) are emitted into the troposphere by natural processes and human activities. The oceans, the cryosphere and land surfaces act as sink to some of these gases. The main sources of pollutants in the northern hemisphere are fossil fuel combustion (energy production, traffic, industry and domestic heating) coupled with some biomass burning; in the southern hemisphere biomass burning is dominating. However modern agricultural intensive practise is also becoming a significant source of emissions. Pollutants are emitted within urban and near-urban areas where they are dispersed over the surrounding countryside and, depending on the atmospheric lifetime of the pollutant or its secondary reaction product(s), are transported around the globe. The pollutants from North America can e.g. be responsible for a significant fraction of tropospheric O_3 in the European summer due to complex transport and transformation mechanisms.

Tropospheric processes as sketched in Fig. 1.4 are well known to exhibit strong variability influenced by meteorology, diurnal as well as seasonal variations in the sources of the emissions and solar illumination. Photolysis of O_3 initiates the production of OH that determines, to a large extent, the oxidative (or cleansing) capacity of the troposphere. The role of the halogen oxides in the boundary layer as oxidants is currently a research matter. Many of the tropospheric trace gases are transformed into acids and other soluble products, which are removed from the atmosphere by precipitation or by uptake on aerosols and subsequent dry deposition on surfaces. The atmospheric oxidation power is of vital importance to the control of radiatively and chemically active pollutants. Therefore, any change in the atmospheric oxidation efficiency directly affects air quality.

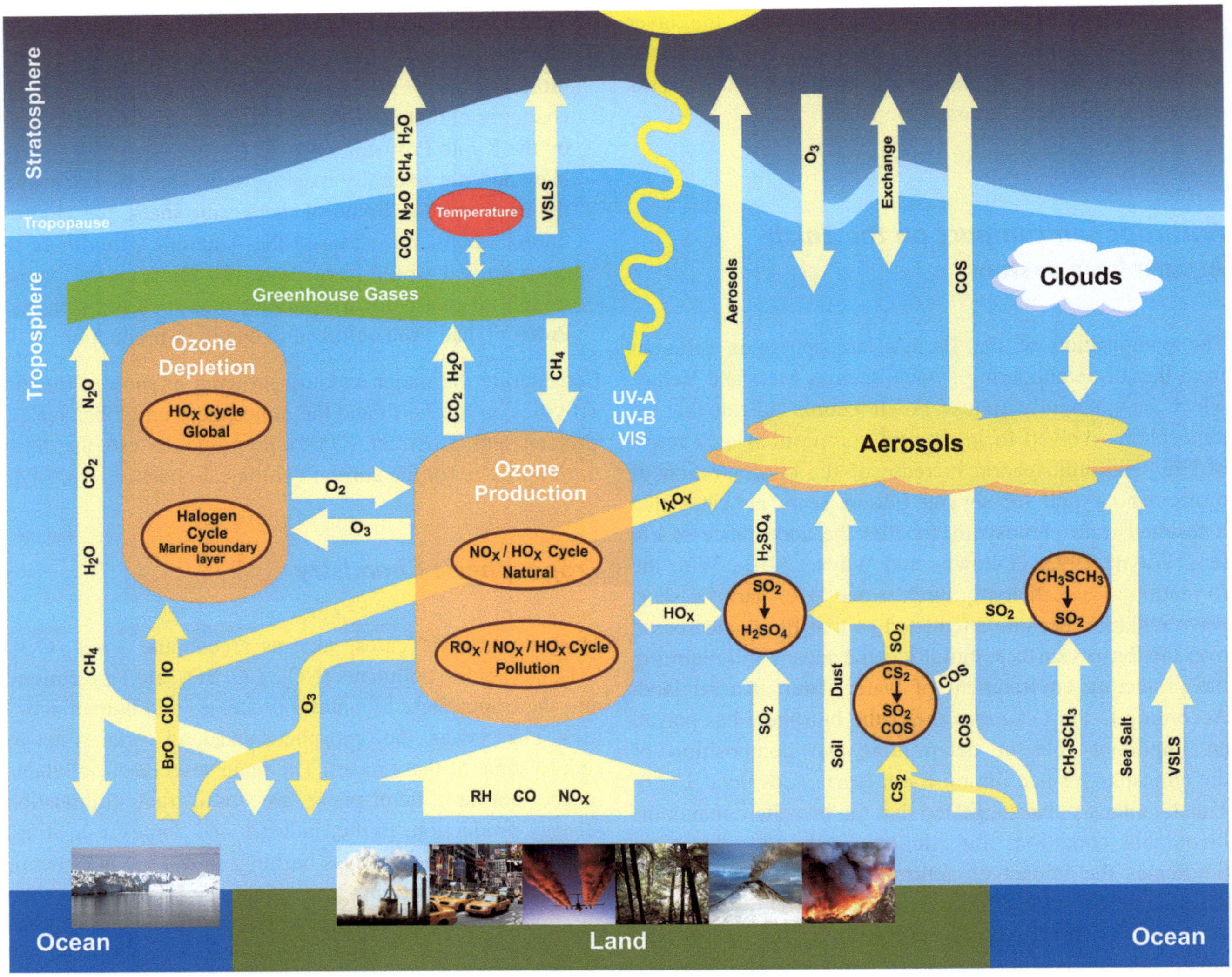

Fig. 1.4 The dominant physical and chemical processes determining the composition of the troposphere (Courtesy: IUP-IFE, University of Bremen).

The lack of information on the temporal and spatial distributions of the relevant species, as well as the source strengths of CO, CH_4 and NO_x, severely limits the quantitative understanding of the processes involved in tropospheric O_3 production and destruction. This is also a prerequisite for quantitative estimates of the hydroxyl radical distribution and thus of the cleansing power of the atmosphere, which is expected to be changing as a result of increasing emissions and resulting concentrations of O_3, CH_4, NO_x, CO, and non-methane hydrocarbons. One of the major challenges facing atmospheric science is to assess, understand and quantify the impact of a changing climate, which has feedbacks with atmospheric composition and air pollution.

SCIAMACHY is designed to determine the amounts of trace gases and aerosol in the lower troposphere, including the planetary boundary layer under cloud free conditions. From SCIAMACHY nadir and limb measurements tropospheric columns of O_3, CO, NO_2, BrO, CH_4, H_2O, SO_2, HCHO, CHOCHO, and IO are retrieved. In addition, surface spectral reflectance, aerosol and cloud parameters – cloud cover and cloud top height – can be obtained. These data are required for studies of the oxidising capacity of the troposphere, photochemical O_3 production and destruction, tropospheric pollution (biomass burning, industrial activities, aircraft), long-range transport of pollutants as well as quantification of natural and human emissions.

Greenhouse Gases: Global Warming and Climate Change

Arrhenius, back in 1896 published his account *On the influence of carbonic acid in the air upon the temperature on the ground*, identifying carbon dioxide and water vapour as important greenhouse gases that change climate (Arrhenius

1896). He calculated, with the help of Langley lunar plots measuring the infrared absorption of the atmosphere, that doubling the atmospheric CO_2 content by burning fossil fuel would lead to an average global temperature rise of 5–6°C, not far from the current IPCC estimates of 2–4.2°C. At the present rate of rise, doubling CO_2 would occur in the last quarter of this century. The list of greenhouse gases now comprises many species including H_2O, CO_2, CH_4, nitrous oxide, CFCs and tropospheric ozone. In addition, clouds and aerosols play an important, yet highly uncertain, role in global warming and cooling. Figure 1.5 summarises their global annual mean radiative forcing contributions. In order to compare greenhouse gases, taking into account their different infrared absorption and atmospheric lifetime, the concept of *global warming potential* over a 100 year time period has proved useful. For example, N_2O has a global warming potential of 296, i.e. causing a global warming contribution 296 times that of the same mass of CO_2.

Governments of many nations, concerned with the potential harmful consequences of global warming, have mandated the Intergovernmental Panel of Climate Change (IPCC) in 1989 to make evaluations aiming to provide policymakers with an assessment of our current understanding of climate change. The increasing evidence that current global warming is to a large extent man-made was documented in the fourth assessment report of the IPCC in 2007. As the concentrations of atmospheric greenhouse gases and their radiative forcing have continued to increase as a result of human activities, global warming and its impact on the Earth–Atmosphere system is predicted to further increase. One of the future challenges is to quantify the complex feedback cycles (see Fig. 1.3) between climate, atmospheric composition, natural and human activity which are driven by global warming. Climate change is expected to result in more frequent dry, hot summer periods in Europe – like the summer of 2003 – with degraded air quality in wide parts of Europe. Stratospheric ozone recovery may be delayed through climate change lowering stratospheric temperatures.

For use in climate research, SCIAMACHY measurements aim to provide the distributions of several important

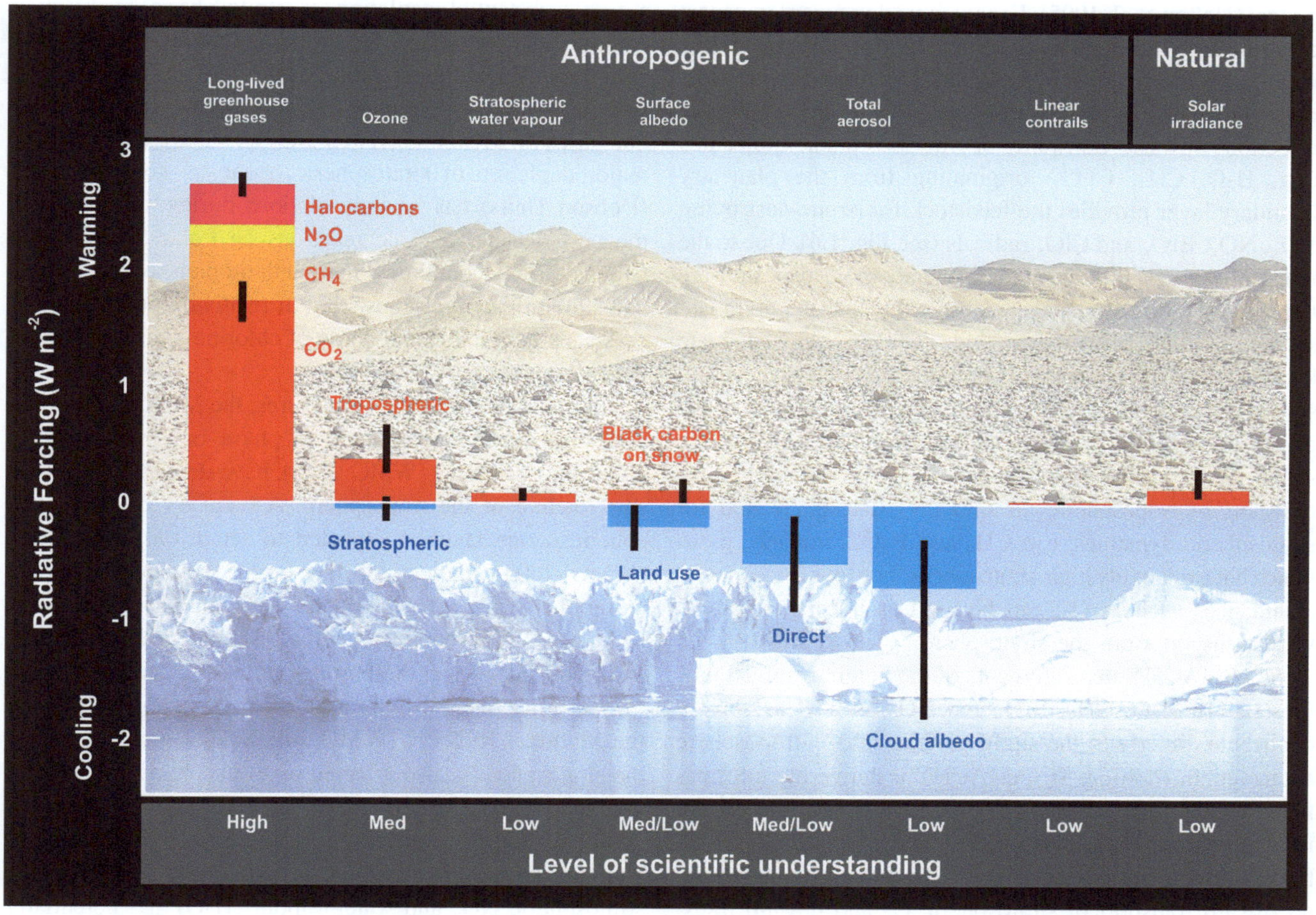

Fig. 1.5 Global, annual mean radiative forcings (W/m^2) due to a number of agents for the period from pre-industrial (1750) to present (2005). The height of each box denotes a central or best estimate value. The vertical bars visualise an estimate of the uncertainty range, for the most part guided by the spread in the published values of the forcing. A 'level of scientific understanding' index is associated to each forcing, with high, medium and low levels, respectively (Courtesy: adapted from IPCC 2007).

greenhouse gases (CH_4, CO_2, H_2O, tropospheric O_3), aerosol and cloud data, surface spectral reflectance (280–2386 nm), the incoming solar spectral irradiance and the outgoing spectral radiance (214–2386 nm). The observation of the greenhouse gases CH_4 and CO_2 will better constrain the quantification of emissions globally, thereby improving the scientific basis of the Kyoto Protocol, which was put into force in spring 2005 and currently ends in 2012. As SCIAMACHY has now made 8 years of observations and is capable of making them for many more years, this long-term data set will not only provide unique information about atmospheric composition but in addition is also of great value for studying solar-terrestrial interactions, e.g. how solar output variations impact climate change.

The Tropopause Region

Exchange of gases and particles between the stratosphere and the troposphere is of importance for the chemical composition of both regions as well as for the atmospheric energy budget (Holton et al. 1995). For example downward transport of stratospheric ozone is a source of tropospheric ozone which, as a precursor of OH radicals, to a large extent determines the oxidising power of the troposphere. In the opposite direction upward transport of the precursor molecules (e.g. H_2O, CH_4, CFCs) originating from the planetary boundary layer provides the feedstock for ozone-destroying HO_x, NO_x, BrO_x and ClO_x radicals (see Fig. 1.6). Due to the long tropospheric lifetime of the precursor molecules, they are transported to the stratosphere, where they are the dominant source of the ozone-destroying radicals. An adequate knowledge of the processes is required that determine strato sphere–troposphere exchange and the distribution of trace gases, especially in the lower stratosphere. Photo-chemically stable gases in the troposphere are useful as tracers for transport of tropospheric air into the stratosphere and for stratospheric dynamics, e.g. CH_4 and H_2O. Similarly, gases which have relatively high stratospheric but low tropospheric abundances such as O_3, can be used as tracers for downward transport from the stratosphere.

SCIAMACHY measurements of the height resolved profiles of O_3, H_2O, CH_4, BrO, and NO_2 as well as aerosol contribute directly to the study of stratosphere–troposphere exchange. In addition, SCIAMACHY is delivering information on the thermodynamical phase of clouds, which are important for the water and energy budget especially in the tropical tropopause region. With these measurements the downward transport of stratospheric O_3 and upward transport of important species (e.g. aerosol, CH_4, H_2O) can be investigated. Across the tropopause, the different measurement modes of SCIAMACHY have different vertical and horizontal resolutions. The solar and lunar occultation mode yields measurements with a vertical resolution of 2.5 km and a horizontal resolution of 30 km across track and extending roughly 400 km along track. For the limb measurements the spatial resolution is approximately 3 km vertically and typically 240 km horizontally across-track and extending roughly 400 km along-track. Therefore studies of relatively small scale features such as tropopause folding at mid-latitudes requiring a high spatial resolution are unlikely to be unambiguously observed by SCIAMACHY, but the larger scale stratosphere–troposphere exchange and phenomena such as volcanic eruptions will be readily observed.

Stratospheric Chemistry and Dynamics

No part of the global environment has been disturbed by human activity as significantly as the stratosphere. In the upper stratosphere and lower mesosphere ozone is removed by catalytic cycles involving halogen oxides. In addition, a very substantial depletion of stratospheric ozone over Antarctica – the ozone hole – has been observed during springtime since the beginning of the 1980s. This depletion is largely due to the emission of industrial chlorofluorocarbon gases (WMO 2003, WMO 2007). Also over the Arctic a major depletion of stratospheric ozone by about 100 DU (Dobson Units) has become obvious during springtime in the past decade. Surface reactions on Polar Stratospheric Clouds (PSC) containing hydrated droplets of nitric acid and sulphuric acid or consisting of relatively pure water ice crystals are – via the activation of chlorine – responsible for these changes.

International regulatory measures, the Montreal Protocol and subsequent amendments to phase out production of chlorofluorocarbons (WMO 1995), have decreased the chlorine loading of the troposphere. A recovery of the stratospheric ozone layer is expected to set in only after 2010 (WMO 2003). Indications that ozone abundances in the extrapolar regions (60°S–60°N) have stabilised in recent years were observed, in agreement with the observed decline in the abundances of anthropogenic ozone-depleting gases in the atmosphere as a result of the implementation of the Montreal Protocol (WMO 2007). The loss of ozone in the stratosphere is also affected by tropospheric emission of greenhouse gases (see Fig. 1.6). For example, the anthropogenic tropospheric concentrations of nitrous oxide and methane are increasing, leading to additional formation of stratospheric NO_x and water vapour (H_2O) and potentially enhancing the probability for formation of PSC. Thus, even though the stratospheric chlorine content is expected to decline at the beginning of the twenty-first century, ozone depletion in the lower stratosphere at higher latitudes may

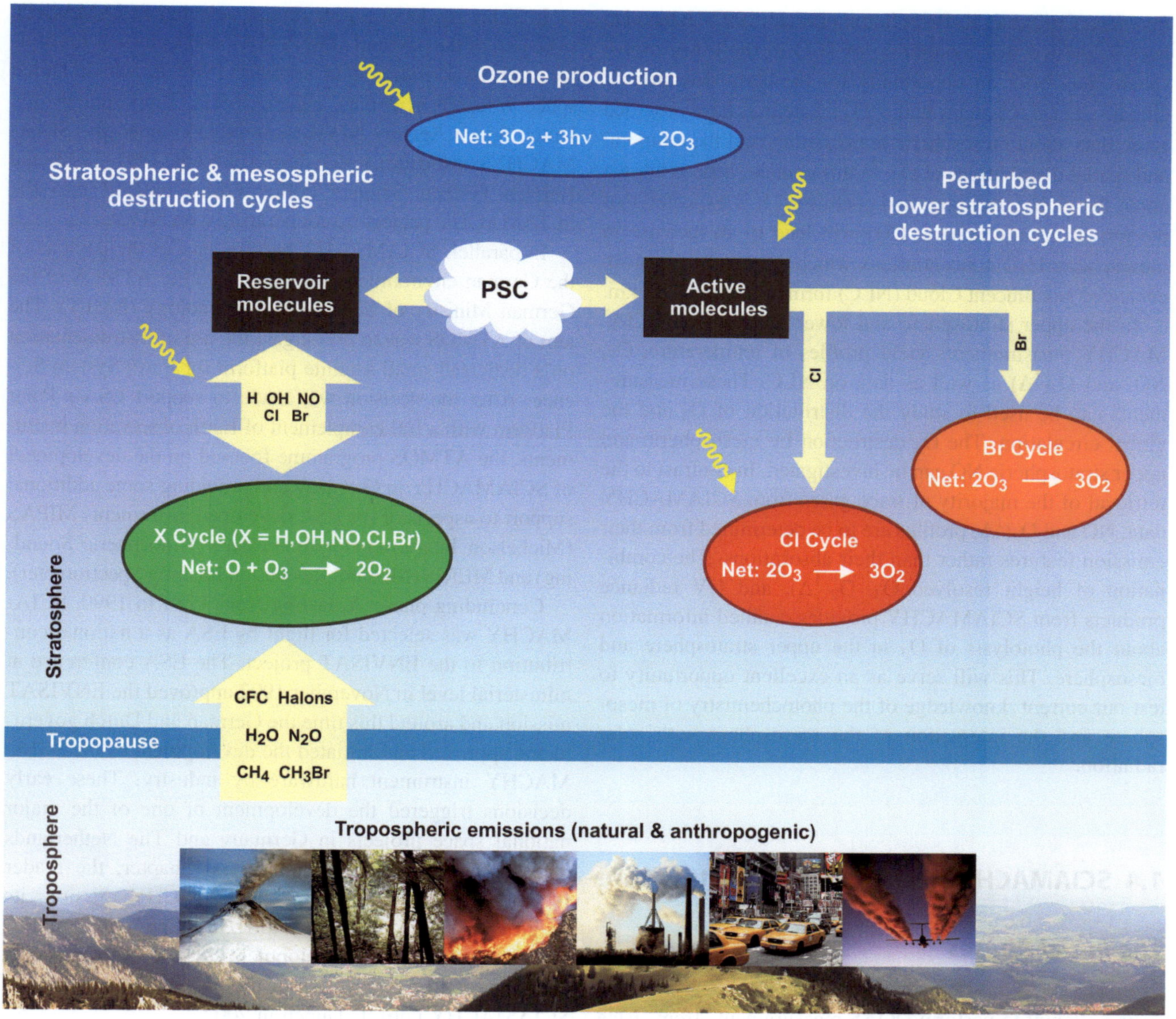

Fig. 1.6 Schematic sketch of the stratospheric ozone chemistry showing production and major destruction cycles (Courtesy: IUP-IFE, University of Bremen).

not. Therefore, continued monitoring of O_3, the sources of catalysts and the catalysts themselves remains mandatory in future.

SCIAMACHY's observations of the stratospheric chemistry utilise the simultaneous retrieval of total columns from nadir measurements and vertical stratospheric profiles from limb and occultation measurements of O_3, NO_2, BrO, H_2O, OClO (under ozone hole conditions), as well as aerosol and stratospheric cloud information. SCIAMACHY is making measurements when the halogen loading of the stratosphere maximised at the beginning of the twenty-first century and SCIAMACHY observations yield detailed information about the development of stratospheric O_3 above the Arctic and Antarctica, the global stratospheric active halogen species (BrO, OClO), nitrogen species (NO_2), which participate in the catalytic destruction of O_3, and the global O_3 budget as a function of the height in the atmosphere. Thus the SCIAMACHY data provide a unique opportunity to test the accuracy of current stratospheric models and their predictive capabilities.

Mesospheric Chemistry and Dynamics

The mesosphere extends from the temperature maximum at the stratopause around 50 km altitude to the atmospheric temperature minimum at the mesopause around 85 km. There has been much discussion of upper stratospheric and mesospheric chemistry in the context of the *Ozone Deficit*

Problem (Crutzen et al. 1995; Summers et al. 1997). It has also been suggested that monitoring of H_2O in the lower mesosphere may offer an opportunity for the early detection of climate change (Chandra et al. 1997). Satellites have provided some data about mesospheric temperatures and the temporal and spatial distributions of O_3. In this context, little is known about the global dynamics and chemistry. It is expected that the growth in atmospheric CH_4 will lead to an increase in mesospheric H_2O concentrations which might also result in enhanced Noctilucent Cloud (NLC) formation around 85 km.

In the upper stratosphere and lower mesosphere, SCIAMACHY measurements yield profiles of temperature, O_3, NO, and $O_2(^1\Delta)$ as well as data on NLC. These measurements can be used to study the distribution of O_3 and the global circulation. The O_3 destruction by mesospheric and upper stratospheric NO can be investigated. In contrast to the retrieval of the majority of trace gases from SCIAMACHY data, NO and $O_2(^1\Delta)$ profiles are to be determined from their emission features rather than their absorptions. The combination of height resolved O_3, $O_2(^1\Delta)$, and UV radiance products from SCIAMACHY provides detailed information about the photolysis of O_3 in the upper stratosphere and mesosphere. This will serve as an excellent opportunity to test our current knowledge of the photochemistry of mesosphere and the interaction of the mesosphere with solar radiation.

1.4 SCIAMACHY's Past and Beyond Its Future

The Initial Phases

A first attempt to perform DOAS measurements from space named MAP (Measurement of Atmospheric Pollution) dates back to 1985. It was submitted by J.P. Burrows, D. Perner, P.J. Crutzen, (MPI for Chemistry, Mainz) in response to an ESA call for research instruments to fly on EURECA, a free-flying European platform carried into space by the Space Shuttle allowing measurements for a period of several months. This proposal was not selected at that time but it was further developed, whilst international scientists joined a growing team. The concept of SCIAMACHY resulted from these endeavours. The SCIAMACHY proposal (Burrows et al. 1988a) was submitted in July 1988 by the SCIAMACHY Science Team – and supported by the German Space Agency DARA GmbH (now DLR) – in response to ESA's call for experiments to fly on-board the Polar Platform, an element of the Columbus Programme. This mission, the Polar Orbiting Earth Observation Mission (POEM-1), finally evolved into the mission now better known as ENVISAT.

In February 1989 a peer review selected SCIAMACHY to be part of the payload of ENVISAT and a phase A feasibility study was initiated in summer 1989 in which Dutch space research scientists and engineers, supported by the Dutch Space Agency NIVR (now NSO), joined the SCIAMACHY consortium. Later, during phase B (1991–1992), the Belgian Federal Science Policy Office entered into the SCIAMACHY project by cooperating with NIVR.

In parallel, SCIAMACHY became part of the payload of the German environmental research satellite ATMOS of the German Ministry of Science and Technology (BMFT). The aim of ATMOS was to investigate the use of instrumentation on a dedicated small satellite platform for Earth System Science. After the decision was made to support ESA's Polar Platform with a full complement of Earth observation instruments, the ATMOS programme focused on the development of SCIAMACHY in Germany and providing some additional support to aspects of the ESA developed instruments MIPAS (Michelson Interferometer for Passive Atmospheric Sounding) and MERIS (Medium Resolution Imaging Spectrometer).

Concluding phase A, lasting from 1989 to 1990, SCIAMACHY was selected for flight by ESA as a national contribution to the ENVISAT project. The ESA conference at ministerial level in November 1992 approved the ENVISAT mission and around this time the German and Dutch governments approved and initiated the development of the SCIAMACHY instrument hardware by industry. These early decisions triggered the development of one of the major national space projects in Germany and The Netherlands in the 1990s. Starting with the next chapter, the reader is invited to learn about the SCIAMACHY mission, its on-ground and in-orbit life. Chapter 3 is of particular interest as the story of the development of the instrument is continued with a description of the phase C/D 'making' of SCIAMACHY until its launch in 2002.

Following the recognition of the observation of the precipitous loss of ozone in austral spring mid of the 1980s, the *ESA User Consultation* meeting in Paris of November 1988 recommended for Europe to develop early independent ozone monitoring capability, rather than awaiting the launch of ENVISAT. ESA therefore announced a call for experiments to measure atmospheric constituents from ERS-2, which was under construction at that time. In response, J.P. Burrows and P. Crutzen proposed the SCIA-mini experiment (Burrows et al. 1988b), a scaled-down version of the original SCIAMACHY instrument concept. This proposal was selected for flight on ERS-2, after being further descoped to fit into the available platform resources, to include nadir observation only and was re-named GOME (Global Ozone Monitoring Experiment). GOME was 'fast tracked' on ERS-2, benefiting from the scientific and industrial development work on SCIAMACHY already performed. Within 4 years, GOME was successfully designed, manufactured, calibrated and integrated into ERS-

2 on time. ERS-2 was launched on 20 April 1995. Currently, GOME is still flying on the ERS-2 satellite and delivering data, however limited due to degradation of the platform and the instrument. Although originally a spin off from SCIAMACHY, GOME has demonstrated the feasibility and performance of both the spaceborne instrument hardware and the ground segment data retrieval software, which in turn has benefited the development of SCIAMACHY with respect to its nadir UV-VIS-NIR mission.

The Next Decades

With the impact of environmental and climate change becoming palpable, as documented for example by the scientific results from GOME and SCIAMACHY, the need for sustained global monitoring of ozone, greenhouse gases and air pollutants has turned into a prime requirement for future satellite missions. The councils of ESA and EUMETSAT approved therefore the EUMETSAT Polar System comprising a series of three METOP satellites. GOME-2 was selected as ozone and atmospheric trace gas payload complement. With the first METOP-A launched in 2006, this series of satellites provides continuous monitoring capability to the year 2020 from polar sun-synchronous low-Earth orbit. SCIAMACHY together with the data from GOME on ERS-2 (1995–2009) and the GOME-2 series on METOP (2006–2020), will therefore deliver a data set about atmospheric composition that is covering more than 25 years.

In parallel, initiatives emerged to realise a comprehensive global Earth observation system. Global Monitoring for Environment and Security (GMES), a joint initiative of the European Commission and ESA started in 1998, aims at bridging the gap between scientific data produced and the provision of reliable and timely information services to government and data users. This initiative resulted in the definition of a series of five GMES Sentinel satellites. Sentinels 4 and 5 are devoted to monitoring the atmosphere and were subsequently approved by EUMETSAT Council and the ESA ministerial conference in 2008. GMES is the European contribution to a wider approach, the Global Earth Observation System of Systems (GEOSS), within the international Global Earth Observation initiative (GEO). GEOSS couples the various Earth observation networks internationally to improve global coverage, whilst ensuring open access, accuracy and consistency of data of different sources. In addition, under auspices of WMO and in collaboration with ESA, the Integrated Global Atmospheric Chemistry Observation (IGACO) strategy was published in 2004 and aimed at integrating space data with ground-based and airborne measurements. IGACO established a basis for atmospheric chemistry instrument requirements for future satellites both for scientific and operational applications (IGACO 2004). These international initiatives were accompanied by new satellite missions and mission concepts.

Based on SCIAMACHY/GOME heritage, the Ozone Monitoring Instrument (OMI) is now flying on the NASA EOS satellite platform AURA, launched in 2004. GOME technology (nadir observations in the UV-VIS) is advanced by employing two-dimensional CCD detector arrays for the first time, improving spatial and temporal resolution in combination with wide global coverage without the need for a mechanical scanning mirror. This concept will be further improved by TROPOMI, as the spectral range will be extended to the NIR and SWIR. TROPOMI is planned for launch in 2014 as the GMES Sentinel–5 Precursor mission, to bridge parts of the data gap between ENVISAT, Sentinel 4 (launch 2019) and Sentinel 5 (launch 2020).

In addition, stimulated by the SCIAMACHY concept, US colleagues proposed and built instruments and missions with elements of SCIAMACHY heritage. The Orbiting Carbon Observatory (OCO) is a mission to quantify natural CO_2 emissions by performing NIR and SWIR nadir measurements similar to SCIAMACHY, but with higher spatial and spectral resolution. Sadly the OCO launch failed early 2009, but a reflight is envisaged for the timeframe 2013/2014. The Ozone Monitoring and Profiling Suite (OMPS) on NPP and NPOESS, operational from 2012 to 2025, with a combined limb (UV-VIS-NIR) and nadir (UV) backscatter spectrometer using one spectrometer for nadir and one for limb, follows the original SCIAMACHY idea.

Unfortunately and ignoring the needs from European GMES users, there will be no European measurements of the dry column of CO_2 after ENVISAT and prior to Sentinel 5. In order to fill this data gap, the CarbonSat project was initiated in 2009 (Bovensmann et al. 2010). CarbonSat shall allow to determine CO_2 and CH_4 emitted by fossil fuel combustion, biomass burning and natural emissions and CO_2 uptake in the terrestrial and ocean biosphere employing inverse modelling techniques in combination with measured data. In comparison to SCIAMACHY, CarbonSat will have a much higher spatial resolution (2 km × 2 km) with a 3–6 days revisiting time. In comparison to OCO, the mission will measure CO_2 and CH_4. Due to its across-track imaging capability, 500 km swath compared to 10 km swath for OCO, imaging of emission hot spots will become feasible. The data shall support monitoring of strong localised anthropogenic emission hot spots and shall investigate the feasibility of independent (post) Kyoto treaty verification.

The missions discussed so far refer to space platforms in low-Earth sun-synchronous polar orbit, i.e. at approximately 800 km altitude, measuring the Earth's atmosphere under fixed illumination conditions, revisited after 1 day or more. The diurnal variation of photo chemically active gases (NO_2, CO, O_3) cannot be captured. However air pollution quantification and forecast require measurement of diurnal variability. From geo-stationary orbit, i.e. at approximately 36000 km altitude, a certain hemisphere of the Earth is permanently in view such that the same scene can be monitored throughout the day at typically hourly temporal and high spatial (<10 km) resolution. In addition increased occurrences of cloud-free conditions, required for troposphere and boundary layer observation, become likely. To address these aspects for the first time, the Geostationary Scanning Imaging Absorption Spectrometer (GeoSCIA) was proposed in 1998. The concept of geostationary air pollution observations was later further enhanced by combining solar backscatter observations with thermal infrared observations in the Geostationary Tropospheric Explorer (GeoTROPE) mission concept, which now in parts will be implemented on the next generation of Europe's geostationary meteorological satellite METEOSAT Third Generation (MTG). The GeoSCIA concept has been adopted by EUMETSAT and ESA to form the concept for the Sentinel 4 UVN instrument as the EU-ESA GMES payload contribution to the MTG programme. Sentinel 4 UVN on MTG shall perform daytime observations of solar backscattered spectra to determine the total and tropospheric column amounts, including the boundary layer, of ozone, ozone precursors and pollutants (NO_2, SO_2, HCHO, CHOCHO) as well as the aerosol optical depth (AOD) over Europe and the surrounding regions with temporal and spatial resolutions of 1 h and better than 10 km × 10 km. On the same MTG platform EUMETSAT will install the meteorological InfraRed Sounder (IRS), complementary to Sentinel 4 UVN. The IRS sensor will focus on meteorological parameters (H_2O, temperature and pressure) but shall also obtain vertical distributions of O_3 and CO during day and night with temporal and spatial resolutions of about 20–60 min and 4 km × 4 km. The launch of MTG with the Sentinel 4 UVN and IRS payload is planned for 2019. In order to accomplish an operational MTG in-orbit phase of at least 15 years, a second platform carrying the same instrumentation shall be launched a few years later.

In retrospect, several Earth Observation missions for atmospheric remote sensing, extending well into the twenty-first century, are of SCIAMACHY heritage. This is by far more than the authors of the original SCIAMACHY proposal of 1988 could have expected.

References

Arrhenius S (1896) On the influence of carbonic acid in the air upon the temperature of the ground. Philos Mag J Sci 41:237

Bovensmann H, Burrows JP, Buchwitz M, Frerick J, Noël S, Rozanov VV, Chance KV, Goede APH (1999) SCIAMACHY: mission objectives and measurement modes. J Atmos Sci 56:127–150

Bovensmann H, Buchwitz M, Burrows JP, Reuter M, Krings T, Gerilowski K, Schneising O, Heymann J, Tretner A, Erzinger J (2010) A remote sensing technique for global monitoring of power plant CO_2 emissions from space and related applications. Atmos Meas Tech 3:781–811

Burrows JP, Chance KV, Crutzen PJ, van Dop H, Geary JC, Johnson TJ, Harris GW, Isaksen ISA, Moortgat GK, Muller C, Perner D, Platt U, Pommereau J-P, Rodhe H, Roeckner E, Schneider W, Simon P, Sundqvist H, Vercheval J (1988a) SCIAMACHY – a European proposal for atmospheric remote sensing from the ESA Polar Platform. Technical Document, Max Planck Institute for Chemistry, Mainz, Germany

Burrows JP, Crutzen PJ, Chance KV, Geary JC, Goutail F, Harris GW, Moortgat GK, Muller C, Perner D, Platt U, Pommereau J-P, Schneider W, Simon P (1988b). SCIAmini. Technical Document, Max Planck Institute for Chemistry, Mainz, Germany

Burrows JP, Weber M, Buchwitz M, Rozanov VV, Ladstädter-Weissenmayer A, Richter A, de Beek R, Hoogen R, Bramstedt K, Eichmann K-U, Eisinger M, Perner D (1999) The Global Ozone Monitoring Experiment (GOME): mission concept and first scientific results. J Atmos Sci 56:151–175

Burrows JP, Platt U, Borrell P (eds.) (2011) The remote sensing of Tropospheric Composition from Space. Springer Heidelberg Dordrecht London New York, ISBN 978-3-642-14790-6, DOI: 10.1007/978-3-642-14791-3

Chandra S, Jackman CH, Fleming EL, Russell JM III (1997) The seasonal and long-term changes in mesospheric water vapor. Geophys Res Lett 24:639–642

Crutzen PJ, Grooß J-U, Brühl C, Müller R, Russell JM III (1995) A reevaluation of the ozone budget with HALOE UARS data: No evidence for the ozone deficit. Science 268:705

Crutzen PJ, Stoermer EF (2000) The anthropocene. IGBP Newsl 41:17–18

Holloway AM, Wayne R (2010) Atmospheric chemistry. Royal Society of Chemistry Publishing, Cambridge, UK. ISBN 978-1-84755-807-7

Holton JR, Haynes PH, McIntyre ME, Douglass AR, Hood RB, Pfister L (1995) Stratosphere–troposphere exchange. Rev Geophys 33:403–439

IGACO (2004) An Integrated Global Atmospheric Chemistry Observations Theme for the IGOS Partnership, IGOS Atmospheric Chemistry Theme Report. ESA-SP1282, WMO GAW Report No. 159, WMO TD No. 1235

Intergovernmental Panel on Climate Change (IPCC) (2007) Climate change 2007 – the physical science basis. In: Solomon S, Qin D, Manning M, Chen Z, Marquis M, Averyt KB, Tignor M, Miller HL (eds) Contribution of working group I to the fourth assessment report of the intergovernmental panel on climate change. Cambridge University Press, Cambridge, UK and New York, USA. Available at http://www.ipcc.ch

Lovelock JE (1979) Gaia: a new look at life on earth. Oxford University Press, Oxford

Munro R, Siddans R, Reburn WJ, Kerridge BJ (1998) Direct measurement of tropospheric ozone distributions from space. Nature 392:168–171

Reber CA, Trevathan CE, McNeal RJ, Luther MR (1993) The Upper Atmosphere Research Satellite (UARS) mission. J Geophys Res 98(D6):10643–10647, DOI:10.1029/92JD02828.

Shindell DT, Rind D, Logan P (1998) Increased polar stratospheric ozone losses and delayed eventual recovery owing to increasing greenhouse-gas concentrations. Nature 392:589–592

Summers ME, Conway RR, Siskind DE, Stevens MH, Offermann D, Riese M, Preusse P, Strobel DF, Russell JM III (1997) Implications of satellite OH observations for middle atmospheric H_2O and ozone. Science 277:1967–1970

United Nations Economic Commission for Europe (2007) Hemispheric transport of air pollution 2007. Air Pollution Studies, 16. United Nations, New York and Geneva, ISBN 978-92-1-116984-3
Wayne RP (1992) Chemistry of atmospheres. Oxford University Press, Clarendon Press, Oxford
Wild M, Gilgen H, Roesch A, Ohmura A, Long CN, Dutton EG, Forgan B, Kallis A, Russak V, Tsvetkov A (2005) From dimming to brightening: Decadal changes in solar radiation at earth's surface. Science 308:847–850
World Meteorological Organization (WMO) (1995) Scientific assessment of ozone depletion 1994. Global Ozone Research and Monitoring Project Report, 37, Geneva
World Meteorological Organization (WMO) (2003) Scientific assessment of ozone depletion 2002. Global Ozone Research and Monitoring Project Report, 47, Geneva
World meteorological organization (WMO) (2007) Scientific assessment of ozone depletion 2006. Global Ozone Research and Monitoring Project Report, 50, Geneva

Chapter 2
ENVISAT – SCIAMACHY's Host

M. Gottwald, F.-J. Diekmann, and T. Fehr

Abstract ENVISAT is Europe's ambitious Earth Observation enterprise to study the many facets of the Earth system. It carries ten remote sensing instruments with SCIAMACHY, MIPAS and GOMOS forming the atmospheric chemistry mission. SCIAMACHY, although provided by national space agencies to ESA, is an integral part of the payload. Orbit and attitude of ENVISAT determine the framework of SCIAMACHY's observing capabilities. As a polar, sun-synchronous satellite, ENVISAT provides a stable platform for orbiting the Earth every 100 min. All instruments share the available on-board resources, particularly on-board data handling capabilities. The ENVISAT ground segment consists of the Flight Operation Segment for platform and instrument control and of the Payload Data Segment for measurement data acquisition, processing, archiving and dissemination. The SCIAMACHY data processing occurs at the LRAC and the D-PAC, depending on whether the data is of type level 0, 1b or 2. Access to SCIAMACHY data follows the general ENVISAT data policy with the exception that the instrument providing agencies receive a separate copy of such data.

Keywords ENVISAT • Orbit • Flight Operation Segment • Payload Data Segment • Product levels

The decision to place SCIAMACHY onto the ENVISAT platform (Figs. 2.1 and 2.2) made it part of ESA's most ambitious Earth observation mission. Being a 'derivative' of the Columbus programme, ENVISAT is composed of the Polar Platform and the payload complement consisting of ten instruments (ESA 1998; ESA 2001a). With an overall length in orbit of more than 25 m, it is the largest satellite ever built by ESA (Table 2.1). The instrument complement was selected to permit studies of many aspects of the complex Earth system (Table 2.2). This system is considered to be driven by interactions of its main 'spheres', i.e. lithosphere/geosphere, hydrosphere, atmosphere, cryosphere and biosphere. The goal of the mission is to obtain insights into local and global processes of our environment. With a current in-orbit mission lifetime of 8 years ENVISAT generates a large data repository for long-term analyses in Earth sciences. Seven instruments of the payload complement have the status of an ESA developed instrument (EDI). SCIAMACHY, AATSR and DORIS are Announcement of Opportunity Instruments (AOI). They were provided by national agencies, developed under their responsibilities. Each AOI is an integral part of the payload following the same guidelines as the EDI. However tasks in instrument operation and data processing are shared between ESA and the AOI providers. Overall the ENVISAT instruments utilise various remote sensing methods. Active microwave sensors (ASAR, RA-2) are supplemented by passive microwave (MWR, DORIS) and optical (AATSR, MERIS, MIPAS, GOMOS, SCIAMACHY, LRR) sensors.

Together with MIPAS (ESA 2000) and GOMOS (ESA 2001b), SCIAMACHY forms ENVISAT's atmospheric mission. The three atmospheric instruments operate in different wavelength ranges using different measurement principles. They complement each other such that synergistic views are generally possible. While SCIAMACHY, as an absorption spectrometer in the UV-SWIR range is requiring sunlight, MIPAS – operating in the thermal infrared – can measure over the complete orbit. This is also the case for GOMOS, where the UV-VIS component of stars in occultation is used to probe the atmosphere. SCIAMACHY is viewing in flight direction (limb mode) and towards the sub-satellite point (nadir mode). MIPAS looks along-track into anti-flight and across-track into anti-sun direction. GOMOS can steer its

M. Gottwald (✉)
Remote Sensing Technology Institute, German Aerospace Center (DLR-IMF), Oberpfaffenhofen, 82234 Wessling, Germany
e-mail: manfred.gottwald@dlr.de

F.-J. Diekmann
European Space Operations Centre (ESOC), European Space Agency (ESA), Robert-Bosch-Str. 5, 64293 Darmstadt, Germany

T. Fehr
European Space Research Institute (ESRIN), European Space Agency (ESA), Via Galileo Galilei, 00044 Frascati (RM), Italy

M. Gottwald and H. Bovensmann (eds.), *SCIAMACHY - Exploring the Changing Earth's Atmosphere*,
DOI 10.1007/978-90-481-9896-2_2, © Springer Science+Business Media B.V. 2011

Fig. 2.1 Artist's impression of ENVISAT in orbit. SCIAMACHY is located at the upper right corner of the payload front panel (Photo: ESA).

Fig. 2.2 ENVISAT in ESTEC's test facility (Photo: ESA).

Table 2.1 ENVISAT characteristics

ENVISAT parameters	
Dimensions	26 m × 10 m × 5 m
Total mass	8140 kg
Payload mass	2050 kg
Launcher	Ariane-5
Launch	1 March 2002

line-of-sight (LoS) towards stars which set between anti-flight and across-track anti-sun direction (Fig. 2.3). Additionally, the optical imaging instruments MERIS and AATSR (Advanced Along Track Scanning Radiometer) deliver data for scientific applications in the fields of clouds, aerosols and water vapour. Their nadir views overlap with SCIAMACHY's nadir geometry thus permitting synergistic analyses.

2.1 ENVISAT Attitude and Orbit

Attitude and orbit of the ENVISAT platform are contributing to the status of the SCIAMACHY line-of-sight. Both are controlled by the Attitude and Orbit Control System (AOCS). It provides attitude determination and attitude and orbit control capabilities (Bargellini et al. 2006). Attitude measurements are performed via gyroscopes, Earth sensors, Sun sensors and star sensors while the control function uses reaction wheels/magnetorquers and hydrazine thrusters. The platform is three-axis stabilised with ENVISAT's local relative orbital reference coordinate frame forming an orthogonal right-hand system with the -Y (roll) axis pointing close to the velocity vector of the platform and the Z (yaw) axis pointing in the direction of the outward local normal of the Earth's reference ellipsoid (Fig. 2.4). For the specification of target directions, azimuth and elevation angles are defined. Azimuth is measured clockwise from the $-Y$ axis around the $-Z$ axis to the projection of the target LoS in the plane perpendicular to the orbital plane containing the velocity vector. Elevation is the angle between that projection and the target LoS (Milligan et al. 2007).

The attitude mode of ENVISAT for nominal operations offering the best pointing performance is the *stellar yaw steering mode* (SYSM). In this mode the position of a star from the uplinked star catalogue as measured by two star trackers is compared with the predicted position to derive prediction updates. Yaw steering is required by the microwave imaging instruments in order to compensate for the rotational velocity of the Earth's surface at the sub-satellite point. It is achieved via small rotations about the roll, pitch and yaw axis, with the transformation around yaw being by far the dominant component. The maximum yaw amplitude of $\pm 3.92°$ occurs when ENVISAT passes the equator at ascending or descending node, the minimum amplitude of $0°$ is reached close to the poles. In total, yaw steering imposes a sinusoidal 'wobble' of the platform around the flight direction. In the illuminated part of the orbit the platform is turned to the right side of the flight direction, in the eclipse part to the left. For measurements with long line-of-sight requiring high spatial resolution, as is the case for SCIAMACHY operating in limb mode, the quality of the scientific results, e.g. altitude profiles, is strongly dependent on the precision of the platform attitude. Even small perturbations in the order of a few 10^{-3} deg can significantly impact the measured altitudes in limb profiles.

The orbit of the hosting platform was selected to be similar to its precursors, ERS-1 and ERS-2. Thus ENVISAT

Table 2.2 The ENVISAT payload complement and applicable components in the Earth system

Instrument	Lithosphere	Hydrosphere	Atmosphere	Cryosphere	Biosphere
ASAR	×	×		×	×
RA-2	×	×		×	
MWR			×		
DORIS	×				
MERIS	×	×	×	×	×
AATSR	×	×	×	×	×
GOMOS			×		
MIPAS			×		
SCIAMACHY		×	×	×	×
LRR	×	×		×	

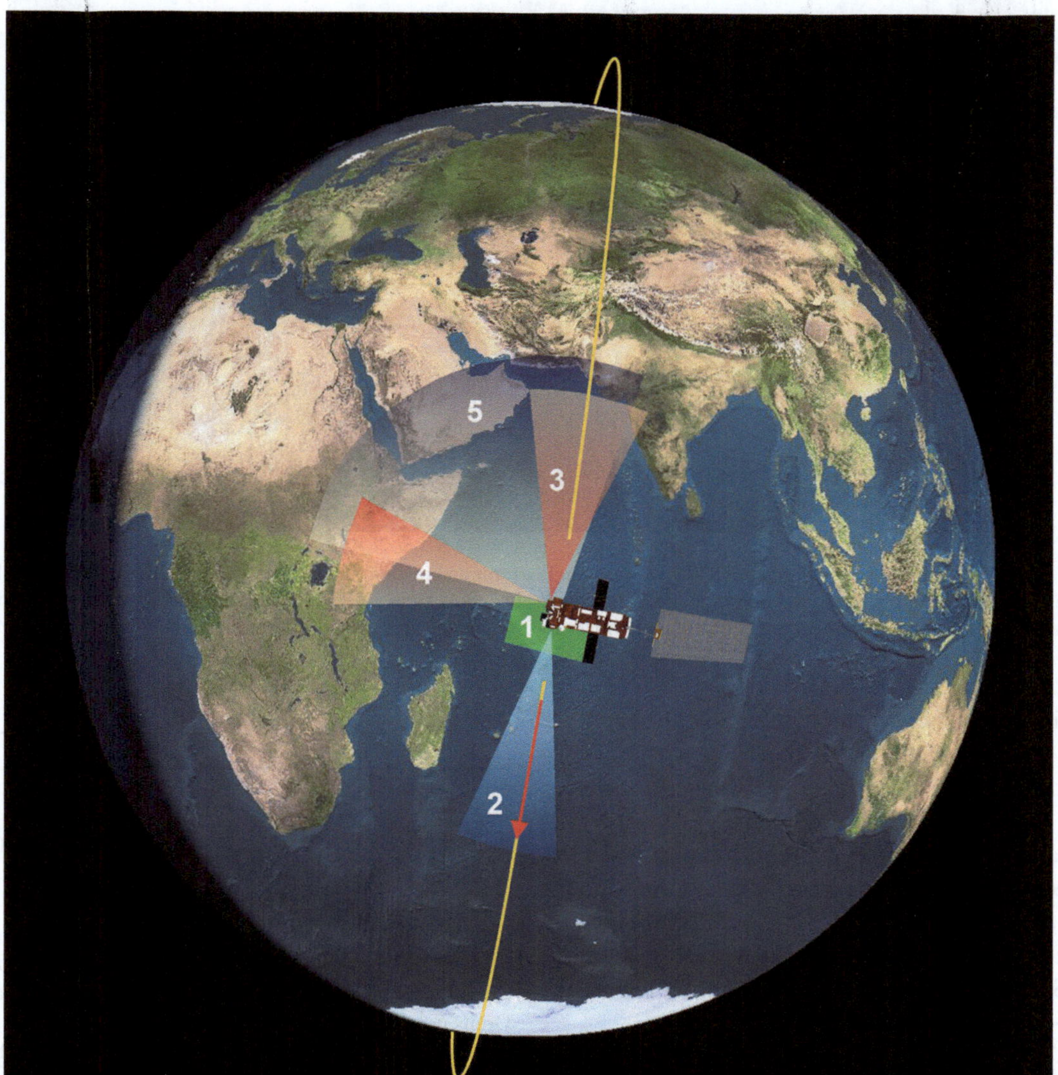

Fig. 2.3 The total field of views of the atmospheric mission instruments. SCIAMACHY viewing comprises nadir (1) and limb in flight direction (2). MIPAS looks into anti-flight direction (3) and sideways (4) for special events monitoring. GOMOS can observe setting stars in anti-flight and sideways direction (5) (Courtesy: DLR-IMF).

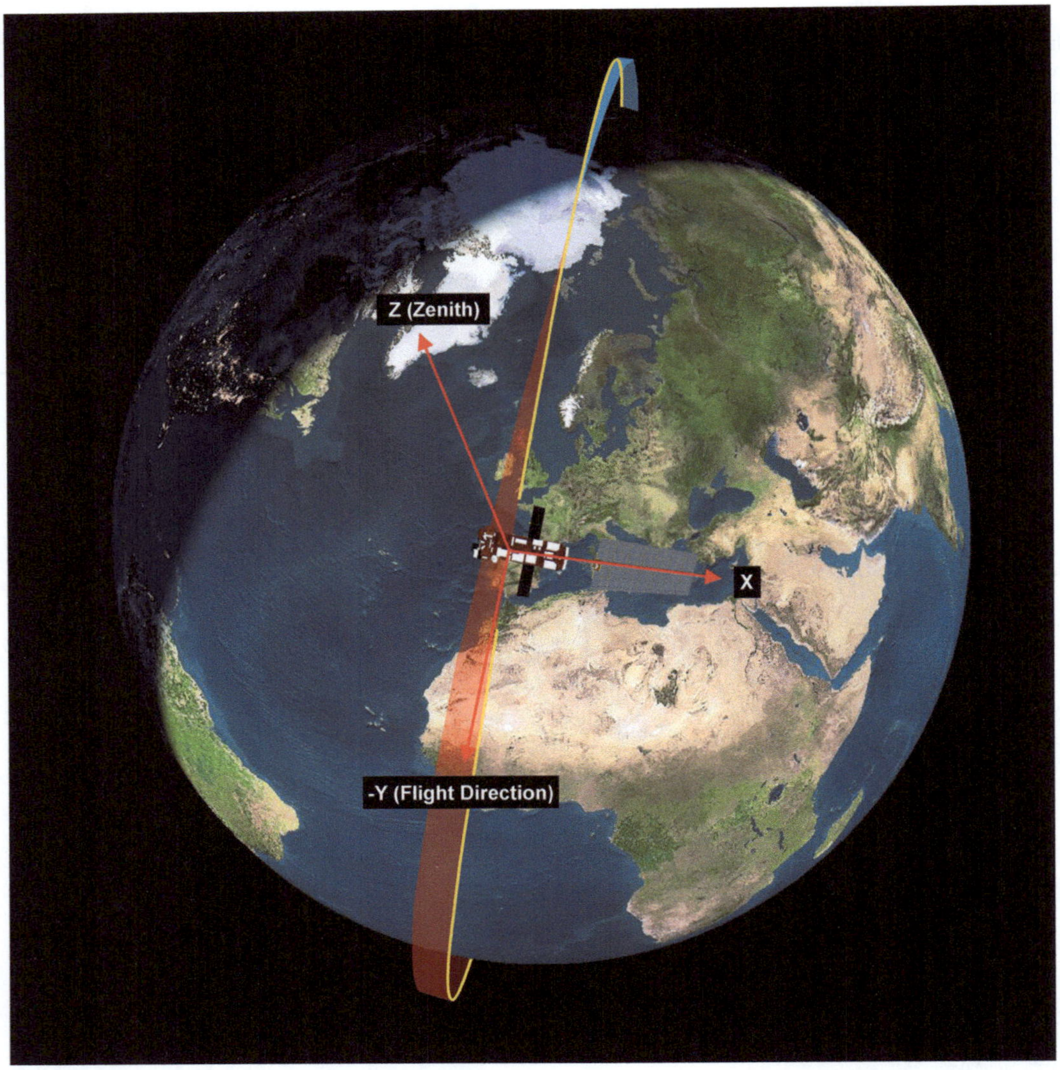

Fig. 2.4 The coordinate system specified in the ENVISAT mission. Platform and instruments use the same definitions. The red/blue areas right and left of the orbit sketch the platform yaw steering range. It gradually increases from pole to equator and decreases from equator to pole. The local -Y axis always points to that side, where Earth rotation moves the surface towards the sub-satellite track (Courtesy: DLR-IMF).

was placed in a polar, sun-synchronous orbit with a morning descending node crossing (DNX) time. Sun-synchronicity yields identical illumination conditions whenever crossing the equator. The selected Mean Local Solar Time (MLST) at DNX of 10 a.m. means that the Sun is always positioned left of the spacecraft, relative to the orbital plane. Detailed orbit parameters are given in Table 2.3. Since SCIAMACHY is located at the upper right corner on the payload module front panel, the Sun can only be seen when pointing the instrument's line-of-sight to the left.

ENVISAT operations have to ensure that the actual ground track does not deviate by more than ±1 km from the reference orbit. However orbit perturbations slightly modify the specified reference orbit with time. Once the 1 km deadband is going to be violated, an orbit control manoeuvre has to adjust the actual orbit. These manoeuvres can either correct for the effect of air drag on altitude (*In-Plane*, Stellar Fine Control Manoeuvre – SFCM) or adjust the inclination which drifts due to solar and lunar gravity perturbations (*Out-of-Plane*, Orbit Control Manoeuvre – OCM). Orbital manoeuvres are

Table 2.3 Nominal ENVISAT orbit

Orbital parameters	
Semi-major axis	7159.50 ± 0.07 km
Inclination	98.55° ± 0.01°
Eccentricity	0.001165 (−0.001165/+0.005)
Argument of perigee	90.0° ± 3°
Mean altitude	799.8 km
Orbital period	100.6 min
Type	Polar, sun-synchronous
Mean local time at DNX	10:00 a.m. ±5 min
Orbits per day	14 11/35
Repeat cycle	35 days (501 orbits)

executed by short firings of the thrusters. Since each firing consumes hydrazine, the available on-board resource 'fuel' diminishes with time. The originally available 314 kg of fuel is reduced by several kg per OCM but only by less than 100 g per SFCM. In addition, non-nominal manoeuvres, such as, e.g. space debris collision avoidance manoeuvres or manoeuvres triggered by platform anomalies further reduce the available fuel. SCIAMACHY continues measurements during In-Plane manoeuvres. While executing an Out-of-Plane manoeuvre, SCIAMACHY measurements are interrupted but the thermal status of the Optical Bench Module (OBM) and the detectors remain unchanged.

2.2 ENVISAT On-board Resources

The total ENVISAT payload shares the same on-board resources. Thus instruments may not interfere with each other. For SCIAMACHY this is particularly important with respect to the allocated data rate (Fig. 2.5). The data rate depends on an instrument and its measurement mode. The Regional Mission of ASAR (Advanced Synthetic Aperture Radar) and MERIS produces data with high and medium rates aiming at acquiring information with high spatial resolution for specific areas. Providing maximum coverage of the Earth is the goal of the Global Mission. It generates data with a lower rate and is established by all other instruments together with ASAR and MERIS in their low rate modes. Global Mission instruments are operated continuously throughout the orbit. Measurement data from the Global Mission instruments are processed on-board via the High Speed Multiplexer (HSM). SCIAMACHY uses HSM resources together with MERIS in Reduced Resolution (RR) mode. As long as MERIS runs in RR mode, only the low rate data can be used by SCIAMACHY. This MERIS mode is operational for solar zenith angles <80°. Although SCIAMACHY is able to generate data with a data rate of 1.8 Mbit/s yielding measurements with high spatial resolution, for most parts of the orbit only the low rate of 400 kbit/s can be achieved. Only when ENVISAT is close to sunrise during each orbit, the solar zenith angle condition is not fulfilled for MERIS and measurements with the higher rate are possible for SCIAMACHY when observing a solar occultation.

ENVISAT measurement data can be transmitted to ground stations via the Artemis data relay satellite using the Ka-band. The antenna on the platform is located on the upper payload panel just above SCIAMACHY's sub-solar port, i.e. the instrument window allowing observations

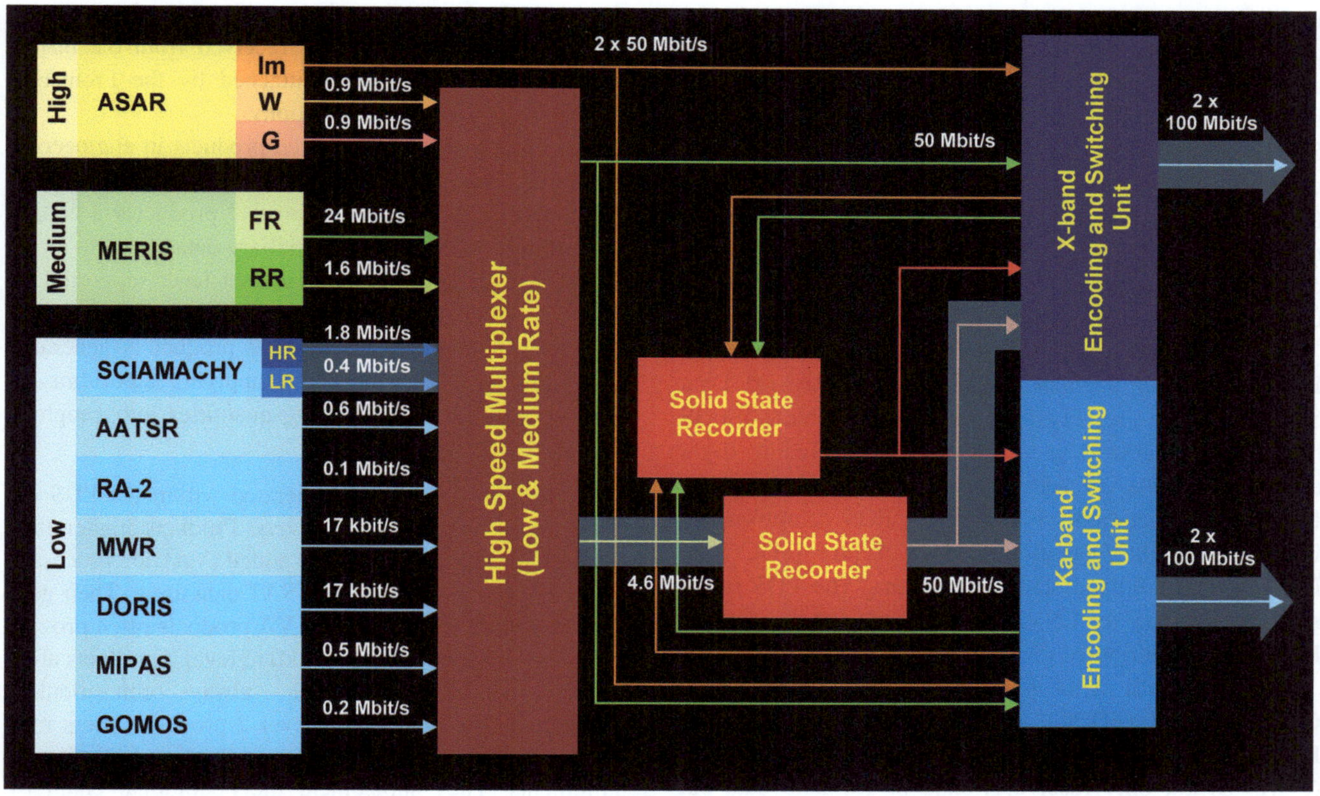

Fig. 2.5 Schematic view of the ENVISAT on-board data handling. SCIAMACHY shares HSM resources with the other instruments generating low rate data. Full orbit data of the Global Mission is stored on one of the solid state recorders and then disseminated to ground via X-band or Ka-band (Courtesy: DLR-IMF).

of the Sun when it has reached highest elevation each orbit. When the antenna dish is deployed in operating position, it vignettes SCIAMACHY's sub-solar window. Therefore sub-solar measurements, necessary to monitor the long-term behaviour of optical components, can only be executed when the Ka-band antenna has been moved to its parking position. This is accomplished via the ENVISAT/SCIAMACHY mission planning process by applying strict planning rules.

2.3 ENVISAT Ground Segment

All information generated by the instruments, i.e. housekeeping (HK) telemetry reporting the instrument status and measurement data, is either downlinked directly via S-band or as part of the science data stream via the X-band or the Ka-band link. The X-band link transmits measurement data to the high latitude receiving ground station at Kiruna while the Ka-band link connects ENVISAT via the Artemis data relay satellite with the receiving station at ESRIN. In the early phase of the mission, when Artemis was not operational, only Kiruna and an additional X-band station at Svalbard were used. In the nominal 'Kiruna-Artemis' scenario, Global Mission data are provided to Kiruna for 8 orbits per day, to ESRIN via Artemis for 6 orbits per day. HK telemetry is mainly sent via the command and control (C&C) S-band link of the Kiruna station and its supplementing Svalbard C&C facility. This telemetry stream is only available for the coverage interval of Kiruna or Svalbard visible orbits. Of the 14.3 orbits per day, 9–10 are orbits with Kiruna coverage intervals amounting up to 12 min (spacecraft elevation >5°). In addition, low elevation coverage may also be used. Each coverage in a daily cycle occurs at a different time relative to ascending node crossing (ANX) depending on the actual location of the orbital track.

ENVISAT and its instruments are controlled via the Flight Operation Segment (FOS) and the Payload Data Segment (PDS) – see Figs. 2.6 and 2.7 and Table 2.4. FOS handles all command and control issues including flight dynamics aspects. In addition, mission planning is performed in cooperation with the PDS. FOS comprises the ENVISAT Flight Operation Control Centre (FOCC) at ESOC and the S-band station at Kiruna. The PDS is the responsible entity for ENVISAT measurement data. It is a Europe-wide distributed ground segment.

Central control of the PDS occurs at the Payload Data Control Centre (PDCC) at ESRIN. The Payload Data Handling Stations in Kiruna (PDHS-K) and Frascati (PDHS-E) not only receive X-band or Ka-band data but also process and disseminate data in near-realtime (NRT). X-band data reception for the Regional Mission is complemented by the Payload Data Acquisition Station (PDAS) in Matera/Italy. For data products in offline (OL) mode processing, archiving and distribution is shared among several facilities. At the Low Rate Reference Archive Centre (LRAC), located at Kiruna, level 0 data are consolidated and archived. From these consolidated products, the Processing and Archiving Centres (PAC) generate, archive and disseminate consolidated level 1b and level 2 products. Also NRT products from the two PDHS are collected at the PAC and can be distributed, if required. Although implemented at national remote sensing facilities, PACs are an integral part of the PDS. They are controlled by the PDCC, which also drives product generation via order mechanisms. In order to maintain a certain product quality throughout the PDS, PDCC executes product quality monitoring.

2.4 ENVISAT Data Products

The ENVISAT ground segment concept defines product levels which fall under the responsibility of the PDS. These are

- Raw data: Instrument raw data is generated from the X- and Ka-band demodulator output. It serves as input for the level 0 data.
- Level 0: Produced from the raw data, level 0 contains instrument source packets as received from the instrument with a small header attached by the Front-End Processor at the receiving stations.
- Level 1b: These are geolocated products in engineering or physical units. Level 1b products are generated by transforming the associated level 0 product via certain algorithms, calibration and auxiliary data.
- Level 2: This is the highest product level supported by the ENVISAT PDS. It represents the final geolocated geophysical product. The level 2 product is generated from the level 1b product by applying algorithms to convert calibrated engineering quantities into geophysical parameters.

Level 0 to level 2 products generated within the PDS are referred to as *operational products*. Products higher than level 2, often referred to as value-added (VA), are processed under the responsibility of ENVISAT data users. Their generation is not a PDS task. While VA products often provide regional or global coverage, level 0 to level 2 products are – with the exception of GOMOS – always orbit oriented. In addition to the operational level 2 products, users may generate scientific level 2 products according to their own requirements. They supplement the operational chain by either applying new retrieval algorithms or concentrating on trace gas species where operational retrieval algorithms still have to be developed.

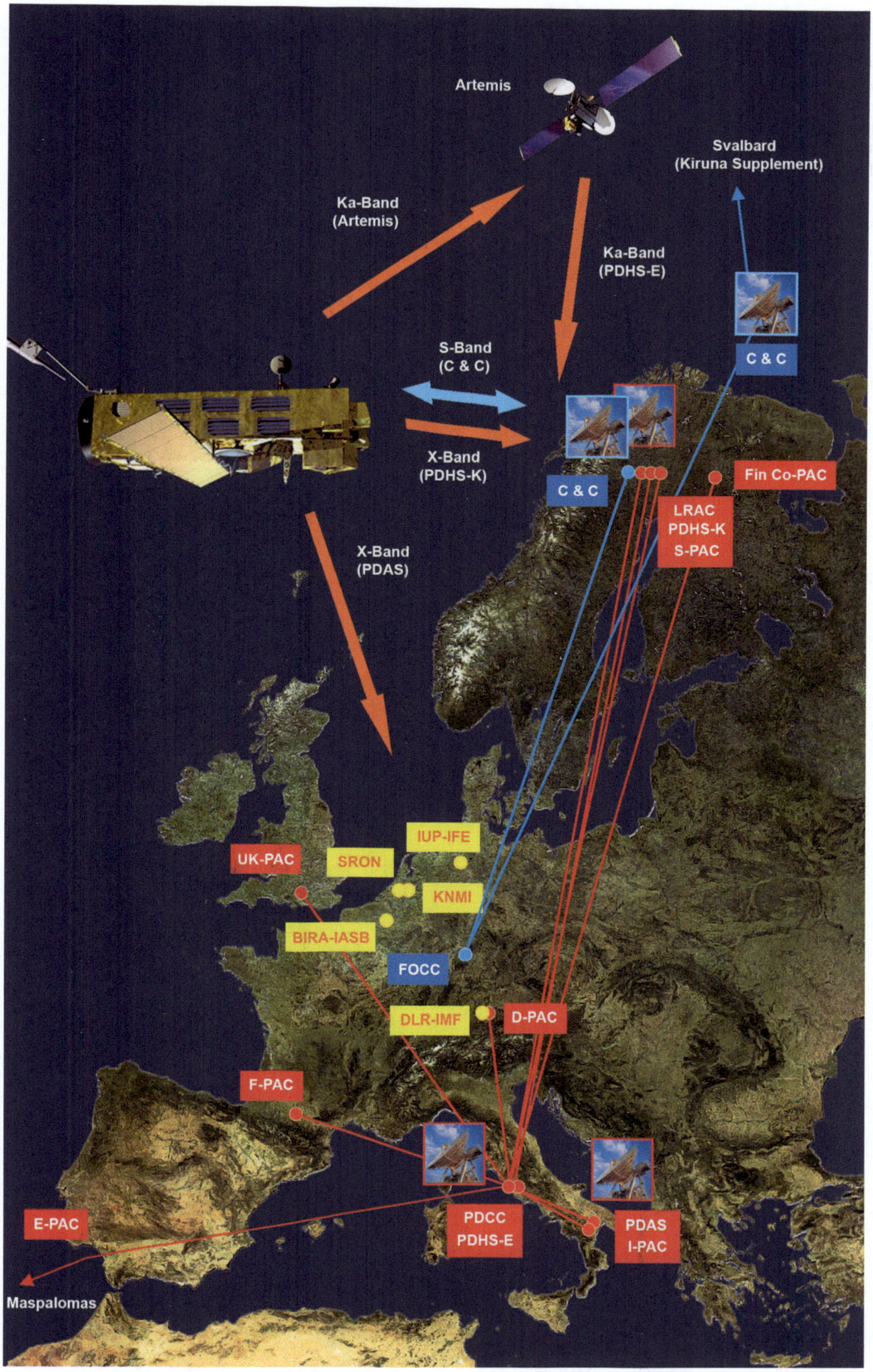

Fig. 2.6 The Europe-wide distributed ENVISAT ground segment (*FOS: blue, PDS: red, national facilities serving the SCIAMACHY mission: yellow*) (Courtesy: DLR-IMF; background map: ESA).

Fig. 2.7 The ESA centres hosting major ENVISAT ground segment facilities with ESOC/ Darmstadt (*top*) and ESRIN/ Frascati (*middle*). The bottom photograph shows ESA's S-band and X-band station at Kiruna-Salmijärvi (Photos: ESA, S. Corvaja).

Table 2.4 ENVISAT ESA ground segment facilities

Ground segment	Location	Facility
Flight operations	Darmstadt (ESOC)	FOCC
	Kiruna-Salmijärvi (ESA)	C&C station
	Svalbard (KSAT)	C&C station
Payload data	Frascati (ESRIN)	PDCC, PDHS-E
	Matera (ESA)	PDAS
	Kiruna-Salmijärvi (ESA)	PDHS-K, LRAC
	Oberpfaffenhofen (DLR)	D-PAC
	Sodankylä (FMI)	Fin Co-PAC
	Kiruna-Salmijärvi (SSC)	S-PAC
	Maspalomas (INTA)	E-PAC
	Toulouse (CNES)	F-PAC
	Matera (ASI)	I-PAC
	Farnborough (NRSC/Infoterra)	UK-PAC

Product levels 1b and 2 can be generated in NRT or OL mode. NRT products are processed and disseminated within a few hours from sensing. Because of stringent time requirements, level 2 NRT data may sometimes not cover the complete data set contained in the level 1b product. When distributed on media, NRT products show a larger time delay from sensing. NRT data are unconsolidated. Unconsolidated products are characterised by using NRT auxiliary information available at time of product generation, e.g. not the most precise orbit state vectors or calibration data. Additionally, NRT products have start/stop times which are defined by data receiving coverage times. They do not cover complete orbits and may overlap.

Consolidated products benefit from a posteriori knowledge of information concerning calibration, auxiliary data and precise orbit. The start/stop times of consolidated products refer to ANX of the corresponding orbit, i.e. a consolidated product is usually generated from two unconsolidated products. Consolidated data show no overlaps. Time gaps are only present when the instrument was not recording data, either planned or unexpectedly due to an anomaly. A consolidated product is the best representation of the scheduled and executed measurements. In the case of SCIAMACHY, a consolidated product begins with the first measurement state (see Chapter 4) starting just after ANX and ends when the last measurement state, which started just before the next ANX, has run to completion.

Product generation in the PDS uses auxiliary data. Auxiliary data may come from instrument calibration and monitoring measurements, the satellite platform itself or sources external to the PDS. For each product, product confidence data is attached providing the results of an evaluation of the overall quality of the product against predetermined thresholds. If a product does not meet the predefined quality level, a flag is set.

For each instrument, algorithm and processor development is controlled by a Quality Working Group (QWG). This QWG also takes care of those aspects of operational instrument performance monitoring which are related to product quality. Each QWG is composed of personnel from scientific and engineering facilities having proven expertise on the particular fields covered by the individual instruments. Additionally ESA operational team members supplement the group. Due to the AOI status, the SCIAMACHY Quality Working Group (SQWG) was established later than the groups for the EDI payload by combining existing activities both on national agencies and ESA side. The SQWG work also benefits largely from the progressing algorithm evolution in the framework of science product generation.

2.5 Data Access

Details of SCIAMACHY product handling in the PDS are listed in Table 2.5. The status shown corresponds to end 2009. As the mission progressed, tasks had been shifted between centres and may still be changed in the future. Changes also apply for product dissemination when technical progress requires modifications for access mechanisms. The PAC serving the atmospheric mission on ENVISAT is the D-PAC, located at the German Remote Sensing Data Center (DFD) at the DLR premises at Oberpfaffenhofen/Germany. D-PAC interfaces with PDHS, LRAC and the FIN-CoPAC to receive products from SCIAMACHY, MIPAS and GOMOS for archiving, further processing and dissemination. Usually NRT products are distributed from the PDHS using the dedicated Data Dissemination System (DDS), which is based on leased EUTELSAT transponder capacity. Besides the PDS internal satellite links, several SCIAMACHY specific DDS terminals had been set up in Germany and The Netherlands. The DDS system is designed to support data transfer of up to dedicated 7 Mbit/s at user level. This is sufficient to receive the NRT level 0 and level 1b products generated each day from SCIAMACHY data. Developed initially only for Commissioning Phase requirements, the DDS has become a robust and stable data transfer mechanism. A specific fast delivery (FD) service has been set up for SCIAMACHY level 2 products at the D-PAC. It provides certain level 2 information within 24 h from sensing.

Guidelines for data access are defined in the ENVISAT data policy. Data utilisation is classified as 'Category 1' (C1) and 'Category 2' (C2). Under C1, research and applications in support of the ENVISAT mission objectives including long-term studies of the Earth system, are summarised. C2 users are all others not falling under C1, i.e. operational and commercial entities. Due to the scientific character of the atmospheric sensors on ENVISAT access to SCIAMACHY, MIPAS and GOMOS data is always as for C1. Identi-

Table 2.5 SCIAMACHY products in the PDS

Level[a]	Dissemination timescale	Source	Task[b]	Medium[c]
cL0[d]	OL	LRAC	G, A	
cL1b	OL	D-PAC	G, A, D	DVD, ftp
cL2	OL	D-PAC	G, A, D	ftp
L0[d]	NRT	PDHS	G, A, D	DDS
L1b	NRT	PDHS	G, D	DDS, ftp
L2	FD[e]	D-PAC	G, D	ftp
L2	NRT	D-PAC	A	

[a] *cL* consolidated level
[b] *G* generation, *A* archiving, *D* dissemination
[c] Only the standard media are listed
[d] Only access for selected users
[e] Fast delivery service (24 h from sensing)

fication of C1 users was originally based on peer review processes, e.g. ESA announcements of opportunity for data exploitation or research and application development programmes of participating member states or European organisations. Because now most of the SCIAMACHY products are systematically available either on Internet or via DDS, and both in near-realtime or offline, a simple registration through ESA's EO Principle Investigator Portal (http://eopi.esa.int/) is sufficient to register and get access to SCIAMACHY data as C1 user. The SCIAMACHY data are cost-free when available on Internet or via DDS, and provided at cost of reproduction when needed on media.

The status of SCIAMACHY as an AOI grants additional data rights. As stipulated in the Flight Operation and Data Plan (FODP), which identifies responsibilities both for ESA as ENVISAT mission provider and DLR/NIVR as SCIAMACHY instrument provider, SCIAMACHY data are delivered at no charge to the instrument providing national agencies.

General help on ENVISAT can be obtained via ESA's EO user service (http://earth.esa.int/services/). It provides information on EO mission documentation, EO data catalogues, software tools for reading and processing EO data and EO data product definitions and content. This website also links to the EO help desk at ESRIN which acts as the central facility for user services in the PDS. EO help is contacted via EOhelp@esa.int. A user may also wish to learn about the actual performance of ENVISAT and its instruments. This request is supported by ESA's dedicated ENVISAT website (http://envisat.esa.int/). Specific websites with detailed information on SCIAMACHY operations and performance as well as science product status do exist on AO provider (AOP) side (see Chapters 4 and 8).

References

Bargellini P, Garcia Matatoros MA, Ventimiglia L, Suen D (2006) ENVISAT attitude and orbit control in-orbit performance: An operational view. Proceedings of 6th international conference on guidance, navigation and control systems, Loutraki, Greece, ESA SP-606

ESA (1998) ENVISAT-1 Mission & System Summary. ESA Brochure, ESA

ESA (2000) Envisat – MIPAS, an instrument for atmospheric chemistry and climate research. ESA Special Publication, ESA SP-1229

ESA (2001a) ENVISAT Special Issue. ESA Bulletin, 106

ESA (2001b) Envisat – GOMOS, An instrument for global atmospheric ozone monitoring. ESA Special Publication, ESA SP-1244

Milligan D, Ventimiglia L, Eren Y, Mesples D, Diekmann FJ, Kuijper D, Garcia-Matatoros M (2007) Envisat platform operations: Providing a platform for science. Proceedings of ENVISAT Symposium 2007, Montreux, Switzerland, ESA SP-636

Chapter 3
The Instrument

M. Gottwald, R. Hoogeveen, C. Chlebek, H. Bovensmann, J. Carpay, G. Lichtenberg, E. Krieg, P. Lützow-Wentzky, and T. Watts

Abstract The objective of the SCIAMACHY mission requires the instrument to be capable of determining concentrations of a large number of trace gas species over the full vertical extent of the atmosphere from the troposphere up to the mesosphere. In addition, aerosol properties as well as pressure and temperature shall be derived. Therefore SCIAMACHY was designed as a passive imaging spectrometer, comprising a scan mirror system, a telescope and a spectrometer, controlled by thermal and electronic subsystems. Scan mirrors, telescope and spectrometer together form the optical assembly. The scan mechanisms permit steering the line-of-sight according to the required viewing geometries. Solar radiance and irradiance are dispersed by the spectrometer into eight channels from the UV to the SWIR range. With signals gained from the calibration unit and the Polarisation Measurement Device, the spectral and radiometric calibration of the science channels can be maintained over the mission lifetime. Thermal stability is ensured via active and passive thermal control systems including the Radiant Cooler assembly. Those units which control the entire instrument and interface electrically with the ENVISAT platform are hosted by the Electronic Assembly. SCIAMACHY was developed in a combined effort of German, Dutch and Belgian space agencies, industry and scientists and ready for the ENVISAT launch in March 2002.

Keywords Imaging absorption spectrometer • Optical Assembly • Detector channels • Scanner • Radiant Cooler • Electronic Assembly

The scientific requirements define the overall concept of the instrument. They comprise detecting all species listed in Chapter 1. This can be achieved by continuously observing the wavelength ranges 214–1773, 1934–2044 and 2259–2386 nm. The retrieval of the amount of constituents depends on the ability to measure their absorptions precisely. Retrieving total column concentrations of minor trace gases with an accuracy of 1–5% – or 5–10% for their profiles – requires measuring intensity changes of 10^{-3}–10^{-4}. An instrument capable of acquiring data with a high signal-to-noise ratio and a good radiometric calibration can accomplish this task (SSAG 1998).

To fulfil the mission objectives with respect to spatial resolution and coverage, it is necessary to observe the scattered and reflected solar photons in nadir and limb direction as well as the light transmitted through the atmosphere in solar and lunar occultation geometry (Burrows and Chance 1991; Bovensmann et al. 1999). For calibration and monitoring purposes the solar and lunar irradiance above the atmosphere has to be determined. As total column amounts and height resolved profiles are required, the spectrometer must alternately observe the atmosphere in limb and nadir viewing. Combining both geometries of a single orbit for the same volume of air allows the study of tropospheric properties. Global coverage has to be obtained within 3 days in limb or nadir mode.

M. Gottwald (✉), G. Lichtenberg, and E. Krieg
Remote Sensing Technology Institute, German Aerospace Center (DLR-IMF), Oberpfaffenhofen, 82234 Wessling, Germany
e-mail: manfred.gottwald@dlr.de

R. Hoogeveen
SRON, Netherlands Institute for Space Research, Sorbonnelaan 2, 3584 CA Utrecht, The Netherlands

C. Chlebek
German Aerospace Centre, Space Agency, Königswinterer Straße 522-524, 53227 Bonn, Germany

H. Bovensmann
Institute of Environmental Physics/Institute of Remote Sensing (IUP-IFE), University of Bremen, Otto-Hahn-Allee 1, 28359 Bremen, Germany

J. Carpay
Netherlands Space Office (NSO), Juliana van Stolberglaan 3, 2595 CA The Hague, The Netherlands

P. Lützow-Wentzky
EADS Astrium GmbH, Claude Dornier Straße, 88090 Immenstaad, Germany

T. Watts
Dutch Space B.V., Newtonweg 1, 2333 CP Leiden, The Netherlands

M. Gottwald and H. Bovensmann (eds.), *SCIAMACHY - Exploring the Changing Earth's Atmosphere*, DOI 10.1007/978-90-481-9896-2_3,

3.1 Instrument Concept

The requirements outlined above, together with the accommodation on the ENVISAT platform, were translated into an instrument concept for SCIAMACHY providing spectroscopic capabilities from the UV via VIS and NIR to SWIR with

- Moderately high spectral resolution
- High radiometric accuracy
- High spectral stability
- High dynamic range
- High signal-to-noise

Possible viewing geometries must include steering the line-of-sight (LoS) towards nadir, limb (in flight direction), sunrise and moonrise direction. Additional observability of the Sun around occurrence of the sub-solar point is needed. Maintaining thermal stability at all operating temperature levels is fundamental for achieving high radiometric and spectral accuracy. This is ensured by sophisticated thermal control systems. Finally, instrument control must be executed continuously in a highly autonomous manner with the ability to react to a wide variety of operations conditions. This includes not only measurement data relevant parameters as, e.g. line-of-sight, signal-to-noise levels and spectral sampling but also the tasks of overall instrument command and control. All SCIAMACHY instrument requirements were documented in the 'SCIAMACHY Instrument Requirements Document' (SIRD, DARA 1998).

Conceptually, SCIAMACHY is a passive imaging spectrometer, comprising a scan mirror system, a telescope and a spectrometer, controlled by thermal and electronic subsystems (Fig. 3.1 and Table 3.1). Functionally, three main blocks, the Optical Assembly (OA), the Radiant Cooler Assembly (SRC) and the Electronic Assembly (EA) can be identified (Fig. 3.2). The instrument is located on the upper right (i.e. starboard, referring to nominal flight direction) corner of the ENVISAT platform with the OA mounted onto the front and the EA mounted onto the top panel. The Radiant Reflectance Unit (RRU) of the SRC points sideways into deep space away from any heat source. Interfaces with the ENVISAT platform exist for the provision of on-board resources. These include power and command interfaces from the platform to the instrument. In the other direction measurement data and housekeeping (HK) telemetry from SCIAMACHY are routed into the overall ENVISAT data stream for downlinking.

Whilst a complex instrument such as SCIAMACHY cannot provide for full redundancy – particularly not for the optical path and the detectors – some essential hardware components do exist twice (Table 3.2). Nominally the instrument is operated using its primary chain of hardware, called *side A*. In case of a malfunction in side A, SCIAMACHY can be switched to *side B* which has identical

Table 3.1 SCIAMACHY instrument physical characteristics

Dimensions	
Optical Assembly	109 cm × 65 cm × 101 cm
Electronic Assembly	82 cm × 90 cm × 28 cm
Radiant Cooler Assembly	51 cm × 91 cm × 62 cm
Total mass	215 kg
Power consumption	140 W

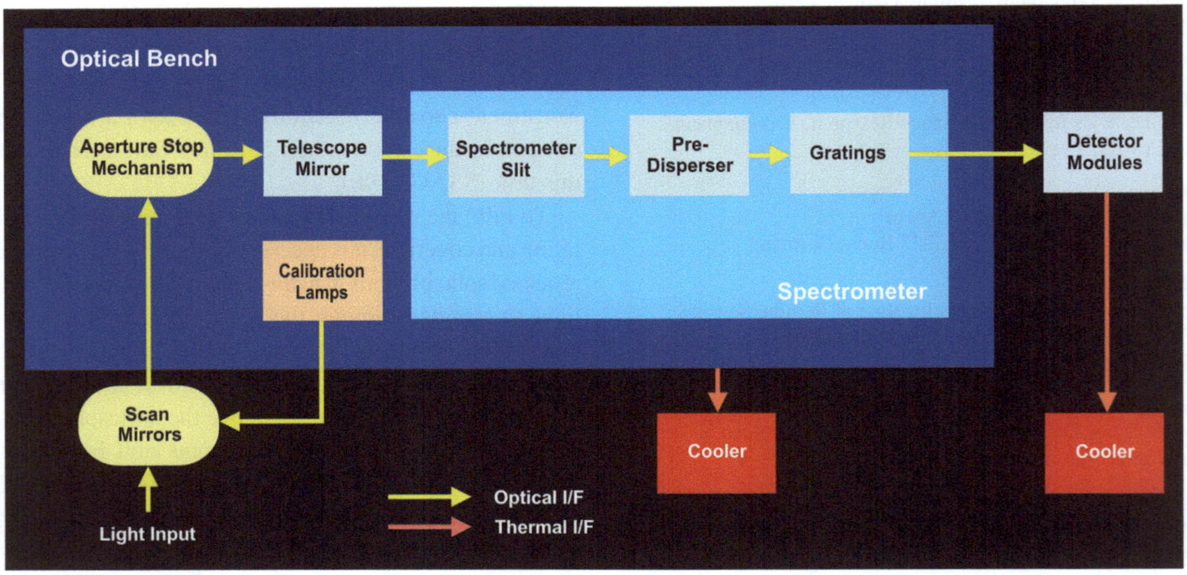

Fig. 3.1 General spectrometer layout illustrating SCIAMACHY's main components (Courtesy: DLR-IMF).

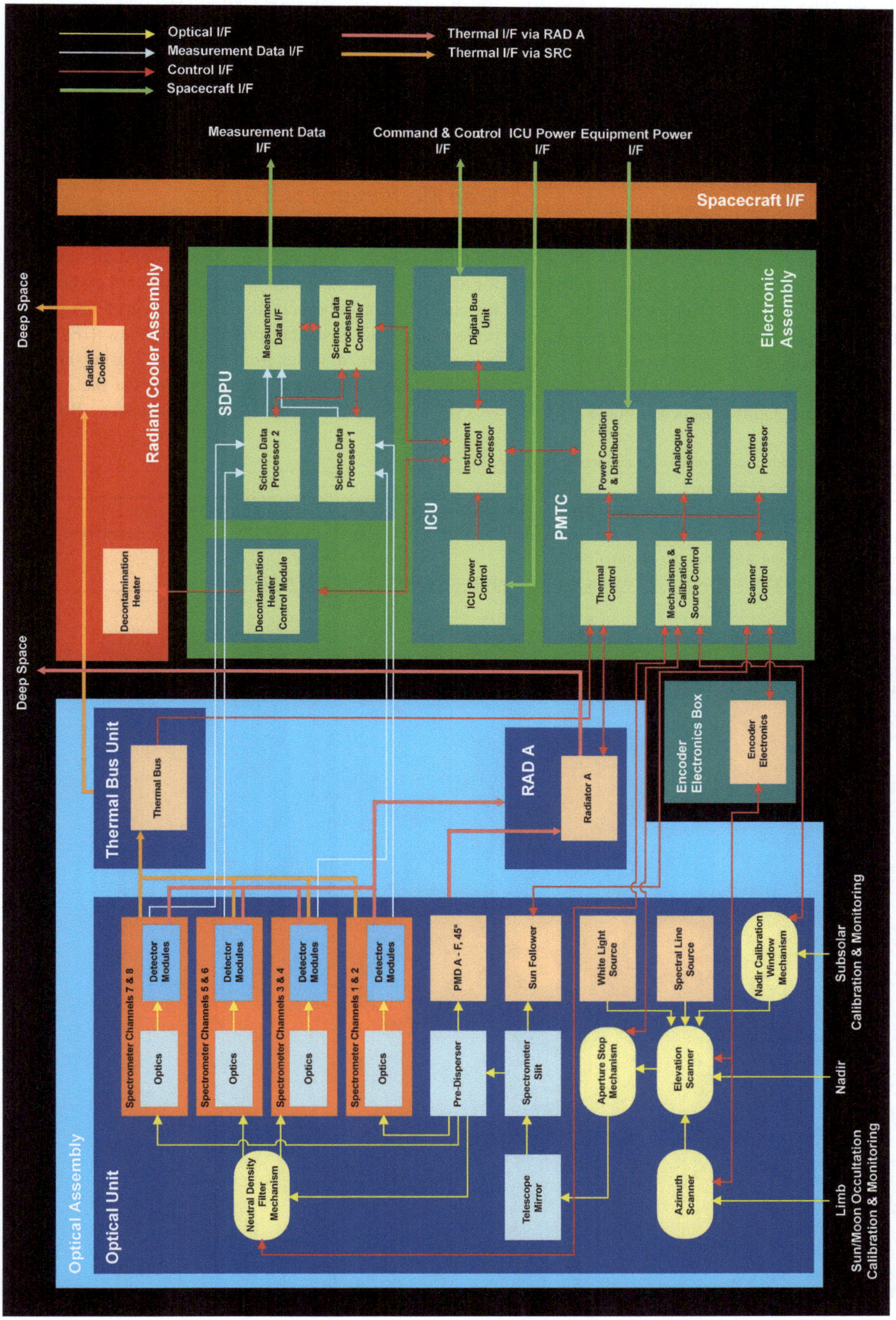

Fig. 3.2 SCIAMACHY functional block diagram with simplified interfaces between assemblies and modules (Courtesy: DLR-IMF; adapted from EADS Astrium 2004).

functions. However, since several modules are shared by both chains, a loss of one of these non-redundant components would result in degraded instrument performance.

3.2 Optical Assembly

The Optical Assembly is the part of the instrument which collects solar radiation as input and generates the spectral information as output. This occurs in the Optical Unit (OU). To maintain the specified thermal conditions, the OA includes Radiator A and the Thermal Bus Unit. The Optical Unit is organised into two levels. Entrance optics, pre-disperser prism, calibration unit and channels 1 and 2 can be found in level 1 facing in flight direction (Fig. 3.3). Channels 3–8 are located in level 2 (Fig. 3.4). All components are mounted onto the Optical Bench Module (OBM) which serves as the structural platform and maintains overall alignment between modules. By combining the optical-components described below, various optical paths ('trains') from external and internal light sources to detectors can be established (Fig. 3.5).

Scan Mechanisms and Baffles

Scanning is required in order to steer the line-of-sight both for executing particular observation geometries and for collecting light not only from the limited size of the ground projection of the Instantaneous Field of View (IFoV, see below) but from a wider ground scene. During nominal measurements light enters the instrument via the azimuth (ASM) or elevation (ESM) scan mechanisms. Both are

Table 3.2 SCIAMACHY redundant components for nominal operations

Component	Redundant hardware
Interfaces	Equipment power (incl. converter), ICU power, DBU (command and control), measurement data
Electronic Assembly	ICU, PMTC mechanisms and calibration source control, PMTC control processor, PMTC ESM scanner control, SDPU
Encoder Electronics Box	ESM and ASM encoder electronics
Optical Assembly	ESM/ASM encoders, ESM motor, redundant equipment can be powered by redundant chain

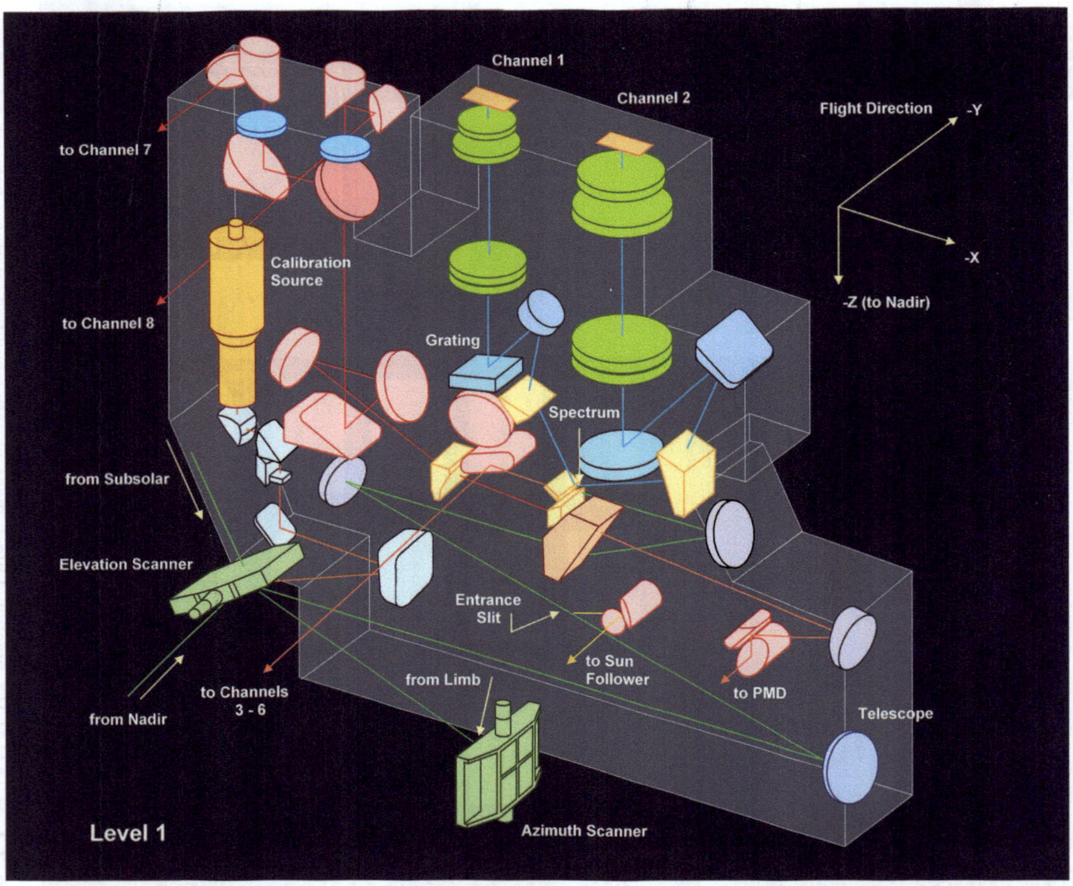

Fig. 3.3 Optical configuration level 1 (Courtesy: DLR-IMF; adapted from SJT 1996).

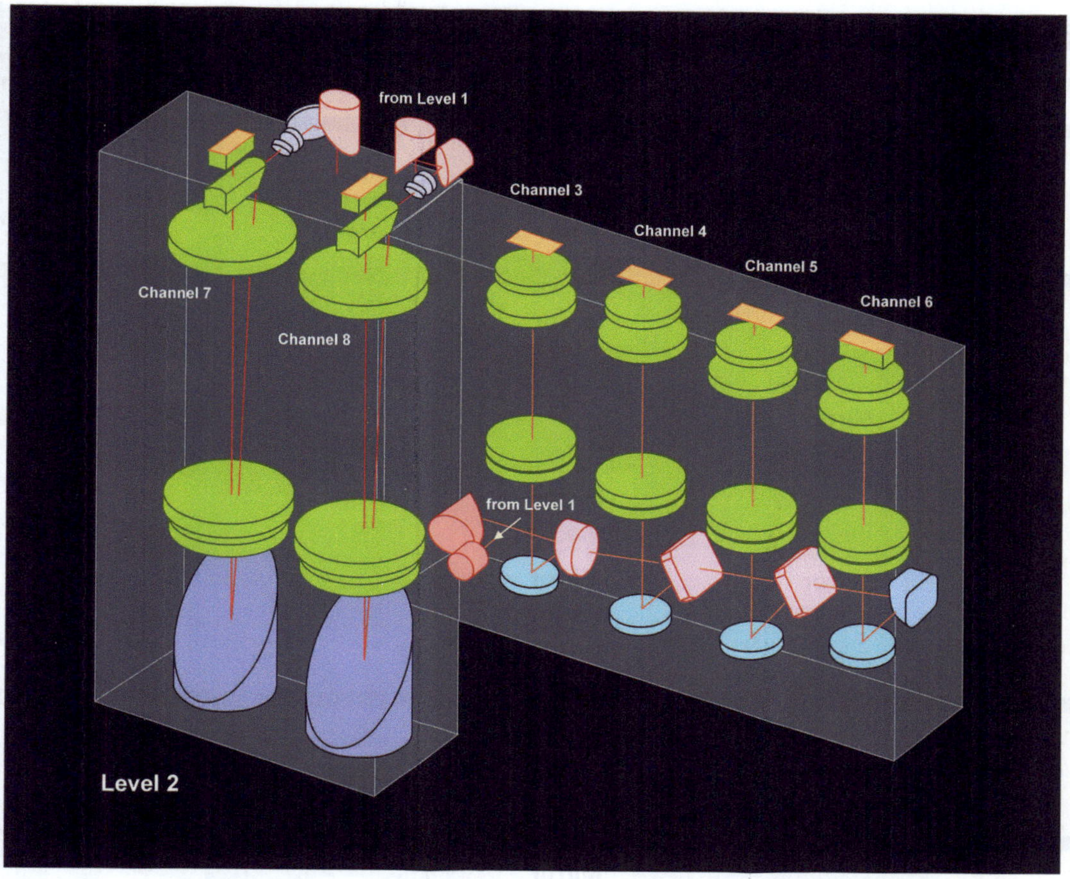

Fig. 3.4 Optical configuration level 2 (Courtesy: DLR-IMF; adapted from SJT 1996).

located below the lower part of level 1 of the OU. Mechanically, each scanner comprises a mirror block, bearings, a drive motor and encoders (Fig. 3.6). Bearings use a special lubrication allowing quasi-continuous in-orbit operation without life-time limitations. The scanning mirrors have uncoated, polished aluminium surfaces with a size of 90 mm × 60 mm for the ESM and 125 mm × 110 mm for the ASM. Whilst the ASM captures radiation coming from regions ahead of the spacecraft, the ESM either views the ASM or the region directly underneath the spacecraft. In limb observations light from the ASM mirror is directed via the ESM mirror into the spectrometer. In nadir observations, only the ESM mirror is used. Both scanners shall ideally be mounted such that their axes are parallel to the platform coordinate system. Because placing the scanners actually resulted in small alignment errors, their pitch, roll and yaw mispointings had been measured on-ground during instrument integration. The corresponding values are stored on-board and are used in scanner control to compensate these misalignments. The LoS pointing knowledge is of particular importance for limb observations where even a small pitch mispointing of, e.g. only 10 mdeg would translate into an altitude error of 570 m thus degrading the scientific value of the information in limb profiles.

Although both scan devices can be rotated by 360°, baffles limit the effective field of view. This results in the Total Clear Field of View (TCFoV) which depends on the observation mode as listed in Table 3.3 and sketched in Fig. 3.7. For the limb and occultation LoS, the baffles provide a symmetric range on either side of the flight direction while vertically they restrict viewing from slightly below the horizon to an altitude of about 380 km, i.e. well above the top of the atmosphere at 100 km. The nadir LoS is limited to an area of about ±32° across track. For a special type of measurement, the rectangular shaped Nadir Calibration Window (NCW) can be opened temporarily allowing sunlight from above to enter the instrument via the ESM mirror. Its elongated TCFoV of 1.7° × 14.8° is designed to view the Sun at high elevation when the spacecraft crosses the orbital subsolar point. In almost all measurements the scanners execute oscillating movements (forward/backward scans) or follow a certain trajectory which is a segment of a circle. Each type of

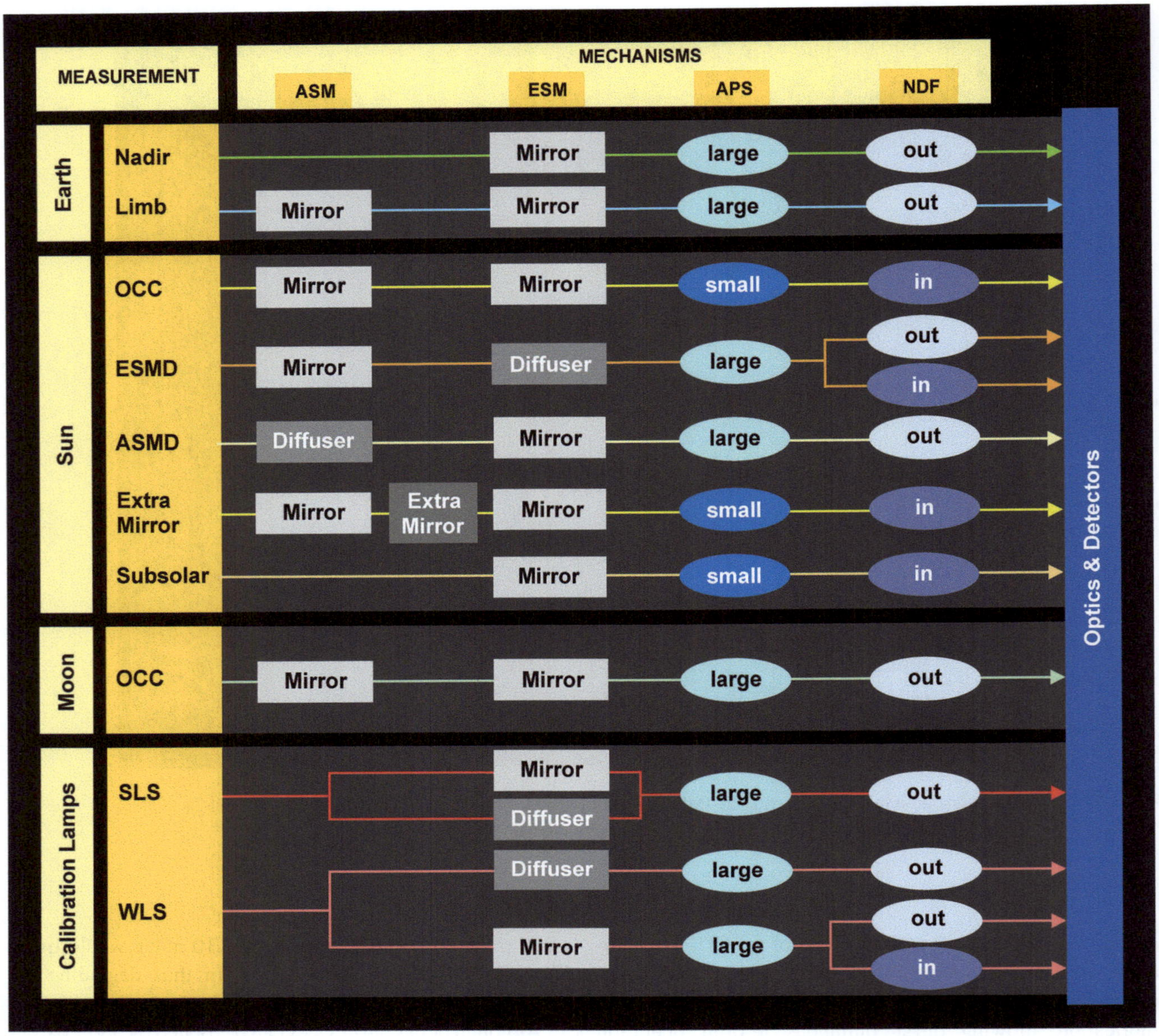

Fig. 3.5 SCIAMACHY optical 'trains'. Each 'train' defines a path for the measured light through the instrument to the detectors. Light sources can be external or internal (Courtesy: DLR-IMF/SRON).

measurement has its specified scanner start positions and scan ranges. These are defined by the orientation of the normal of the ASM and ESM scan mirror (Figs. 3.8 and 3.9).

Executing various LoS trajectories and pointing to the Sun or Moon requires sophisticated scanner control functions, particularly when the movement of both mechanisms has to be synchronised. The scanner control tasks are programmed in on-board software with supporting information being generated by the Sun Follower (SF) in the case of solar and lunar observations. Nominally, each scanner is operated separately in feedback control using measurements of the rotation angle by an incremental optical encoder. Angular scan trajectories are assembled from preprogrammed basic and relative scan profiles for offset and motion generation. Since precise LoS steering to the Earth's limb or celestial targets depends on various scanner internal or external parameters, the selected trajectory can be corrected correspondingly. In limb measurements the horizontal scans through the atmosphere maintain a constant altitude by applying a correction which takes into account the varying curvature of the Earth (WGS84 model) along the orbit. Further corrections provide for the yaw steering attitude mode of the ENVISAT platform and the known misalignment of the instrument reference frame relative to the spacecraft frame. Sun and Moon observations

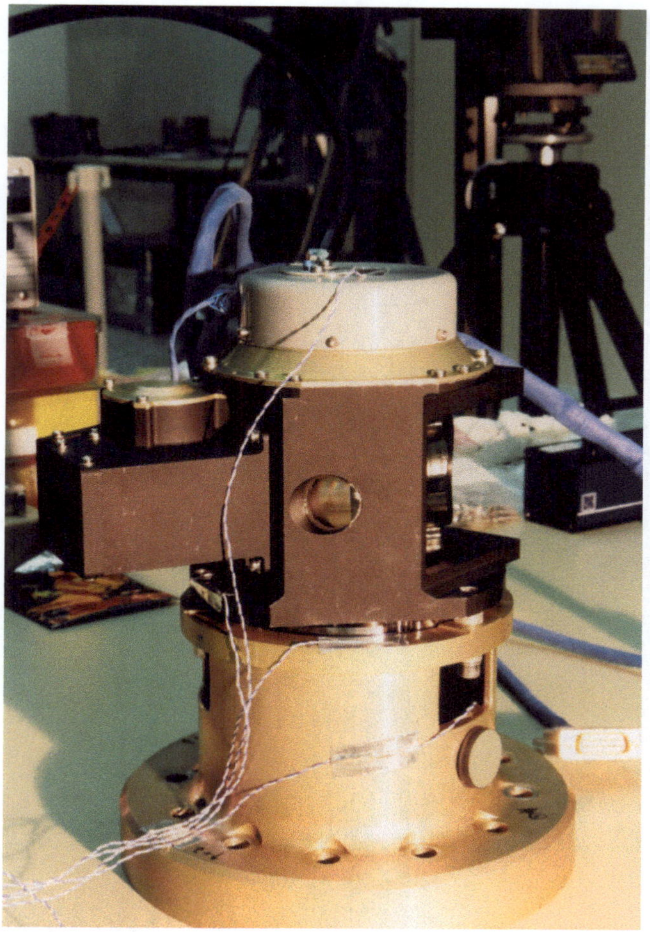

Fig. 3.6 ESM on vibration test adaptor. The leftmost rectangular opening is the Nadir Calibration Window (Photo: EADS Astrium).

Table 3.3 SCIAMACHY Total Clear Field of View

Observing geometry	Total Clear Field of View (TCFoV)
Nadir	32°/31° (ESM, across-track left/right)
Limb, occultation	88° (ASM, azimuth 316°–44°) × 8° (ESM, elevation 19.5°–27.5° from X-Y plane downwards)
Sub-solar	1.7° (azimuth) × 14.8° (ESM, elevation 53.7°–68.5° from X–Y plane upwards)

require the LoS to be centred onto the target. Analytical control algorithms cannot always ensure this. Therefore, information derived from the readout of the four quadrants of the SF is fed into the control loop to steer the scanner motors such that the mirrors – either the ASM or the ESM or both – lock onto the central part of the intensity distribution and follow the trajectory of Sun or Moon after successful acquisition. The SF receives light which is reflected from the polished blades of the spectrometer entrance slit. It is able to detect the Sun or Moon in a 2.2° × 2.2° wide field.

For obtaining the solar irradiance, the Sun has to be measured via a diffuser. Two aluminium diffusers are mounted on SCIAMACHY: one on the backside of the ESM mirror, the other on the backside of the ASM mirror. Originally the ESM diffuser was the only one in the instrument. During calibration it had turned out that this type of diffuser exhibits spectral features which would have jeopardised successful retrieval of some trace gas species. Thus very late in the development of phase C/D an ASM fitted with an additional diffuser was integrated. Its surface was ground in a different way to yield optimised diffuser properties.

Telescope and Spectrometer

The ESM reflects light towards the telescope mirror, which has a diameter of 32 mm. From the telescope mirror the light path continues to the spectrometer entrance slit. With linear dimensions of 7.7 mm × 0.19 mm (cross-dispersion × dispersion) the entrance slit defines an Instantaneous Field of View (IFoV) of 1.8° × 0.045°. This corresponds to a ground pixel size of 25 km × 0.6 km at the sub-satellite point (nadir) and of 103 km × 2.6 km at the Earth's horizon (limb). For solar observations, the IFoV can be reduced to 0.72° × 0.045° by the Aperture Stop Mechanism (APSM), which is located between the ESM and telescope mirror. The APSM reduces the aperture area and thus the intensity level. Channel dependent effects lead to a reduction by a factor of about 11000 for channels 1 and 2, 5000 for channels 3–5 and 2500 for channels 6–8.

The overall spectrometer design is based on a two stage dispersion concept: The pre-disperser prism, located behind the entrance slit, serves two purposes. It weakly disperses the light and directs fully polarised light for further processing to the Polarisation Measurement Device (PMD). Small pick-off prisms and subsequent di-chroic mirrors direct the intermediate spectrum to the 8 science channels where the light is further dispersed by individual gratings. The selected approach has the advantage of reducing stray light in the channels with low light levels in the UV and NIR-SWIR part of the spectrum. It also effectively prevents the various spectral orders from one grating overlapping with the other parts of the spectrum (Goede et al. 1991). In the light path routed to channels 3–6 the Neutral Density Filter Mechanism (NDFM) can move a filter into the beam. With a filter transmission of 25% it can be used, in conjunction with the APSM, to even further reduce light levels during solar measurements.

Detector Modules

The full resolution spectral information is generated in 8 science channels (Fig. 3.10 and Table 3.4). These employ two types of detectors (Fig. 3.11). For the UV-VIS-NIR range covered by channels 1–5, standard Silicon photodiodes (RL

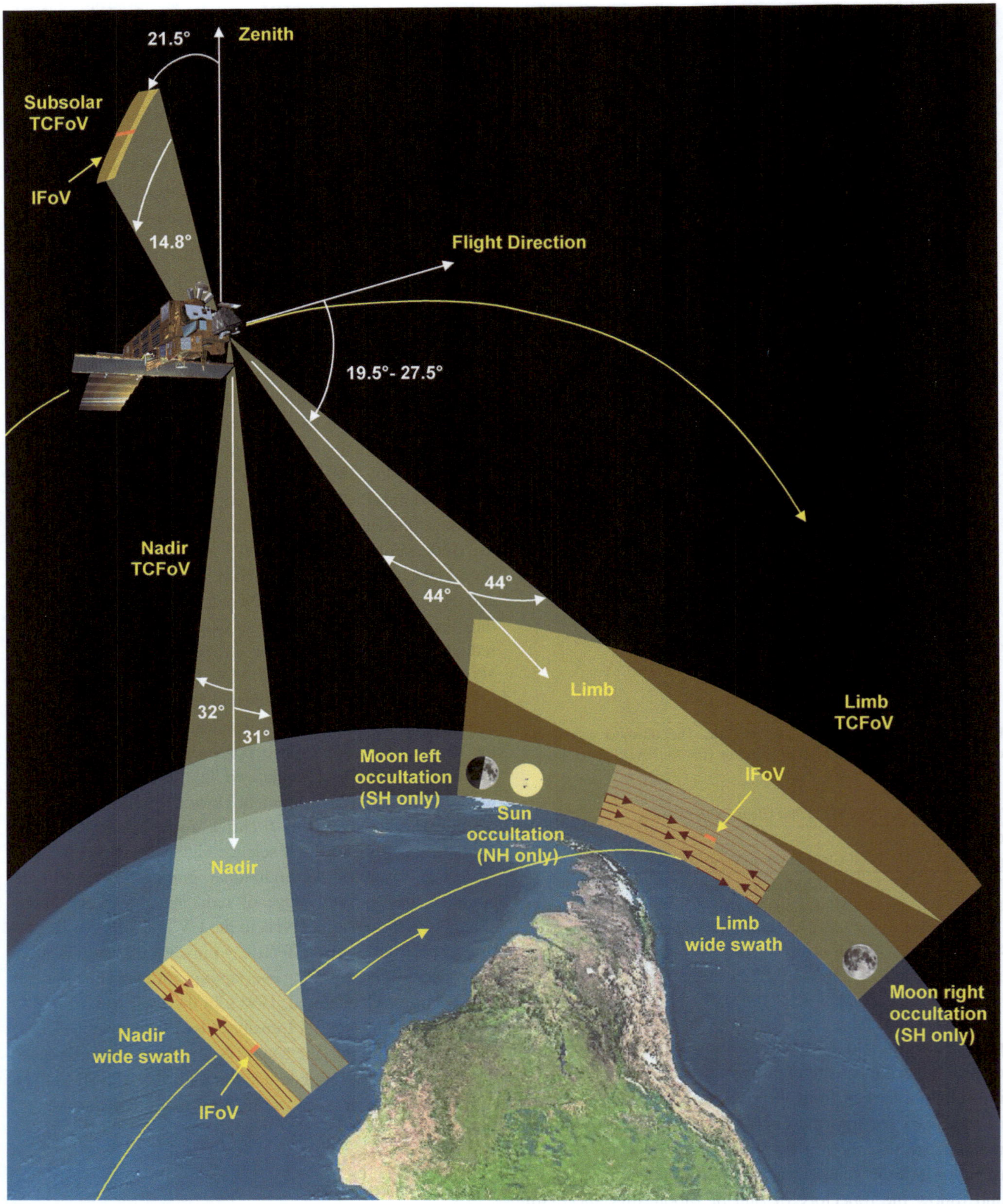

Fig. 3.7 Sketch of SCIAMACHY's TCFoV and observation geometries (Courtesy: DLR-IMF).

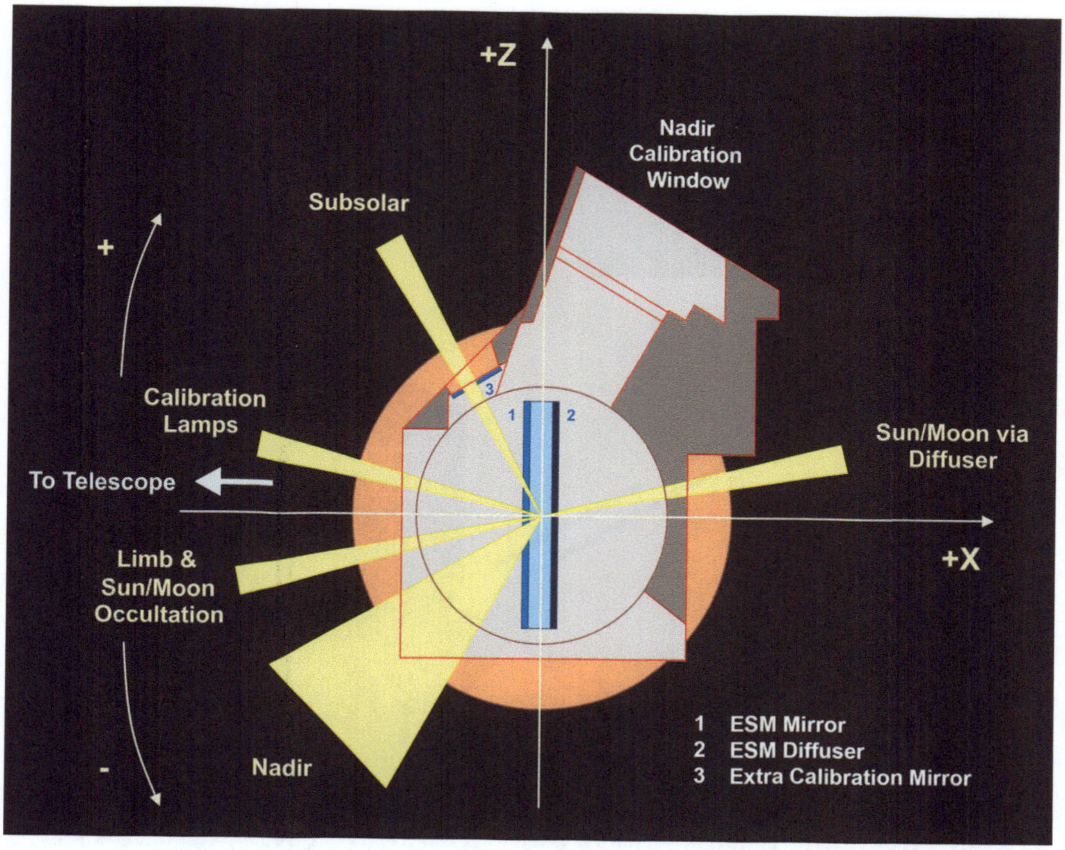

Fig. 3.8 ESM mirror's normal during specified observing geometries (Courtesy: DLR-IMF).

1024 SR, EG&G RETICON, California) with 1024 pixels are used which are sequentially read out. Additionally, UV channels 1 and 2 are electronically divided into two virtual bands 1a/1b and 2a/2b, which can be configured separately. The SWIR channels 6–8 use Indium Gallium Arsenide detectors (InGaAs by EPITAXX, New Jersey) specifically developed and qualified for SCIAMACHY (Hoogeveen et al. 2001). In the SWIR channels all pixels are read out in parallel. In order to be sensitive to wavelengths beyond 1700 nm, the detector material in the upper part of channel 6 above pixel number 794 (named 'channel 6+') and channels 7 and 8 were grown with higher amounts of Indium. All channels have to be cooled to achieve the specified signal-to-noise performance. The operational temperature range is channel dependent and lowest for the SWIR channels 7 and 8.

The pixel readout sequence in channel 2 is reversed in wavelength to avoid spatial aliasing in the overlap region of channels 1 and 2 and channels 2 and 3. Spatial aliasing occurs due to the fact that for channels 1–5 the detector pixels are read out sequentially with a time delay between the first and the last pixel of about 28.75 ms. Therefore pixels that are read out at a different time see a somewhat different ground scene because during the sequential readout the platform and the scan mirrors continue to move. The size of the wavelength dependent spectral bias depends on the variability of the ground scene. Reversing the readout of channel 2 ensures that the channel overlaps observe the same ground scene.

In the UV-VIS channels the minimum Pixel Exposure Time (PET) amounts to 31.25 ms. In order to avoid saturation in the SWIR channels, exposures in channels 6–8 can be set as short as 28 μs. Since the movement of the scanners and the readout of the detectors are synchronised by pulses of 62.5 ms, the maximum data transfer rate is 16 Hz. Therefore, for readouts with exposure times shorter than 62.5 ms the readout is taken only once for the specified PET duration within the 62.5 ms interval and transmitted to ground. However, such cases occur only for calibration and monitoring measurements and not when executing nadir or limb observations.

Trace gas features are distributed non-uniformly over the spectrum. The limited total data rate would therefore

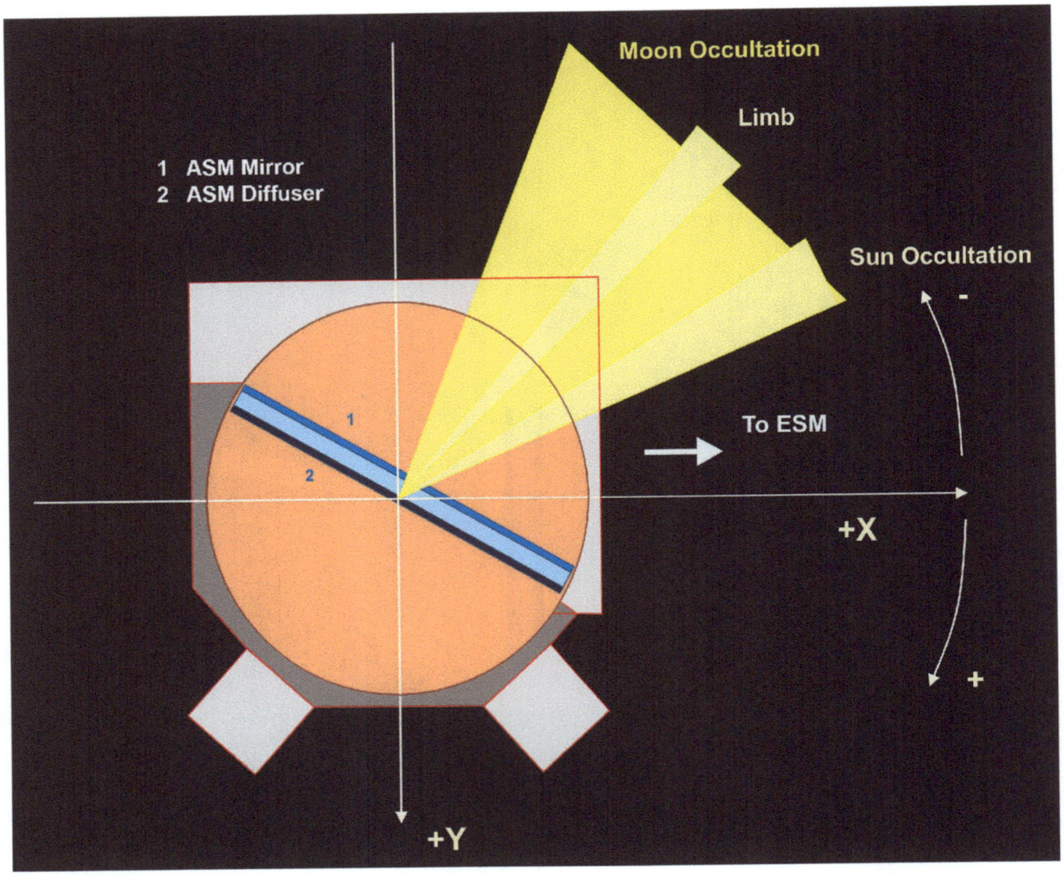

Fig. 3.9 Same as Fig. 3.8 but for the ASM (Courtesy: DLR-IMF).

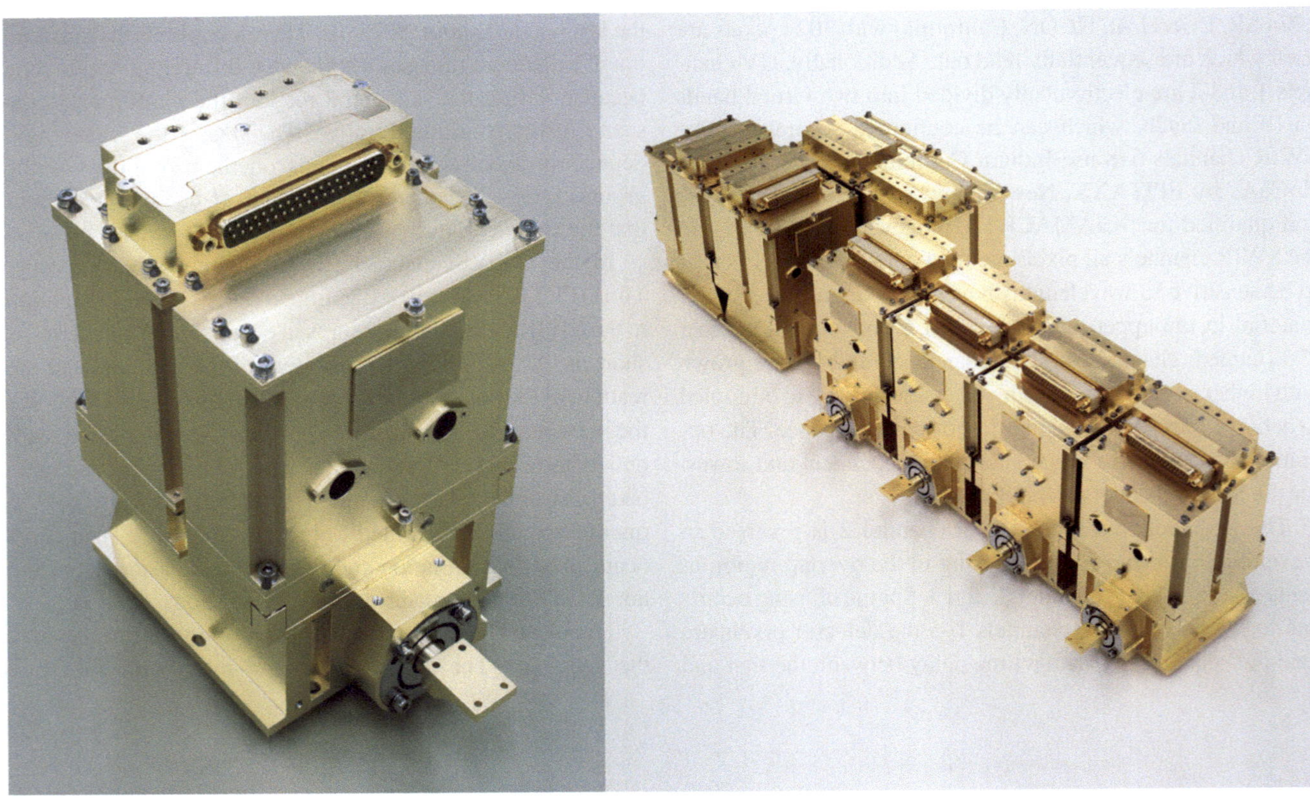

Fig. 3.10 Single SCIAMACHY detector module (*left*) and the full complement of 8 detector modules (*right*) (Photo: SRON).

Table 3.4 SCIAMACHY science channels (1 and 2 = UV, 3 and 4 = VIS, 5 = NIR, 6–8 = SWIR)

Channel	Spectral range (nm)	Resolution (nm)	Stability (nm/100 min)	Temperature range (K)
1	214–334	0.24	0.003	204.5–210.5
2	300–412	0.26	0.003	204.0–210.0
3	383–628	0.44	0.004	221.8–227.8
4	595–812	0.48	0.005	222.9–224.3
5	773–1063	0.54	0.005	221.4–222.4
6	971–1773	1.48	0.015	197.0–203.8
7	1934–2044	0.22	0.003	145.9–155.9
8	2259–2386	0.26	0.003	143.5–150.0

Fig. 3.11 The RETICON (*top*) and EPITAXX (*bottom*) linear detector arrays (Photo: SRON).

prohibit the detailed sampling of those ranges of interest if the full spectrum had to be downlinked as one block. SCIAMACHY avoids this situation by using spectral clusters and co-adding. The 1024 pixels per channel can be subdivided into a number of *clusters* identifying regions where trace gas retrieval will take place. Each cluster can be sampled by on-board data processing applying *co-adding* factors f_{coadd} to the readout of the pixels of this cluster. This results in an integration time (IT)

$$IT = PET \times f_{coadd}$$

which defines how many subsequent readouts of each pixel of a cluster are added to generate one measurement data readout. By appropriately setting the integration time, high or low temporal resolution – equivalent to high or low spatial resolution – can be selected. Thus, depending on the executed measurement states, variable ground pixel sizes as a function of spectral region, i.e. trace gas features, are achieved. Efficient setting of co-adding factors is also required to ensure that the volume of the generated data does not violate the assigned nominal data rate limit of 400 kbit/s. The measurement data stream finally consists of cluster sequences representing different wavelength regions read out at different rates.

Overall the detector performance is characterised by low noise and high instrument throughput. This allows measuring the incoming light with the required very high signal-to-noise ratio (Fig. 3.12), a prerequisite for the retrieval of the targeted geophysical parameters.

Calibration Unit

The requirement to maintain high spectral stability and high relative radiometric accuracy over the mission's lifetime is ensured via an on-board calibration unit. It consists of two calibration lamps, one for white light and one for spectral lines. The White Light Source (WLS) is a 5 W UV-optimised Tungsten–Halogen lamp with an equivalent blackbody temperature of about 3000 K. Its signal is used to verify the pixel-to-pixel signal stability and to monitor the etalon effect (see section "Operational Level 0-1b Processing" of Chapter 8). The Spectral Line Source (SLS) is a Neon filled hollow Pt/Cr cathode lamp. Its operation allows the determination of the pixel-to-wavelength relation. The calibration unit is located close to the ESM. By rotating the ESM mirror into specific positions it is possible to reflect light from the WLS respectively the SLS towards the telescope mirror and thus onto the entrance slit.

An extra calibration mirror near the ESM can be used for an additional reflection of the incoming light onto the ESM mirror. Due to its protected position well within the instrument it is assumed that this extra mirror will not degrade throughout the mission. Thus it can be used as a further means for monitoring optical performance (see section "Optical Performance Monitoring" of Chapter 5).

Polarisation Measurement Device

The sensitivity of the spectrometer depends on the polarisation state of the incoming light. Therefore SCIAMACHY is equipped with a Polarisation Measurement Device. Six of its

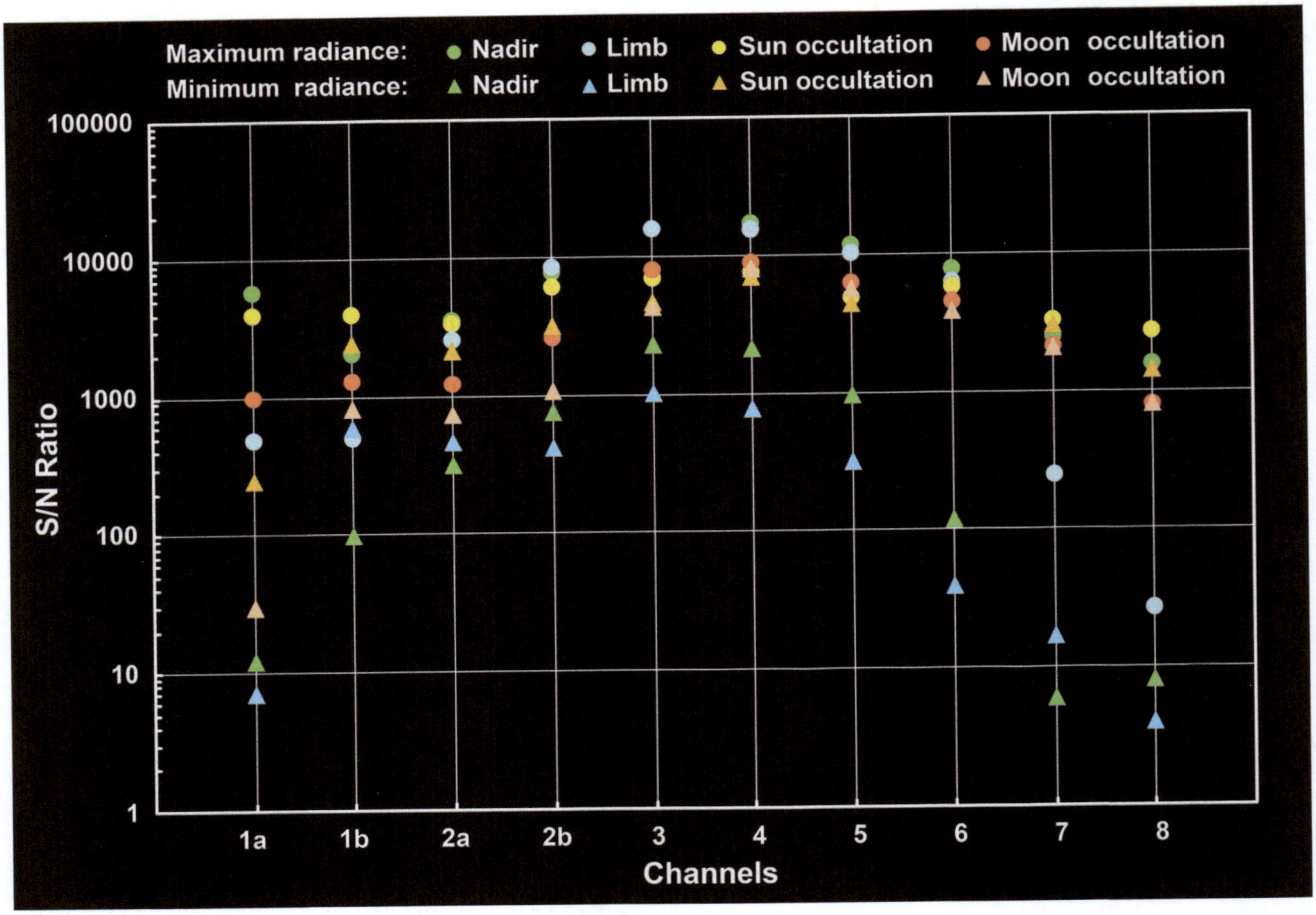

Fig. 3.12 The signal-to-noise ratios for all channels as obtained during OPTEC-5 tests. The ratios were determined for minimum and maximum radiance signals (Courtesy: DLR-IMF).

Table 3.5 SCIAMACHY PMD. The wavelength ranges are defined to contain 80% of the total signal in the respective PMD channel

Channel	Spectral range (nm)	Detector Temperature (°C)
A	310–365	−18
B	455–515	−18
C	610–690	−18
D	800–900	−18
E	1500–1635	−18
F	2280–2400	−18
45°	800–900	−18

channels (PMD A-F) measure light polarised perpendicular to the SCIAMACHY optical plane, generated by a Brewster angle reflection at the second face of the pre-dispersing prism. This polarised beam is split into six different spectral bands (Table 3.5). The spectral bands are quite broad and overlap with spectral regions of channels 2–6 and 8. The PMD and the light path to the array detectors, including the detectors, have different polarisation responses. Consequently, with the appropriate combination of PMD data, array detector data and on-ground polarisation calibration data the polarisation of the incoming light from the nadir measurements can be determined. For atmospheric limb measurements, where both mirrors are used, the light is outside the optical plane of the spectrometer. This requires measurements of additional polarisation information of the incoming light. A seventh PMD channel measures the 45° component of the light extracted from the channels 3–6 light path. All PMD channels are non-integrating and are read out every 1/40 s. They observe the same atmospheric volume as science channels 1–8.

Radiator A and Active Thermal Control

The OBM needs to be operated in orbit at a constant temperature to preserve the validity and accuracy of the on-ground calibration and characterisation. Additionally, a low temperature level is required to keep the thermal radiation of the instrument itself at a minimum in order not to enhance the background in the SWIR channels 7 and 8. Therefore a dedicated radiator, RAD A, is used to cool the OBM and the detector module electronics to between −17.6 and −18.2°C. Its location on the -X side of the instrument avoids direct solar illumination. Heat pipes are used to transfer heat from the OBM and the detector module electronics to the radiator (Fig. 3.13).

While the RAD A provides cooling capacity, thermal stability of the OBM needs to be established via a closed loop Active Thermal Control (ATC) system. It consists of three control loops with heater circuits and thermistors. The heating is controlled by the Power Mechanism & Thermal Control Unit (PMTC) based on measurements by the thermistors. Once ATC settings have been selected, the system maintains the OBM temperature to high precision at the specified level (see section "Thermal Performance" of Chapter 6). When heater control reaches a limit of the specified power range, e.g. due to in-orbit thermal degradation, the OBM temperature can no longer be kept stable over the whole orbit. By commanding appropriately modified ATC parameters the required ATC performance can then be re-established.

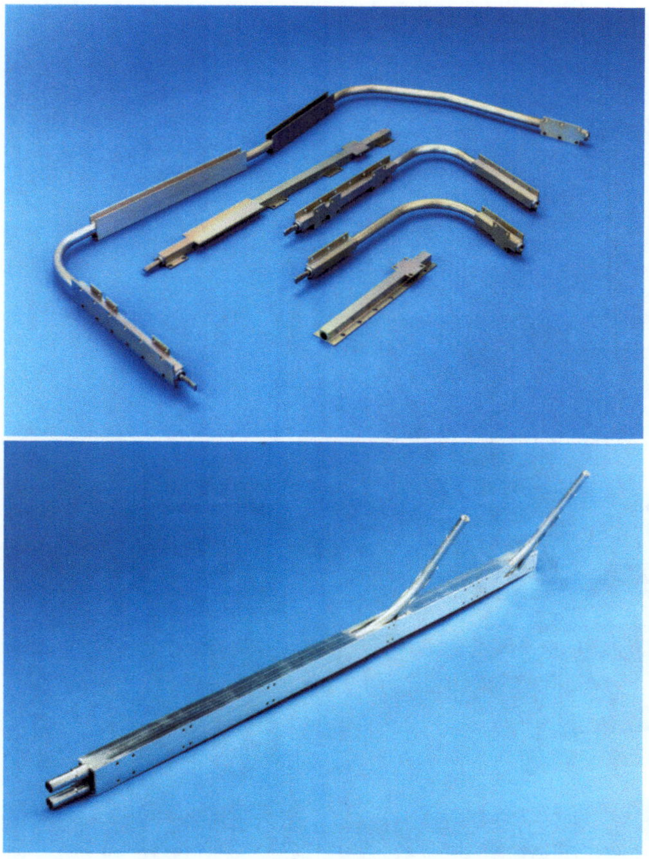

Fig. 3.13 Heat pipes from the optical bench to RAD A (*top*) and cryogenic heat pipes from detector modules 7 and 8 to the SRC (*bottom*) (Photo: EADS Astrium).

Thermal Bus

In-orbit operating temperatures of the detectors are well below ambient. The detectors are cooled via the Radiant Reflector Unit (RRU) of the Radiant Cooler (SRC) Assembly. The Thermal Bus connects the detector modules thermally with the reflector. Heat from detectors 1–6 is transported via an aluminium thermal conductor, from detectors 7 and 8 via two methane filled cryogenic heat pipes (Fig. 3.13). The heat pipes provide an efficient heat transfer in the temperature range 100–160 K.

Since the cooling efficiency of the Radiant Cooler is designed to provide sufficient cooling capacity until the end of the mission, a Thermal Control (TC) system is part of the Thermal Bus. It prevents the detector modules from becoming too cold by counter heating using three trim heaters. The TC system uses open loop heater control. Whenever drifting temperatures of the detectors reach their limits, the power settings of the trim heaters are adjusted by ground command bringing the temperatures back into the specified range.

3.3 Radiant Cooler Assembly

SCIAMACHY's Radiant Cooler dissipates heat generated in the detector modules to deep space to permit cooling of the detector arrays to in-orbit operating temperatures. The reflecting unit and the detectors are connected via the Thermal Bus of the OA. As for RAD A, the RRU points in the -X direction away from the Sun. Earthshine and sunshine are blocked from the radiating surface of the SRC to gain maximum cooling efficiency. Cold temperatures are obtained using a two stage process. An intermediate stage in the Radiator Unit lowers temperatures of detectors 1–6, while the cold stage, fitted with a parabolic reflector, yields temperatures around 150 K for detectors 7 and 8.

Due to its low temperature, the RRU surface is expected to attract contaminants from the in-orbit environment, particularly from ENVISAT itself. This would degrade the performance of the Radiant Cooler leading to reduced cooling efficiency. In order to clean the Radiant Cooler, the cold stage and the reflector are equipped with decontamination heaters. Turning the decontamination heaters 'on' raises the temperatures of the RRU, contaminating substances are removed through evaporation and the cooling performance is re-established. Whether contaminants begin to degrade the SRC performance can be determined from the power settings of the TC trim heaters. Degraded cooling efficiency is equivalent to higher radiator temperatures, i.e. higher detector temperatures. Consequently the trim heaters require less power when the SRC efficiency degrades because of contamination.

3.4 Electronic Assembly

SCIAMACHY's 'brain' resides in the Electronic Assembly. It provides the processing and formatting link of the detectors generating the primary science data with the spacecraft platform transmitting the digitised science data to ground. In addition, the EA houses all electrical and software functions required for autonomous operation of the whole instrument.

Functionally the EA consists of the primary processor called Instrument Control Unit (ICU) and the secondary processors, the PMTC and the Science Data Processing Unit (SDPU) as illustrated in Fig. 3.14. Serial data lines for the transmission of commands and data connect these three units. The EA is supplemented by the Decontamination Heater Control Module (DHCM) for operating the decontamination heaters on the SRC and the Digital Bus Unit (DBU) providing the instrument's command and control communication front-end interface to the ENVISAT platform.

Instrument Control Unit

Central control of all SCIAMACHY equipment in response to commands from ground and autonomous in-orbit operations of the instrument is the task of the ICU. It ensures

- Reception, verification and execution of macrocommands (MCMD) – and potential software updates – relayed by ENVISAT's Payload Management Computer (PMC)
- Autonomous instrument control as required by instrument mode, instrument states and parameters
- Monitoring of instrument HK telemetry data to verify instrument health
- Detection of anomalies and execution of autonomous corrective actions
- Acquisition and formatting of HK telemetry data from the secondary processors and the ICU itself for transmission to the PMC
- Maintaining a History Area to record significant instrument events

Fig. 3.14 The Simplified Engineering Model of the EA with ICU (1), PMTC (2), SDPU (3) and DBU (4) (Photo: EADS Astrium).

All time information is derived from the ICU on-board clock providing the SCIAMACHY On-Board Time (OBT). OBT and ENVISAT's PMC master clock are synchronised. The internal clocks of the secondary processors are not synchronised with the ICU clock. Datation of the scanners and detectors relies on an internal 16 Hz Broadcast Pulse (BCPS) which is generated in the ICU. For many aspects of command execution however, a time resolution of 62.5 ms is too low. Therefore another rate of 256 Hz defines the time unit *Count* – equivalent to 3.9 ms – to synchronise instrument internal control functions. Scanner control operations are driven by a dedicated PMTC internal 1 kHz clock.

The ICU's control logic implements the operations concept. This is not only true for the scientifically oriented (see Chapter 4) but also for engineering related activities. Tables configure ICU functions with the content of the tables being interpreted by ICU software. Operations modifications can thus be implemented by changes of table parameters, loadable via MCMD. Engineering tasks particularly concern the monitoring of the instrument health and safety and autonomous handling of anomalies. Up to 255 HK parameters can be monitored simultaneously. As long as the monitoring function does not report any anomaly, the instrument continues operations. Each anomaly detected triggers a Corrective Action (CA). The relation between anomalies and CA is again defined in tables. Some anomalies result in a CA which does not interrupt operations but is just recorded in the ICU's history area. This area is regularly downloaded via telemetry for inspection. More severe errors cause an immediate stop of ongoing measurements and the transition of the instrument into a safe configuration. After careful analysis of the anomaly and eliminating its cause, the instrument can be commanded from safe configuration back to nominal operations.

Secondary Processors

The SDPU controls and acquires science data from all 8 detector modules and auxiliary information from the PMD, the Sun Follower and the PMTC. On-board data preprocessing in this unit occurs prior to formatting and transfer to ENVISAT's High Speed Multiplexer (HSM) via the measurement data interface. The PMTC receives power from the platform and supplies the various modules in the OA. Additionally it controls the thermal status of the OA and detectors as well as the operations of mechanisms, including scanners and calibration sources.

Modes

Various operational instrument configurations, called 'modes', exist. They are divided into support modes and operations modes. The latter are those where SCIAMACHY can fulfil its measurement objectives. Two operations modes are defined: *Measurement* (see Chapter 4) and *Decontamination*. The *Measurement* mode may either be *Timeline* or *Idle*. *Measurement* corresponds to periods when the instrument collects science data. Once in *Idle*, the instrument has finished a particular measurement and waits until the next measurement will be started. The purpose of support modes is to achieve or maintain full operational conditions. Some of them are the response to an ICU or platform detected anomaly. Depending on the severity of the anomaly the instrument equipment and EA processors may be transferred to restricted activity levels. Other modes represent intermediate steps in the sequence from a lower mode up to *Measurement* mode. Predefined procedures transfer the instrument from one mode to another, both for nominal operations and anomaly cases.

3.5 The Making of SCIAMACHY

Following the selection of SCIAMACHY as an AO instrument aboard ENVISAT in 1989, the space agencies of Germany and The Netherlands initiated the development of the instrument aiming at an ENVISAT launch date before the end of the last century (for details about the early phase of the SCIAMACHY programme see section "SCIAMACHY's Past and Beyond its Future" of Chapter 1). Between 1990 and 1992 conceptual studies were carried out and predevelopment of innovative technology occurred at EADS Astrium (former Dornier Satellitensysteme) in collaboration with TNO-TPD and SRON in The Netherlands. This Phase B ended with a Baseline Design Review in November 1992. Phase C/D work started then in 1993 as a bilateral programme between Germany and The Netherlands and ended with the final delivery of the instrument to ESA in March 2000. DLR and NIVR agreed to a 'Memorandum of Understanding' which outlined the commitments of both partners. Finally Belgium joined the programme by funding the Polarisation Measurement Device, originally part of the Dutch contribution. For the management of the industrial contracts, DLR and NIVR set up DNPM, the common project management entity acting as the interface to the selected industrial contractors. To ensure continuous scientific oversight during instrument development and successive operations, participating institutes formed the SCIAMACHY Science Advisory Group (SSAG), chaired by the Principal Investigator (PI), under the auspices of DLR, NIVR and ESA. This group not only gave advice in scientific data product matters but also hosted the calibration and the validation subgroups with the lead assigned to the Netherlands Institute for Space Research (SRON) and the Royal Dutch Meteorological Institute (KNMI), respectively.

The Phase C/D contractors EADS Astrium and Dutch Space (former Fokker Space) were jointly responsible for the instrument development up to delivery to and acceptance by the agencies at the ENVISAT Launch Readiness Review. EADS Astrium and Dutch Space concluded an agreement between both companies and became co-prime contractors on an equal level. In order to establish a counterpart to the DNPM, the co-primes created the Integrated System Team (IST) acting as the entity to ensure a coherent and harmonised industrial programme. Having two co-primes meant establishing joint plans for many aspects of project development, including Project Management Plan, Product Assurance Plan and Design, Development and Verification Plan.

Each of the two co-prime contractors led an industrial consortium with a well defined share of tasks between the German and Dutch partners (Table 3.6). The German consortium consisted of

- EADS Astrium, Friedrichshafen (lead)
- Jena Optronik, Jena
- OHB System, Bremen
- Carl Zeiss, Oberkochen
- Kayser Threde, München
- SGI, Altenstadt
- EADS Astrium (former Matra Marconi Space), Toulouse
- Austrian Aerospace (former ORS), Wien
- Kongsberg, Oslo
- Turbinegarden, Kopenhagen

Members of the Dutch consortium were

- Dutch Space, Leiden (lead)
- SRON, Utrecht
- TNO-TPD, Delft
- Delft Sensor Systems (OIP), Oudenaarde/Belgium

Dutch Space, SRON and TNO-TPD formed the SCIAMACHY Joint Team (SJT), with its key personnel co-located at TNO-TPD premises in Delft. Development of the OA, with a few exceptions, and the SRC was attributed to the Dutch consortium. German responsibility included EA, Encoder Electronics and instrument harness.

For ground verification and testing, a joint approach was pursued, since final performance verification occurred in many cases at instrument level. The thermal balance/thermal vacuum (TB/TV) and electromagnetic compatibility (EMC RE/RS) instrument tests were performed in the IABG facilities at Ottobrunn near Munich. Instrument level tests included five 'cold' campaigns with the OA being cooled inside the vacuum chamber of the Optical Test Facility (OPTEC) at the Dutch Space Amsterdam/Schiphol premises, designated OPTEC-1 through OPTEC-5. Each OPTEC campaign comprised a sequence of qualification, verification, functional, performance and calibration tests with the Proto Flight Model (PFM) and included also in total three PI Periods. The latter were time slots assigned to scientific measurements to collect representative spectra of trace gases with the instrument. The PI Periods also provided very valuable additional information on the performance of SCIAMACHY.

Table 3.6 Instrument development responsibilities (D = Germany, NL = The Netherlands, B = Belgium)

Item	Responsibility
Optical Assembly	NL
Optical Unit/Optical Bench	NL
Detector modules	NL
PMD and Sun Follower	B
Scan mechanisms (ASM, ESM)	D
Calibration lamps	NL
OA heat pipes	D
Mounting Support Structure	D
Radiator A	NL
Multilayer Insulation (MLI)	NL
Electronic Assembly	D
PMTC	D
ICU	D
SDPU	D
DBU	Customer furnished item
DHCM	D
Base Plate	D
MLI	D
Radiant Cooler Assembly	NL
Radiant Reflector Unit	NL
Thermal Bus Unit	NL
Cryogenic heat pipe	D
Mechanical and thermal accessories	NL
Encoder Electronics	D
Instrument Harness	D

In the course of the development several models were built in support of the instrument and ENVISAT integration, verification and testing requirements. The model finally delivered to ESA for integration onto the ENVISAT platform is referred to as the PFM. Prior to PFM the Simplified Engineering Model (SEM) was required to functionally represent SCIAMACHY in the ENVISAT Engineering Model programme. The SEM consisted of the EA and Instrument Harness while the OA and SRC were represented by electrical dummy loads. Structural Models (STM) were built for the OA and SRC and were used for mechanical testing of the ENVISAT structure (Fig. 3.15).

Further frameworks for the successful development of the SCIAMACHY mission consisted of the 'Instrument Mission Implementation Agreement' (IMIA) and the 'Flight Operation and Data Plan' (FODP), which specified the responsibility split between DLR/NIVR on instrument and ESA on ENVISAT mission provider side. Based on these definitions, DLR created in 1996 the national SCIAMACHY Operations Support Team (SOST) with personnel from DLR at DFD (later at the Remote Sensing Technology Institute – IMF) and at the Institute of Environmental Physics/Institute of

Fig. 3.15 Structural Model of the OA showing the SCIAMACHY design with detector modules (1), ESM with Nadir Calibration Window (2), limb baffle with cover (3), RAD A (4), SRC (5) and Thermal Bus (6) (Photo: ESA).

Fig. 3.17 ENVISAT integration on the ESTEC Hydra vibration test facility. The instruments are covered with MLI. SCIAMACHY is the only instrument located on the front panel (*top right corner*) (Photo: ESA).

Fig. 3.16 Integration of the OA (Photo: Dutch Space).

Remote Sensing (IUP-IFE), University of Bremen. Its objective was to develop the infrastructure and interfaces required for the execution of the in-flight measurements. In addition, SOST supported both industry and agencies in all other aspects of scientific instrument operations. Algorithm and processor development for operational measurement data analysis was performed as part of the implementation of the ENVISAT ground segment. Since the ground segment concept assigned tasks of operational SCIAMACHY product generation to the German Processing and Archiving Centre (D-PAC) located at DLR-Oberpfaffenhofen, a large fraction of the required algorithm and processor development work was put under the responsibility of the data processing group at DFD and IMF.

As not uncommon in the development of complex satellite hard- and software, unexpected difficulties both at management and technical levels introduced delays which had to be accommodated. This also occurred in the SCIAMACHY programme. Fortunately the targeted ENVISAT launch in the late 1990s had to slip as well so that finally the instrument could be delivered on time (Figs. 3.16 and 3.17).

The EA PFM passed its Consent-to-Ship Board successfully in February 1999 when it was handed over to ENVISAT. The OA PFM required investigation of a few deficiencies discovered in late characterisation and calibration data before being accepted by ENVISAT in March 2000. When the flight hardware had been integrated onto the platform, the ESM diffuser was found not to be best suited for the envisaged in-flight calibration strategy. In an extra effort, an additional diffuser was manufactured, tested and mounted on the back side of the spare ASM mirror. ESA accepted the replacement of the current ASM with the modified and retested spare ASM. This exchange operation at spacecraft level took place in March 2001. Only 2 months later, ENVISAT was shipped to the launch site in Kourou where it arrived 15 May on-board an Antonov-124 heavy lifter aircraft. Further months of testing and waiting for the finalisation of the ARIANE-5 requalification programme passed until ENVISAT could finally be launched into orbit on 1 March 2002. This completed SCIAMACHY's on-ground phase and started its in-orbit life.

References

EADS Astrium GmbH SCIAMACHY Team (2004) SCIAMACHY Phase C/D Final Report (RP-SCIA-0000DO/07). Technical Document, EADS Astrium

Bovensmann H, Burrows JP, Buchwitz M, Frerick J, Noël S, Rozanov VV, Chance KV, Goede APH (1999) SCIAMACHY: Mission objectives and measurement modes. J Atmos Sci 56:127–150

Burrows JP, Chance KV (1991) Scanning imaging absorption spectrometer for atmospheric chartography. Proc SPIE 1490:146–155

DARA (1998) SCIAMACHY Instrument Requirements Document (PO-RS-DAR-EP-0001). Technical Document, DARA

Goede APH, Aarts HJM, van Baren C, Burrows JP, Chance KV, Hoekstra R, Hölzle E, Pitz W, Schneider W, Smorenburg C, Visser H, de Vries J (1991) SCIAMACHY instrument design. Adv Space Res 11:243–246

Hoogeveen RWM, van der A R, Goede APH (2001) Extended wavelength InGaAs infrared (1.0–2.4 μm) detector arrays on SCIAMACHY for space-based spectrometry of the earth atmosphere. Infrared Phys Tech 42:1–16

SCIAMACHY Joint Team (SJT), FS-TPD-SRON (1996) Optical and radiant cooler assemblies design description (RP-SCIA-1000FO/025). Technical Document, SJT

SCIAMACHY Science Advisory Group (SSAG) (1998) SCIAMACHY scientific requirements document. Technical Document, SSAG

Chapter 4
Instrument Operations

M. Gottwald, A. Moore, S. Noël, E. Krieg, R. Mager, and H. Kröger

Abstract SCIAMACHY operations have to be highly autonomous. All instrument activities are stored on-board and activated by time-tagged commands. They execute measurements in all required viewing geometries including the unique limb/nadir matching. Observations of the Sun in occultation over the northern hemisphere and the Moon in occultation over the southern hemisphere occur at regular intervals, ranging from each orbit to each month. They are complemented by sub-solar observations. The lunar measurements need particularly accurate preparation due to the orbital motions of the Moon and ENVISAT around Earth. Both Sun and Moon occultations have to take refraction into account for the period when both targets are viewed through the lower atmosphere. Timing of SCIAMACHY measurement operations is entirely Sun or Moon related. The specified mission scenarios combine nadir, limb and eclipse measurements between Sun or Moon related events along the orbit. The smallest functional sequences are states, controlled via various parameter tables. When combining individual states, timelines are generated. Sets of states and timelines are stored on-board. Each timeline is activated by a single macrocommand. Both states and timelines can be modified from ground. Since SCIAMACHY has been provided by national agencies, instrument operations is a shared enterprise between DLR/NIVR and ESA.

Keywords Limb/nadir matching • Sun occultation • Moon occultation • Mission scenarios • States • Timelines • SOST

SCIAMACHY's measurement operations concept required combining the properties of the ENVISAT operational environment – both space and ground segment, the challenging scientific needs and the characteristics of the sensor. Finally a concept of autonomously executing functional sequences, all stored on-board, was developed. This permitted fully exploiting the instrument's flexibility without using too many resources of the ENVISAT mission or violating the requirements of the atmospheric science user community (EADS Astrium 2003).

The characteristics of a polar orbiting platform with short telemetry coverage at the high latitude station Kiruna or via the Ka-band Artemis link demanded highly autonomous on-board operations. This comprises not only on-board anomaly detection and initiating corrective actions as part of the instrument control but also the ability to configure the instrument status and to execute measurements without direct manual intervention from ground. The goal is to store the pre-planned measurement schedule on-board and execute it as a time-tagged sequence of measurement activities.

Scientific requirements include viewing geometries for atmospheric measurements of nadir, limb, Sun occultation and Moon occultation (Fig. 4.1). In addition, external (e.g. dark current, Sun reference) and internal (calibration lamps) observations supplement the measurement schedule. One of SCIAMACHY's main objectives is to measure the same atmospheric volume both in nadir and limb within one orbit, i.e. achieving limb/nadir matching of the geolocation of limb states with associated nadir states. It allows collecting scientific information for the same volume of air from two different measurement modes. The combination of nadir and limb measurements is especially useful for deriving tropospheric information.

M. Gottwald (✉) and E. Krieg
Remote Sensing Technology Institute, German Aerospace Centre (DLR-IMF), Oberpfaffenhofen, 82234 Wessling, Germany
e-mail: manfred.gottwald@dlr.de

A. Moore
Vega IT GmbH, c/o ESA/ESOC, Robert-Bosch-Str. 5, 64293 Darmstadt, Germany

S. Noël
Institute of Environmental Physics/Institute of Remote Sensing (IUP-IFE), University of Bremen, Otto-Hahn-Allee 1, 28359 Bremen, Germany

R. Mager
EADS Astrium GmbH, Claude Dornier Straße, 88090 Immenstaad, Germany

H. Kröger
Bichlmairstraße 17, 83703 Gmund am Tegernsee, Germany

M. Gottwald and H. Bovensmann (eds.), *SCIAMACHY - Exploring the Changing Earth's Atmosphere*,
DOI 10.1007/978-90-481-9896-2_4, © Springer Science+Business Media B.V. 2011

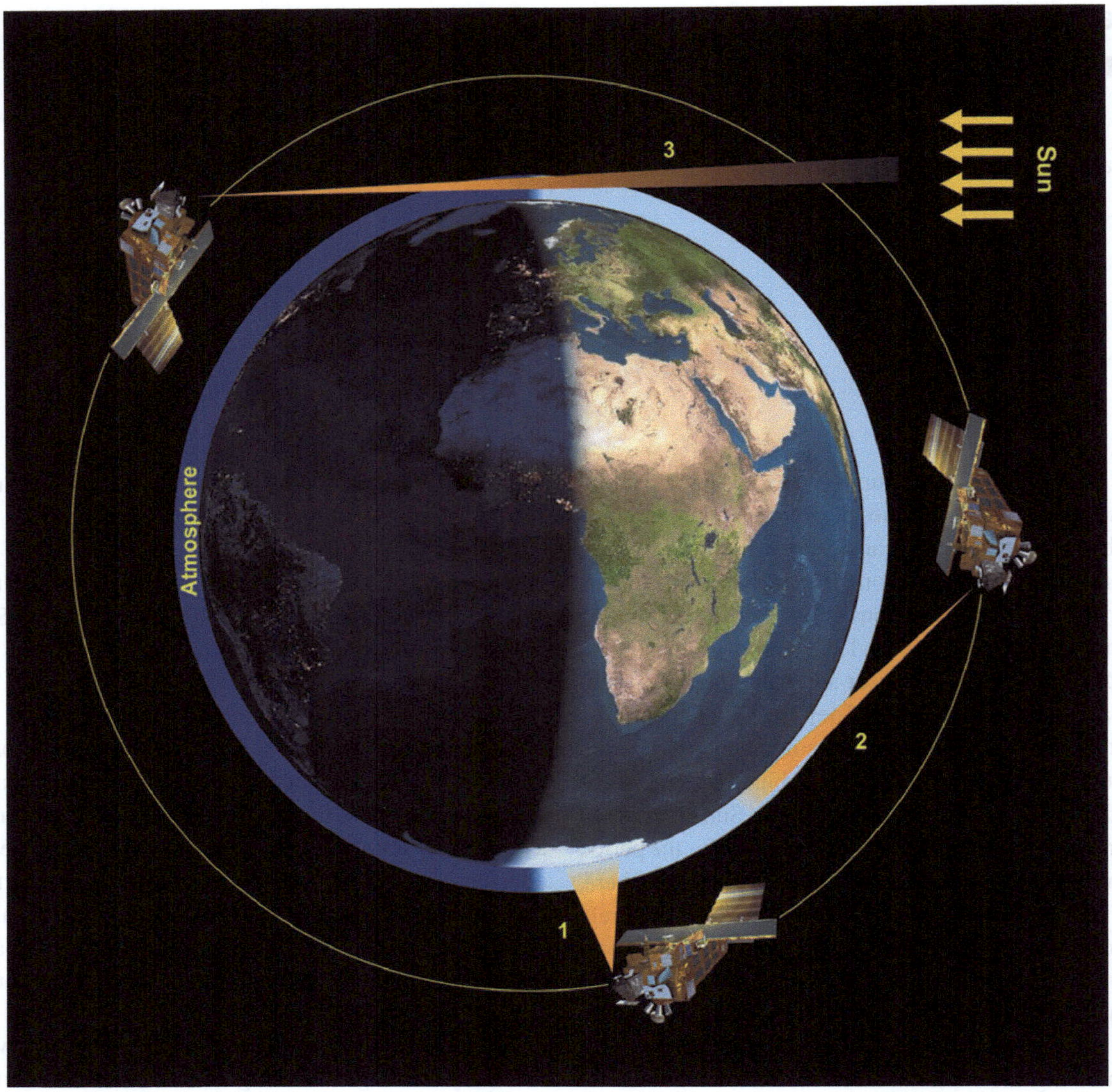

Fig. 4.1 SCIAMACHY's scientific observation modes: 1 = nadir, 2 = limb, 3 = occultation (Courtesy: DLR-IMF).

When matching limb and nadir states, SCIAMACHY first observes an atmospheric volume at the horizon by looking slightly off the flight direction towards Earth's rotation. Later in orbit, after a time interval $\Delta t = 430$ s, the same volume of air crosses the subsatellite point and can be observed under nadir conditions (Fig. 4.2). The interval of $\Delta t = 430$ s is defined by the altitude and angular velocity of the spacecraft platform. In limb mode, SCIAMACHY observes the horizon 3280 km ahead of the instrument close to flight direction. Because the spacecraft's steering law is determined by the latitude-dependent Earth's rotational velocity at the instantaneous subsatellite point, the line-of-sight does not intercept the horizon at a point where the Earth's and spacecraft's angular velocities would lead to limb/nadir matching. Therefore, an instrument yaw steering correction is implemented in SCIAMACHY's on-board software to compensate for the phase shift between local yaw steering and instrument line-of-sight in limb observations. It reflects the angular difference of approx. 27° in latitude between local subsatellite point during limb measurement and line-of-sight interception at the Earth's horizon (Fig. 4.3).

4.1 Sun and Moon Observation

All measurement activities are planned relative to solar and lunar constellations. SCIAMACHY operations are either Sun or Moon fixed, but never Earth fixed (SOST 1996). When viewed from ENVISAT, solar and lunar constellations are determined by the spacecraft propagating along the orbit, the orientation of the orbital plane, the lunar motion around Earth and the Earth orbiting the Sun. Thus the observing conditions can be completely predicted by careful orbit analysis. These predicted conditions are then translated into configurable instrument states and timelines.

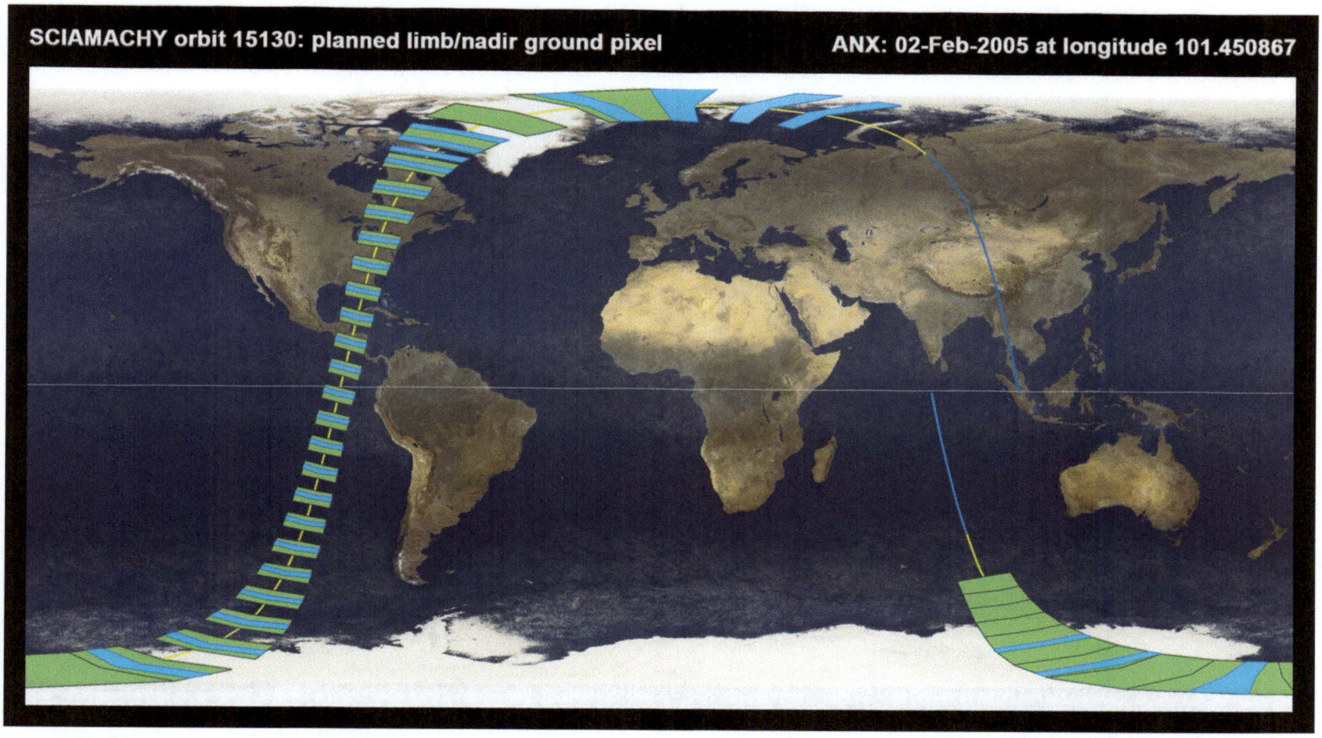

Fig. 4.2 An orbit with planned limb/nadir matching on the dayside of the orbit. The sequence of nadir and limb states in a timeline is arranged such that limb ground pixels (*blue*), defined by the line-of-sight tangent point, fall right into a nadir ground pixel (*green*) in the illuminated part of the orbit (Courtesy: DLR-IMF/SRON).

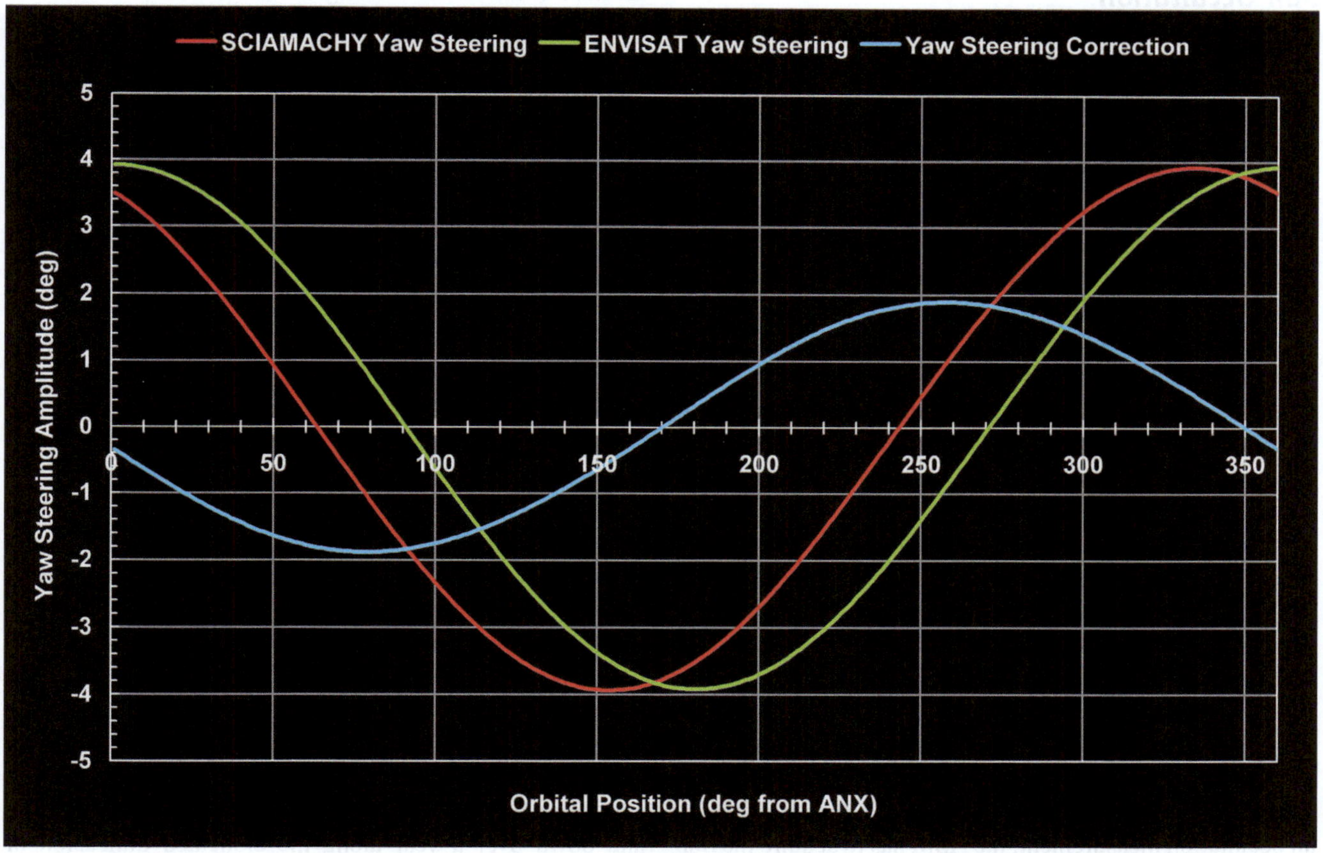

Fig. 4.3 ENVISAT's yaw steering, the yaw steering correction of limb states and the resulting SCIAMACHY yaw steering. Between ENVISAT and SCIAMACHY yaw steering an orbital shift of 27° exists which reflects the observation geometry in elevation when looking to the horizon in flight direction (Courtesy: DLR-IMF).

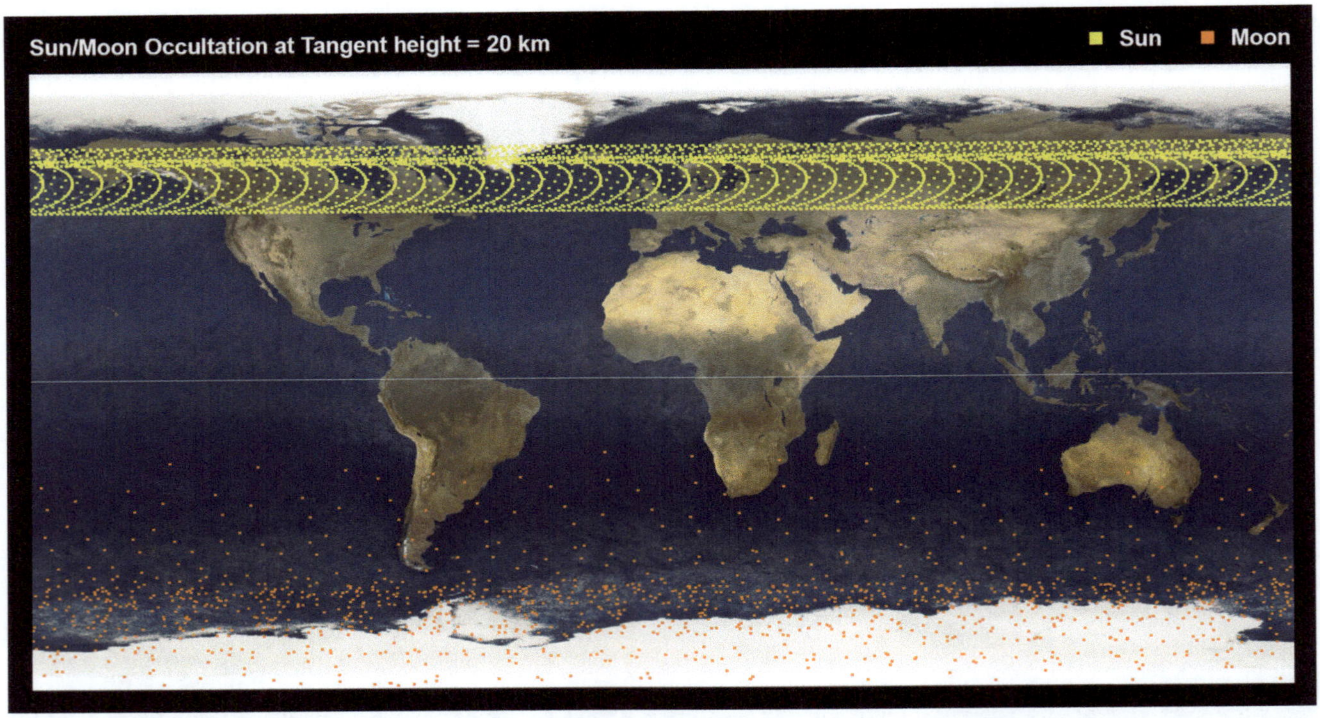

Fig. 4.4 Tangent point geolocation at an altitude of 20 km for all solar and lunar occultation measurements in the year 2009 (Courtesy: DLR-IMF/SRON).

Sun Occultation

The sun-synchronous orbit Mean Local Solar Time (MLST) of 10 a.m. during descending node crossing leaves the Sun always to the left of the flight direction, i.e. at azimuth angles >180°. The elevation of the Sun varies between approx. −70° and +70°. Sunrise occurs after ascending node crossing when ENVISAT moves towards the North Pole on the eclipse side of the orbit. The Sun becomes visible at the Earth's limb left of the flight direction at medium to high geographic latitudes (Fig. 4.4). The exact latitude is dependent on the actual position of the *true* Sun relative to the Earth (*true* Sun reflects the actual annual orbital motion of the Earth contrary to the *mean* Sun which is characterised by assuming a constant Earth orbital velocity). In summer, Sun occultation measurements start when the spacecraft has reached geographic latitudes of about 27° north while in winter the subsatellite point moves up to about 75° north. At sunrise the solar elevation is identical to the elevation of the Earth's limb, i.e. approx. 27.2°. The azimuth angle at sunrise has a mean value of 330°, corresponding to the MLST at descending node crossing of 10 a.m. and changes over a year due to the apparent motion of the true Sun. Caused by the orbital motion of ENVISAT, the Sun rises almost vertically through SCIAMACHY's limb TCFoV. In an occultation measurement, the ASM has to acquire the Sun at an angle of about 330° and to follow the slightly changing azimuth as the Sun moves higher. In the elevation direction the Sun must be tracked by the ESM up to the maximum elevation angle of 19.5°, limited by the TCFoV. From the Earth's limb up to an elevation angle of 25.2°, corresponding to an altitude of 100 km, the sunlight is transmitted through the Earth's atmosphere and partially absorbed. Thus the Sun serves both as a target for probing the atmospheric trace gas constituents (altitude <100 km) and for calibration and monitoring measurements (altitude >100 km). Therefore the total time of the sunrise in the limb TCFoV is referred to as *Sun Occultation & Calibration (SO&C)* window.

Sub-solar Observations

The sub-solar port above the ESM provides additional access to the Sun above the atmosphere. Because Sun viewing in this configuration does not involve the ASM, these measurements, when used in conjunction with data from other light paths, permit monitoring the characteristics of the ASM mirror. The Sun is visible through the sub-solar port when the Sun has reached its highest elevation. This occurs at an azimuth angle of 270°. The sub-solar elevation angle changes continuously with season. Therefore the Sun moves up and down over a year along the elongated dimension of the sub-solar TCFoV when passing through the window.

The duration of a sub-solar measurement is defined by the time it takes the Sun to pass through the azimuth dimension of the sub-solar port, reduced by the small aperture stop to only 0.72°. It amounts to 21 s, with the Sun being fully visible for a short period of only 3.5 s.

Moon Occultation

Individual observations of the Moon follow the same principles as described for the Sun. The lunar disk is acquired by the ASM and ESM and tracked as the moon rises through the limb TCFoV. As in the case of the Sun, the Moon acts both as a target for scientific and for calibration and monitoring measurements. The corresponding time interval is the *Moon Occultation & Calibration (MO&C)* window. Predicting lunar occultation measurements requires analyses of the viewing conditions as a function of the monthly lunar motion. For a full orbit the Moon moves in *direct motion* within a *synodic month* of 29.53 days. The Moon's orbital plane is inclined by 5.1° to the ecliptic. Thus, to a first order, the Moon's orbital plane lies perpendicular to ENVISAT's orbital plane. Whenever the lunar orbit crosses the limb TCFoV of SCIAMACHY, moonrise can be observed (Fig. 4.5). With a total azimuth size of 2 × 44° for the

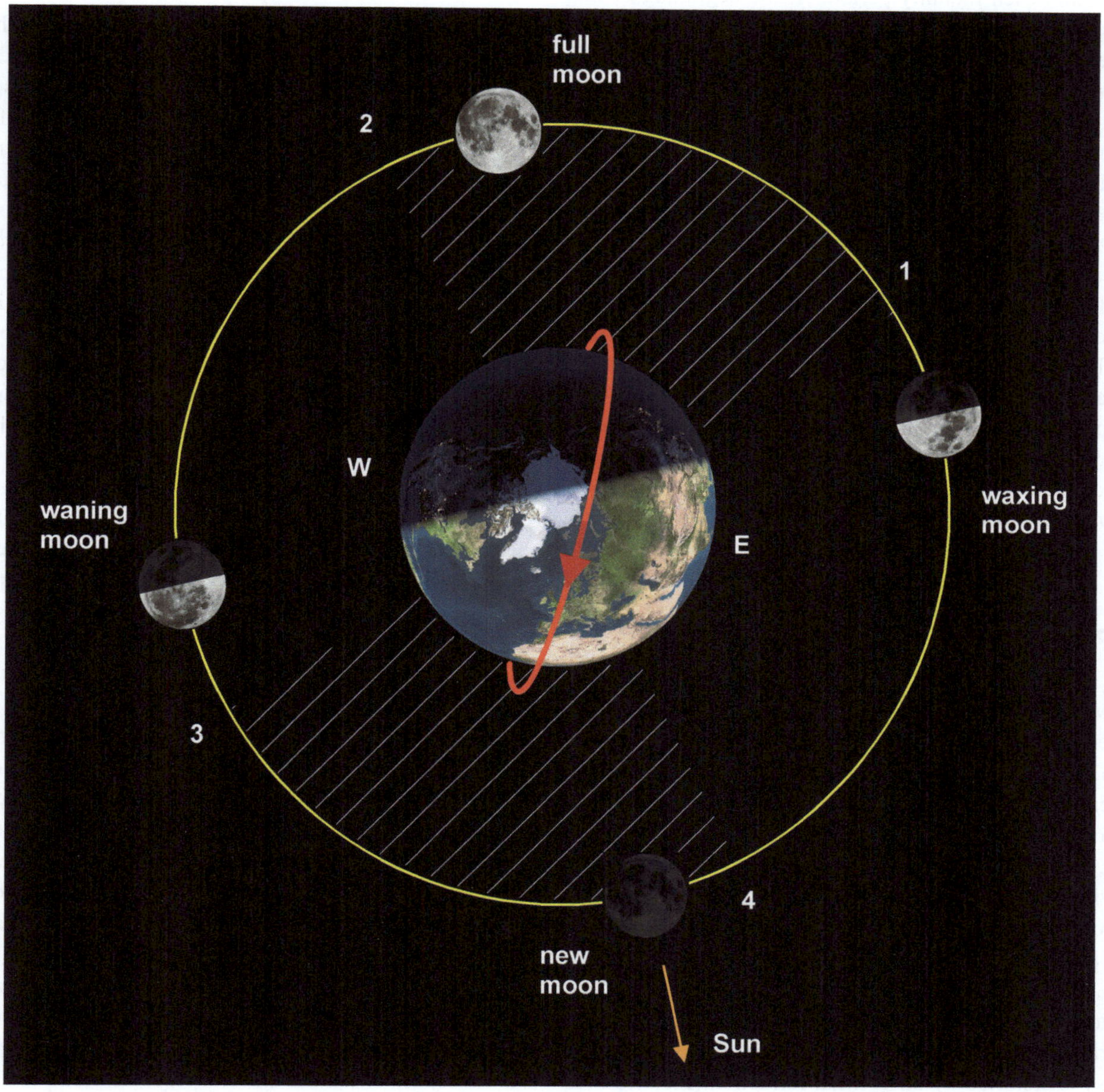

Fig. 4.5 SCIAMACHY's monthly lunar visibility occurs between 1 and 2 over the southern hemisphere (lunar phase >0.5). The hatched area illustrates the limb TCFoV of 88°. Visibility at smaller lunar phases over the northern hemisphere between 3 and 4 is not used because it coincides with solar occultations (Courtesy: DLR-IMF).

limb TCFoV, the monthly time interval when lunar measurements can be executed is

$$\Delta_{\text{moon}} = \frac{88^\circ}{360^\circ} \times 29.5\,\text{days} \approx 7.2\,\text{days}$$

The duration of each monthly observation opportunity displays a seasonal variation of between 5.5–8 days. Due to the lunar orbital motion, the first moonrise in a monthly period occurs on the left side of the limb TCFoV (azimuth = 316°). Each orbit moonrise progresses with an azimuth rate of about 1°/orbit to the right side of the limb TCFoV. At the end of the monthly visibility, the Moon has reached the right edge of the TCFoV (azimuth = 44°). The lunar phase changes continuously within each monthly period. At the beginning of the visibility, the phase amounts to about 0.6–0.7. Full Moon can be observed close to the end of the monthly cycle because the 10 a.m. descending node crossing time criterion of ENVISAT's sun-synchronous orbit only allows full illumination of the lunar disk when the Moon rises at an azimuth of about 30°. Moonrise occurs over a large range of geographic latitudes (see Fig. 4.4). Different to sunrise, where the geographic latitude of the subsatellite point – and thus also the geolocation of the tangent point where the atmosphere is probed – changes slowly over a year, the latitude of the subsatellite point at moonrise varies significantly within a monthly period. This fact is particularly interesting for the occultation measurements because it allows studying the atmosphere at various latitudes. Moonrise at lunar phases >0.5 can only be observed over the southern hemisphere. This complements solar occultations which are restricted to the northern hemisphere.

Refraction

Observation of sunrise and moonrise from a spacecraft is affected by the refractive properties of the Earth's atmosphere. Because the Sun and Moon have almost the same apparent size of 31.5 arcmin the effect of refraction is identical for both. The refraction angle depends on the Earth's radius, the scale height of the exponentially decreasing refractivity profile, the refractivity and the height of the tangent point of the incident rays. For visible light the refraction angle amounts to approx. 70 arcmin at the horizon (h = 0 km), i.e. SCIAMACHY can observe the first solar/lunar photons when the Sun/Moon is still well below the geometric horizon. As the Sun or Moon rises, the refracted image of the disk is distorted by differential refraction (Fig. 4.6). At an altitude of h ≈ 17 km refraction has become so small that refracted image and the celestial position of the solar or lunar disk overlap. Below this height the angular rate of the rising Sun or Moon as defined by the moving spacecraft is larger than the variable rate of their refracted images. At low altitudes, measuring the Sun or the Moon can be difficult due to obscuration by or reflected radiation from clouds. Onboard control of the scan mirrors during occultation uses the Sun Follower with its relatively wide field of view of 2.2° × 2.2°. Since the lunar disk can be outshined by an illuminated cloudy atmosphere, moon occultation measurements are only possible when the Moon rises on the night side beyond the terminator.

4.2 Reference Measurement Orbit

A typical SCIAMACHY orbit starts above the northern hemisphere with an observation of the rising Sun. In order to acquire also light from the sparsely illuminated atmosphere at the limb in the direction of the rising Sun, a sequence of limb measurements precedes each Sun occultation measurement. Once the Sun has risen, it is tracked by the ESM for the complete pass through the SO&C window. After about 175 s the Sun leaves the limb TCFoV at the upper edge. In order to fully exploit the high spatial resolution during occultation, measurement data readout with a high rate is required in the SO&C window. Until the passage of the sub-solar point, a series of matching limb/nadir observations are executed. At the sub-solar point the Sun, generally close to descending node crossing, has reached its highest elevation relative to ENVISAT. A sub-solar measurement is only executed when a sub-solar calibration opportunity has been assigned by ENVISAT. Because the Ka-band antenna in its operational position vignettes the sub-solar TCFoV, only 3 orbits per day with sub-solar opportunities are possible. Again a sequence of matching limb/nadir measurements follows. Above the southern hemisphere, the Moon becomes visible during the monthly lunar visibility period, otherwise matching limb/nadir observations continue. The rising Moon is observed similarly to the rising Sun from bottom to top of the limb TCFoV. A series of limb/nadir observations concludes the illuminated part of the SCIAMACHY orbit. Because the instrument is still viewing sunlight while the projected ground-track in the flight direction will already have seen sunset, the final measurements in this phase are only of the nadir type. When ENVISAT enters the eclipsed part of the orbit, dedicated eclipse observations can be executed until SCIAMACHY moves towards another sunrise and the orbit sequence starts again (Fig. 4.7). Ascending node crossing occurs always in eclipse phase. The reference orbit is entirely based on the *Sun/Moon fixed* concept. While *Sun fixed* events show a relatively stable temporal behaviour over a year, orbital segments related to Moon occultation measurements do not. They exhibit strong variability both within a monthly lunar observation period and over a year.

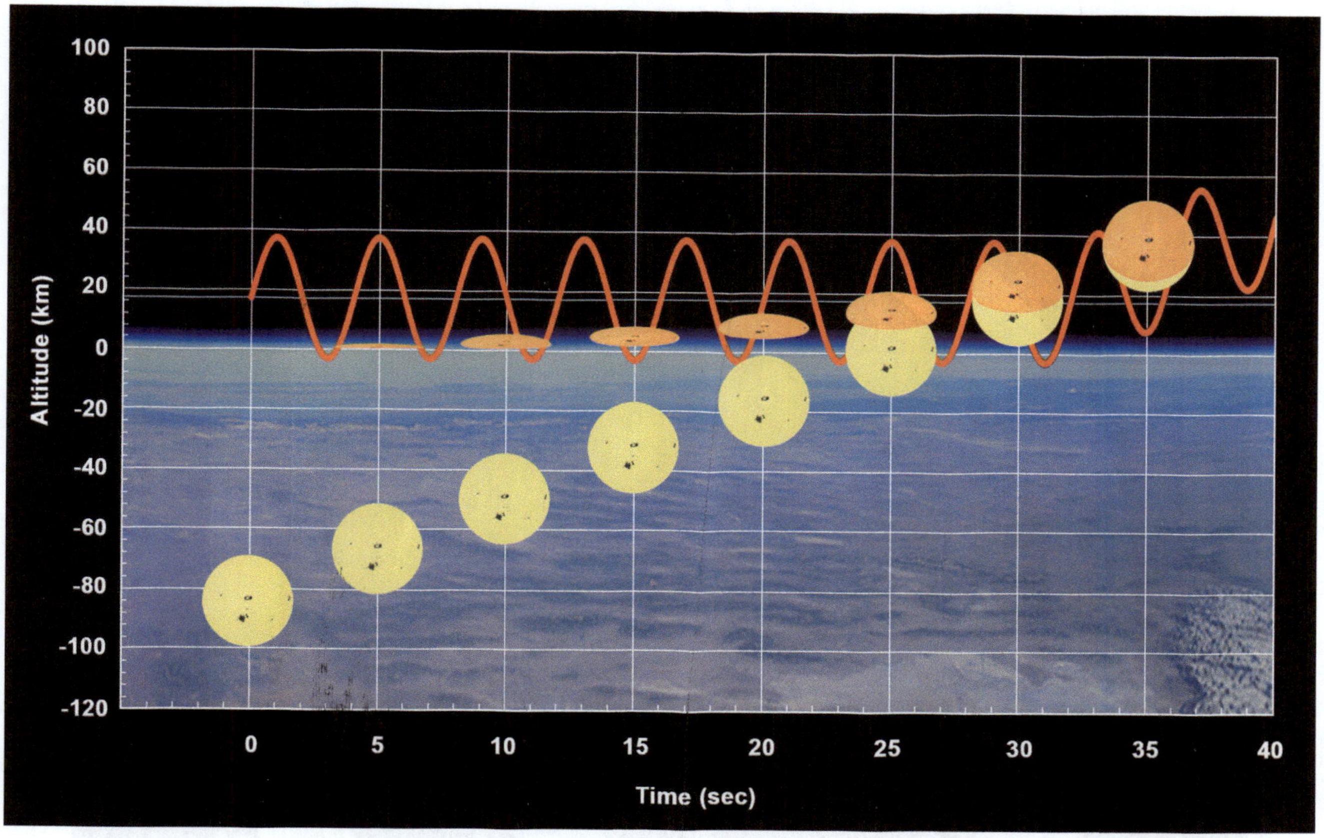

Fig. 4.6 The effect of refraction during sunrise. The position of the Sun is indicated by the yellow disk while the distorted refracted image appears above the horizon. About 30 s after begin of sunrise the position of the true and refracted disk overlap and refraction becomes negligible. The sinusoidal curve illustrates the elevation scan of the ESM at the start of the Sun occultation states (Courtesy: DLR-IMF; background photo: NASA Johnson Space Center and Image Science & Analysis Laboratory).

4.3 Mission Scenarios

It has proven useful to structure SCIAMACHY operations in the form of *mission scenarios* (SOST 2001a). These define in a top-down approach objectives on various planning levels. The mission scenarios lead finally to a scenario for a single orbit, i.e. an orbital mission scenario that is the basis for continuous mission planning and scheduling. For nominal operations the orbital mission scenario is defined by:

- A swath width of ±480 km relative to ground track in nadir and limb scans for global coverage within 6 days (taking the alternating limb/nadir measurements into account – see Fig. 4.8)
- Matching limb/nadir measurements in the illuminated part of the orbit
- Sun occultation measurements each orbit
- Moon occultation measurements whenever possible (moonrise on nightside of Earth)
- Mesosphere/lower thermosphere measurements in eclipse each orbit, intermittent with calibration and monitoring measurements
- The equivalent of 2 days per month with mesosphere/lower thermosphere measurements in the illuminated part of the orbit
- Calibration and monitoring measurements on a daily (every 14th orbit), weekly (every 100th orbit) and monthly basis

The simplest orbital mission scenario is executed whenever the Moon is not visible and no regular calibration and monitoring tasks have to be performed. This scenario occurs about 90% of the time during a month and can be accommodated by 4 timelines. The most complex scenario is defined when implementing monthly calibration and monitoring requirements. This consists of a series of 5 consecutive orbits filled with calibration and monitoring activities.

When defining mission scenarios the accumulated activations of SCIAMACHY's *life limited items* (LLI) have to be taken into account. These are the mechanical components such as NDFM, APSM and the NCWM, the mechanism to open the Nadir Calibration Window, internal calibration lamps WLS and SLS and cryogenic heat pipes Operating LLI according to agreed mission scenarios must

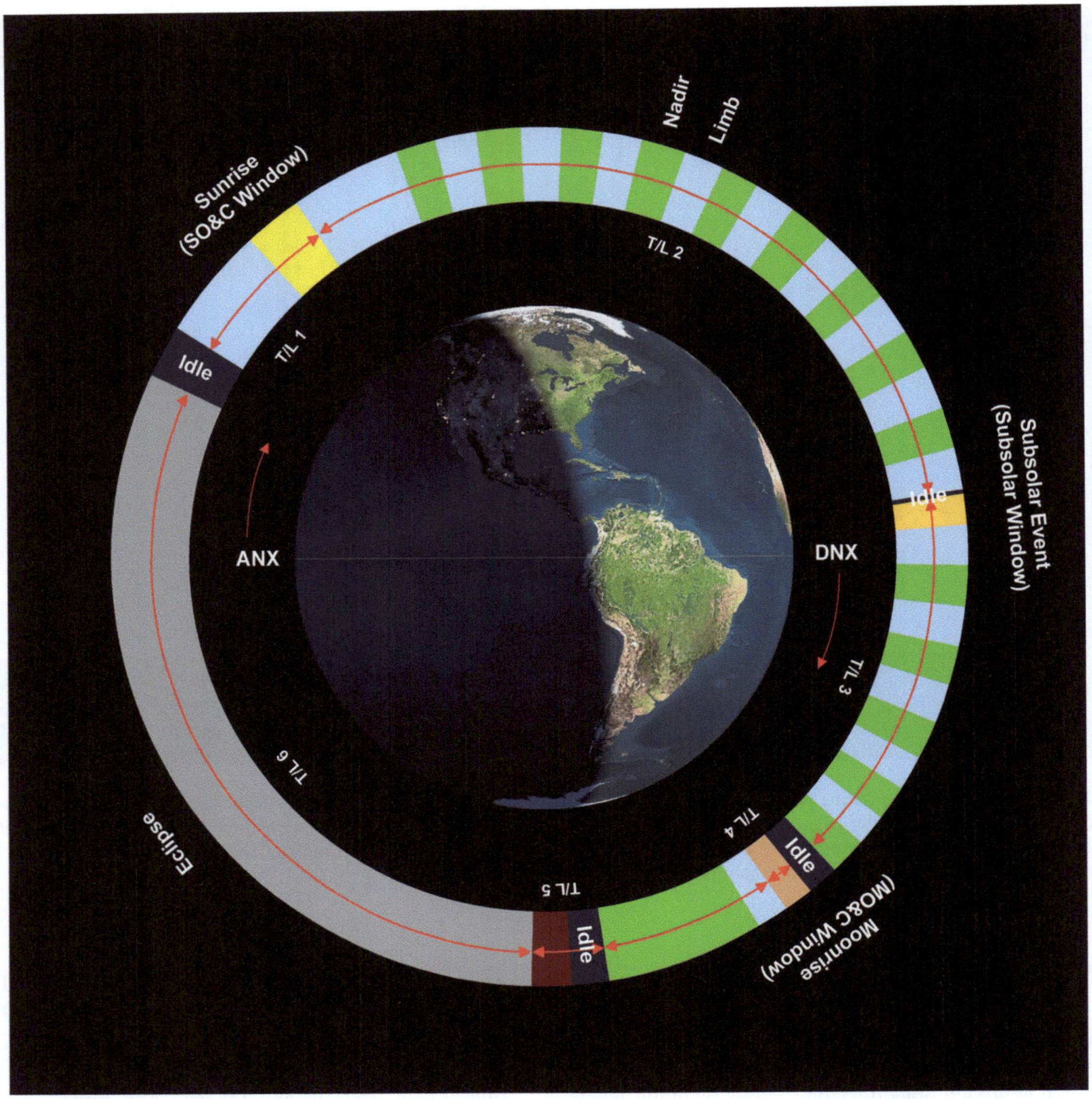

Fig. 4.7 SCIAMACHY reference orbit with Sun/Moon fixed events along the orbit. The events define orbital segments which are filled with timelines. State duration is not to scale (Courtesy: DLR-IMF).

not exceed the specified number of switches, cycles or 'on' times, i.e. individual in-flight LLI budgets, over the mission lifetime.

4.4 Parameter Tables

Utilising the high degree of flexibility in the instrument design can only be accomplished through parameterisation of on-board operations. For the execution of scientific and calibration and monitoring measurements it means that associated functions must be predefined and stored on-board with ground control having the capabilities to modify the instrument configuration. Changing the instrument status includes software updates via patching as well as parameter settings via commanding. Those sets of parameters which are associated with basic instrument properties, e.g. scanner, thermal and mechanism control definitions are termed *engineering* parameters. More than 4600 engineering parameters exist. Most of them have been defined prior to launch and were verified during the Commissioning Phase. During routine operations, engineering parameters are usually not subject to modifications. Parameters relating to the configuration of the spectrometer while acquiring data, the *measurement* parameters, have also been defined prior to launch.

Fig. 4.8 Nadir coverage after 1 day of routine measurements. At the equator consecutive nadir ground pixels are separated by about 1600 km. This gap is filled within 3 days in case of nadir only state sequences. With successive limb/nadir states full nadir coverage at the equator requires 6 days (Courtesy: DLR-IMF/ SRON).

However, changing scientific requirements and following lessons learnt about actual instrument behaviour during the Commissioning Phase, have made it necessary to update measurement parameters occasionally. Both types of parameters were burnt into the EEPROM (Electrical Erasable Programmable Read Only Memory) unit during instrument development. Whenever the ICU is initialised or reset, the EEPROM content is loaded and expanded into the ICU RAM (Random Access Memory). A copy of this content is then further loaded into the work area of the RAM and this part of the ICU RAM can be accessed by macrocommands in order to update parameters as required. Engineering and measurement parameters are organised in a series of parameter tables defining common functional tasks.

More than 25000 measurement parameters are needed to execute all the required measurements. Spectrometer settings, both for the detectors and the scanners, are provided via measurement parameters (Table 4.1). Tables usually relate to individual measurement states (see below) but 'common' tables serve all states. Particularly the values of the pixel exposure times and the co-adding factors ensure, in conjunction with the definition of wavelength clusters, optimised signal-to-noise ratios over the complete orbit. The product of PET and co-adding factor yields the integration time, which defines the readout frequency (see section "Optical Assembly" of Chapter 3). In the case where SWIR channels run into saturation, the *Hot Mode* parameters

Table 4.1 Measurement parameter tables

Type	Table	Number of specified parameters
State	Scanner State	10080
	Pixel exposure time	1400
	Hot Mode	420
	State Index	280
	State duration	420
	Co-adding	4480
	Detector Cmd Words	35
	DME Enable	8
	State RTCS Index	140
Common	Basic Scan Profile	60
	Relative Scan Profile	8652
	Cluster per Channel	40
	Cluster Definition	464

allow decreasing the exposure times in channels 6–8 below the minimum PET of 31.25 ms. The characteristics of the ASM/ESM rotations are defined via the Scanner State, Basic Scan Profile and Relative Scan Profile parameters. The Basic Scan Profile specifies the underlying standard motion of the respective mirror whereas the Relative Scan Profile is a mirror movement superimposed onto the Basic Scan Profile, e.g. fast upward/downward scans in elevation while the ESM is basically slowly moving in elevation.

4.5 Measurement States

The different configurations of the individual functions to operate SCIAMACHY in measurement modes are defined as *states* (SOST 2003). A state controls a sequence of activities to execute a particular measurement task, e.g. nadir observations with certain pixel exposure times, Sun occultation with a certain scan geometry, etc. In total 70 states can be defined on-board. There they are stored as part of the Relative Time Command Sequence (RTCS) table which is also contained in the EEPROM and transferred to the ICU RAM. The entries in the RTCS table determine the execution of each state as a series of *primitive commands* which are activated sequentially once the first primitive command has been started. The parameters in the measurement tables control the execution of the primitive commands, i.e. of the state.

State Definition

SCIAMACHY's scientific measurement objectives and requirements have necessitated the definition of the 70

Table 4.2 Measurement state definition

State ID	State	Measurement type	Remark
1–7	Nadir 960 km swath	Science	All orbital positions
8, 26, 46, 63, 67	Dark current	Calibration	Pointing at 250 km
9–15	Nadir 120 km swath	Science	all orbital positions
16	NDF monitoring, NDF out	Monitoring	
17–21	Sun ASM diffuser	Calibration	Sun above atmosphere
22	Sun ASM diffuser atmosphere	Monitoring	Various azimuth angles
23–25, 42–45	Nadir pointing	Science	All orbital positions
27	Limb mesosphere	Science	Scanning 150–80 km
28–33	Limb 960 km swath	Science	All orbital positions
34–37, 40, 41	Limb no swath	Science	All orbital positions
38	Nadir pointing left	Monitoring	
39	Dark current Hot Mode	Calibration	
47	SO&C scanning/pointing	Science, calibration	Sun through and above atmosphere
48	NDF monitoring, NDF in	Monitoring	
49	SO&C nominal scanning, long duration	Science, calibration	Sun through and above atmosphere
50	SO&C fast sweep scanning	Calibration	
51	SO&C pointing	Science, calibration	Sun through and above atmosphere
52	Sun ESM diffuser, NDF out	Calibration	Sun above atmosphere
53	Sub-solar pointing	Calibration	
54	Moon nominal scanning	Calibration	Moon above atmosphere
55	Limb mesosphere-thermosphere	Science	Scanning 150–60 km
56	Moon pointing	Science, calibration	Moon through atmosphere
57	Moon pointing, long duration	Science, calibration	Moon through and above atmosphere
58	Sub-solar pointing/nominal scanning	Calibration	
59	SLS	Calibration	
60	Sub-solar fast sweep scanning	Calibration	
61	WLS	Calibration	
62	Sun ESM diffuser, NDF in	Calibration	Sun above atmosphere
64	Sun extra mirror pointing	Calibration	Sun above atmosphere
65	ADC, scanner maintenance	Calibration	
66	Sun extra mirror nominal scanning	Calibration	Sun above atmosphere
68	Sun extra mirror fast sweep scanning	Calibration	Sun above atmosphere
69	SLS ESM diffuser	Calibration	
70	WLS ESM diffuser	Calibration	

individual states as listed in Table 4.2. Thirty-five states implement scientific observation requirements, 26 are for the purpose of in-flight calibration, 4 for in-flight monitoring and the data from 5 states can be used for scientific and calibration analyses. The high number of calibration and monitoring states is the result of the thorough and complex in-flight calibration and monitoring concept.

The definition of each state requires the translation of scientific requirements into instrument functions, configurable via parameter settings. Once a state has been specified, its configuration is frozen thus establishing the state *final flight status*. The corresponding parameter tables are uploaded on-board such that routine operations can execute the associated measurements whenever required. Changes to final flight definitions require strict configuration control and follow the Operation Change Request (OCR) procedure.

Nadir and Limb States

Continuous observations of the illuminated atmosphere occur in a sequence of matching limb/nadir states. The standard setting requires a wide swath geometry both in nadir and limb viewing. This enables global coverage within 3 days in nadir only mode or 6 days when running matching limb/nadir sequences. At medium to high latitudes, complete coverage is achieved earlier.

For a nadir state, the instrument Instantaneous Line-of-Sight (ILoS) is pointed, via the ESM, towards the subsatellite point. Starting left of the flight direction, the ESM is then moved during 4 s across-track to the right with an angular rate yielding the required swath width on the ground. Having reached the turnaround on the right side, the ESM returns in 1 s across-track back to the left side. For each integration time, a channel dependent readout of the spectrum occurs, synchronised with the scan. Depending on the measurement duration of the nadir state, a series of forward/backward scans covers a ground scene of typically 400 km along-track. Within the ground scene, individual ground pixels have a size which is defined by the selected integration time. They vary from about 26 km × 30 km (along-track × across-track) up to 32 km × 930 km for wide swath settings (see Fig. 4.9).

While a nadir state requires operation of the ESM only, a limb state has to acquire light from the Earth's limb via the ASM. This light is reflected onto the ESM and then further into the spectrometer. Routine limb observations start with a horizontal scan 3 km below the horizon from the right to the left side. A vertical step of 3 km moves the ILoS upwards and another horizontal scan of the selected width is executed in opposite direction. In a sequence of vertical steps and

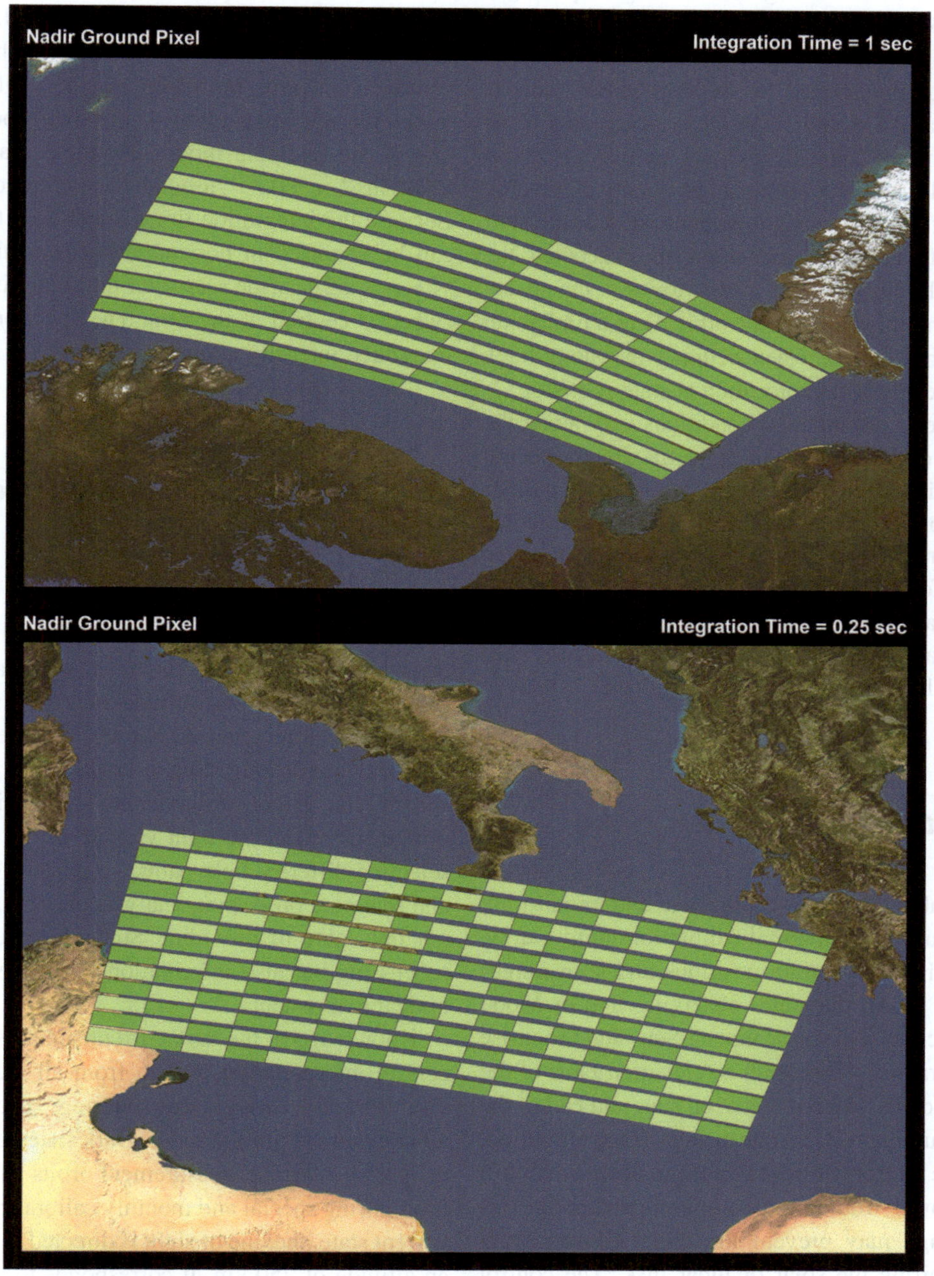

Fig. 4.9 The pattern of ground pixels in a nadir measurement for an integration time of 1 s (*top*) and 250 ms (*bottom*). Only the forward scans are shown. This causes the along-track gaps between consecutive scans which vary in width due to a projection effect. Across-track extent is defined by the integration time while along-track the size reflects the dimension of the IFoV with only a small contribution of the integration time (Courtesy: DLR-IMF; background maps: NASA).

horizontal scans the full range of the atmosphere is observed nearly up to its top at about 93 km. When the last horizontal scan has finished, the ILoS is moved to an altitude of 250 km to obtain a short dark current measurement whilst pointing into open space. The IFoV at the distance of the Earth's limb corresponds to a ground pixel size of 103 × 2.6 km (across-track × height). As for nadir states, the integration time determines the area covered by the readout, i.e. by the scanning IFoV. Largest limb ground pixels span 1060 km × 3.6 km (across-track × height, including step height error), smallest 230 km × 2.6 km across-track × height).

A limb state executes one of the most complex measurement functions defined for SCIAMACHY. This is due to the fact that ASM and ESM movements have to be synchronised with the channel readouts in the horizontal scans. The ground scene of a limb scan is usually defined by the geolocation of the line-of-sight tangent point at the start and end of the state. Due to the elevation steps executed, the tangent point of the line-of-sight moves towards the spacecraft as the platform moves along the orbit. Thus a limb pixel appears to be rather narrow when defining the along-track extent via its tangent height geolocation at state start/stop.

Since atmospheric illumination conditions vary along the orbit as a function of the solar zenith angle, maximum spectral signal-to-noise ratios require selection of pixel exposure times as a function of orbital position. Therefore science oriented states do exist for different orbital positions. In total seven nadir states, each for wide and small swath settings execute the same scanning trajectory but with different pixel exposure times. The same applies to limb states. Here six states each for wide and small swath cover all orbit positions.

Occultation States

While nadir, limb and those calibration and monitoring states which are using the internal calibration units, can be defined by selecting appropriate positions and scan ranges for the ESM and ASM, the states observing Sun or Moon require dynamic control of the scanners. Sun- and moonrise is affected by refraction. This leads to elevation rates of the rising solar/lunar disks different between start and end of the occultation measurement. One rate results from refraction, the other from the platform's orbital motion. Scanner control via the Sun Follower would be able to compensate for this but cloud coverage may prevent the Sun Follower from successfully acquiring the solar or lunar disk. The control loop via the ICU works with a single scan rate only and is unsuitable for tracking the rising Sun/Moon in the early occultation phase. Therefore the corresponding states implement a dedicated Sun/Moon occultation procedure based on the fact that at an altitude of about 17 km, refracted image and true position overlap well. From then on the Sun/Moon rises with an almost constant rate. At the beginning of the state the ESM is rotated to point at an elevation corresponding to 17 km and performs continuous vertical scans of 2 s each with a vertical range of ±0.33°. The ASM is rotated to an azimuth angle which ensures that the Sun or Moon is within the field of view of the Sun Follower when their refracted disk appears at the limb. Because of the scan motion of the ESM, the object is detected when reaching an altitude between the horizon and 17 km. Once the Sun/Moon has arrived at an altitude of 17 km above the horizon, control is switched either to the SF or ICU. In the latter case a constant rate for the ESM is sufficient because from that elevation onwards refraction can be neglected. The ESM tracks the upward motion of the Sun/Moon in pointing or one of the scanning modes. Nominal scanning moves the ESM in 2 s ±0.33° around the centre of Sun or Moon. Because the integration times are shorter than 2 s, the light can be analysed in horizontal slices of the disk. The fast sweep is a 2.1° wide scan over the solar disk in 0.125 s. The sweep is centred at the Sun. The spectrometer records the integrated intensity at one sweep over the full disk.

Calibration and Monitoring States

Usually, calibration and monitoring states operate either the internal calibration lamps SLS and WLS, measure the dark signal from deep space or observe Sun and Moon. As long as the LoS during solar or lunar sunrise traverses the atmosphere, i.e. below an altitude of 100 km, the data serve scientific requirements. Above 100 km, they support calibration and monitoring. Sun measurements above the atmosphere can either observe the solar disk via the scan mirrors or reflect the light via one of the two diffusers. By selecting different light paths – e.g. using the extra mirror – and scanning properties, analysis of solar and lunar states is not only able of providing Sun reference spectra for data processing but also information about the status of various optical components.

Knowledge of the dark current signal is required for successful interpretation of data from all measurement states. Therefore five dark current states are specified which cover all relevant integration times. Dark states are executed on the eclipse side during measurement orbits and along the whole orbit during special and monthly calibration orbits. In a dark current state, the line of sight is directed to and maintained at an altitude of 250 km. It corresponds to pointing into deep space well above the atmosphere. No earthshine light is expected at this altitude and only the detector dark signal should be recorded.

SLS and WLS states are required to derive further pixel-dependent detector properties and to monitor the instrument's stability. Whenever one of these states is operated,

Calibration & Monitoring Timescale	Calibration & Monitoring Measurement																								
	Sun nominal scan	Sun nominal scan/pointing	Sun fast sweep scan	Sun over ESM diffuser (NDFM in)	Sun over ESM diffuser (NDFM out)	Sun over ASM diffuser	Sun over extra mirror fast sweep scan	Sun over extra mirror pointing	Sun over extra mirror nominal scan	Sun over ASM diffuser through atmosphere	Sub-solar fast sweep scan	Sub-solar pointing/nominal scan	Sub-solar pointing	Moon pointing	Moon nominal scan	Dark current	WLS	SLS	WLS over ESM diffuser	SLS over ESM diffuser	NDFM monitoring (NDFM in)	NDFM monitoring (NDFM out)	Nadir pointing left	Nadir pointing	ADC
Window	SO&C										Sub-solar			MO&C											
Orbital																									
each orbit	49													57		8,26,46,63,67									65
Daily																									
1st orbit		47	50	62										56	54	8,26,46,63,67									65
1st orbit (every 3rd day)		47	50	62							53			56	54	8,26,46,63,67									65
2nd orbit		47				17-21	68							57		8,26,46,63,67									65
Weekly																									
1st orbit		47	50	62										56	54	8,26,46,63,67	61	59							65
2nd orbit		47				17-21	68							57		8,26,46,63,67									65
Monthly																									
1st orbit		47	50	62							53			56	54	8,26,46,63,67	61	59							65
2nd orbit		47			52		68					58				8,26,46,63,67			70						65
3rd orbit		47				17-21		64					60			8,26,46,63,67				69	48	16			65
4th orbit		47							66							8,26,46,63,67							38		65
5th orbit										22						8,26,46,63,67								23-25,42-45	65

Fig. 4.10 Calibration and monitoring scenarios from orbital to monthly timescales. In the top row the individual measurements and their targets, e.g. Sun, Moon, lamps, are listed. The states used in each calibration orbit refer to the definitions in Table 4.2. All states unrelated to the Sun or the Moon can be executed several times at any position along the orbit (Courtesy: DLR-IMF).

the ESM is rotated to the position where its mirror reflects light from the lamps onto the entrance slit of the spectrometer. Orbital variations may be detected by running SLS or WLS states several times during an orbit. Since each lamp dissipates heat when operated, thermal perturbations have to be kept to a minimum.

The mission scenarios described in section "Mission Scenarios" specify the execution of calibration and monitoring states on orbital, daily, weekly and monthly timescales. How frequently these are scheduled depends on their specific measurement goals. It has to be ensured that the entire calibration and monitoring measurements cover all aspects of the in-flight instrument characterisation. Figure 4.10 depicts the associated measurements on orbital, daily, weekly and monthly timescales with state ID referring to the definitions in Table 4.2.

4.6 Timelines

Timeline Concept

The execution of single states is possible. However, with an average state duration of about 60 s, commanding a complete orbit, state by state, is cumbersome, both for the ENVISAT ground segment and for operations planning. Therefore, SCIAMACHY allows execution of predefined state sequences. These sequences are called *timelines* (SOST 2001b). A total of 63 timelines can be stored onboard in the Timeline table of the ICU RAM.

Each timeline is characterised by the chronological sequence of states and its total duration. Once the timelines are stored in the Timeline table, they can be started via MCMD. This MCMD provides the scheduled timeline start time. If the timeline includes a state executing a Sun or Moon measurement, e.g. sunrise at a given altitude, subsolar event or moonrise at a given altitude, additional position parameters – aspect and nadir angle together with nadir rate – specifying the solar or lunar celestial positions, are uploaded with the MCMD. They are used by the instrument to correctly position the scan mirrors at the beginning of the particular state to initially acquire the target. Execution of the timeline is a complex interaction between various parameter tables. From triggering the first timeline related commands in the RTCS table until the sequence of states has finally run to completion, information is extracted from tables and used to control instrument measurement activities. It is a prerequisite for successful operations that the content of the associated tables is consistent and free of conflicts. The definition of the timeline header information ensures that Sun or Moon states observe their target at the right time and location to meet the scientific requirements.

Since the start timeline MCMD can provide only one set of position parameters for one solar or lunar event, there can only exist one Sun or Moon related state in a timeline. Thus timelines including a Sun or Moon state are fixed in time and are called *Sun fixed* or *Moon fixed*. All other timelines without a Sun or Moon state are scheduled relative to Sun or Moon fixed timelines.

Timeline Definition

Based on the objectives of each orbital mission scenario and the occurrence of Sun and Moon fixed events along the orbit, timelines can be built from the set of 70 states. Each timeline corresponds to an orbit interval with start/stop being related to a Sun or Moon fixed event. Timelines can be assigned to the following orbit intervals:

- SO&C window
- MO&C window
- Start to end of eclipse
- End of SO&C window to start of eclipse
- End of SO&C window to start of sub-solar window
- End of sub-solar window to start of eclipse
- End of SO&C window to start of MO&C window
- End of sub-solar window to start of MO&C window
- End of MO&C window to start of eclipse

A complete orbital mission scenario is implemented by assembling a sequence of timelines which covers the full orbit and executes those states required in the scenario. This is an efficient building block approach which reduces the command load drastically. The most frequent scenario executes 4 timelines only – a SO&C timeline, followed by a long limb/nadir sequence and two calibration timelines in eclipse. Because the sequence of limb/nadir states generates a ring-like pattern of nadir and limb ground pixels, it has been decided to switch between two limb/nadir sequences in consecutive orbits. At latitudes where nadir ground pixels exist in one orbit, limb ground pixels are generated in the following orbit. The result is a chessboard type pattern better suited for global value-added data processing. In order to obtain earthshine spectra with a signal-to-noise ratio as high as possible, the sequence of nadir and limb states in a timeline for execution on the illuminated side of the orbit needs to reflect the relation between pixel exposure times and orbital position. For a selected swath setting, usually all defined nadir and limb pixels are required to cover the interval from sunrise to sunset.

All timelines starting or ending with the MO&C window have to accommodate the strong temporal variability of lunar events within a monthly visibility period. Therefore several versions of Moon related timelines with different lengths do exist for the same segment. Triggered by mission planning, they are exchanged on-board whenever required by lunar position. This is different from timelines allocated to Sun related orbit segments which require only single instances due to the moderate seasonal changes (Fig. 4.11).

As in the case of states, a final flight configuration exists which reflects the currently agreed and verified set of 63 timelines. Its configuration controlled status can be modified via the same OCR procedure as for states.

4.7 SCIAMACHY Operations Setup

Operations of the instrument include mission planning, configuration control of the on-board measurement status and instrument monitoring. Due to its status as an AO instrument, operational responsibilities for SCIAMACHY were split between ESA and DLR/NIVR. Agreements define that FOCC executes daily SCIAMACHY flight operations as for all other ENVISAT instruments but based on input from the AOP, whereas operational offline tasks are assigned to DLR/NIVR. On the AOP side, the SCIAMACHY Operations Support Team (see section "The Making of SCIAMACHY" of Chapter 3) interfaces with ESA, particularly FOCC, to fulfil these functions in order to accomplish safe operations and generation of high quality measurement data.

SCIAMACHY mission planning has the task to generate and plan timelines around solar and lunar events in each orbit. Since the occurrence of such events depends purely on orbital and celestial mechanics it is predictable. Well in advance of actual operations, orbit analysis determines the properties of Sun and Moon related orbit segments. Timelines are then generated ensuring that the complete orbit is covered with measurements – scientific on the illuminated side of the orbit and calibration and monitoring intermittent with specific scientific in the eclipse phase when the Earth's atmosphere appears dark. Regular mission planning consists of defining the measurement programme for time slices of several weeks by SCIAMACHY Operations Support. This programme is translated into timeline sequences for each orbit and provided to ENVISAT's mission planning system (MPS). The sequences must yield maximum orbital coverage, permit conflict-free instrument operations and comply with the overall ENVISAT flight operation rules. ENVISAT MPS allocates absolute timeline start times. In order to inform validation users well in advance of actual operations about planned measurements, Operations Support determines the geolocation of planned nadir and limb ground pixels based on an orbital simulation of the schedule.

Configuration control of the on-board measurement status is responsible for the final flight status of states and timelines. A database of engineering parameters, measurement

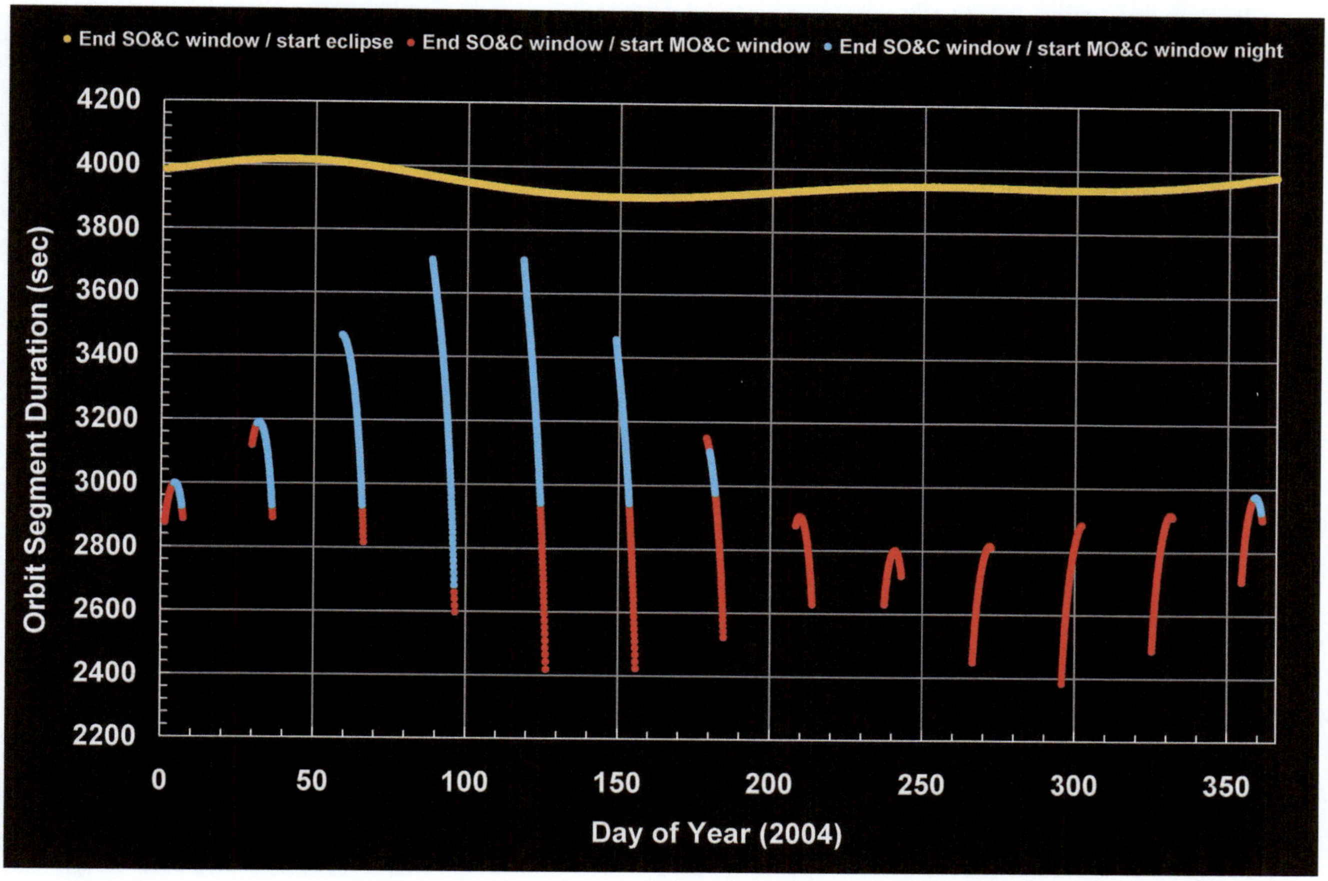

Fig. 4.11 Example of the seasonal temporal variability of orbital segments for the year 2004. The time interval between end of SO&C window and start of eclipse varies only slightly over a year (*yellow*). In the monthly lunar visibility periods, the time between end of MO&C window and start of eclipse shows a much higher variation (*red curves*). The blue segments indicate lunar visibility phases where moonrise occurs on the nightside, i.e. those which can be used for occultations (Courtesy: DLR-IMF).

parameters and timelines is maintained on SCIAMACHY side by Operations Support. Whenever required, modified tables are generated, translated into the MCMD format and sent to FOCC. FOCC integrates SCIAMACHY's input into the overall ENVISAT command database and uploads it according to the specified validity time.

Instrument monitoring includes several operations related aspects. Both the monitoring at FOCC and SCIAMACHY Operations Support uses HK telemetry and report formats to compare executed with planned operations for detecting anomalies. While FOCC focuses on the day-by-day operations following the procedures specified in the Instrument Operation Manual (IOM), Operations Support monitoring covers also long-term aspects including failure investigations in case of anomalies. Particularly important is the monitoring of subsystems which directly impact data quality, i.e. the reliability of the retrieved atmospheric parameters. This concerns the thermal systems ATC and TC. The regular weekly checking of detector and OBM temperatures verifies that the temperatures remain within specified limits. In case of limit exceedings appropriate countermeasures are elaborated by SOST, translated into SCIAMACHY Operations Requests (SOR) and submitted to FOCC for implementation. Additionally the accuracy of the LoS during measurements requires attention. It is determined by the interaction of several components such as, e.g. platform attitude, extra-mispointing or scanners. Those parameters which are relevant for specifying their performance are systematically analysed. SOST provides relevant findings not only to flight operations but also to the SQWG ensuring that the LoS knowledge in data processing does not degrade.

The past, present and planned status of SCIAMACHY mission planning, instrument configuration and long-term monitoring results are reported by SCIAMACHY Operations Support via its dedicated website at http://atmos.caf.dlr.de/projects/scops/. This site includes not only dynamic information but informs also about orbit properties and the overall operations and mission planning concept. A comprehensive description of the final flight states, the corresponding valid parameter settings, a list of timelines and how consolidated level 0 products reflect the SCIAMACHY measurement status supplement the website.

References

EADS Astrium GmbH SCIAMACHY Team (2003) Instrument operation manual (IOM – MA-SCIA-0000DO/01). Technical Document, EADS Astrium

SOST-DLR (1996) SCIAMACHY Orbit Analysis (PO-TN-DLR-SH-0002). Technical Document, DLR

SOST-DLR (2001a) SCIAMACHY operations concept I. Mission scenarios (PO-TN-DLR-SH-0001/1). Technical Document, DLR

SOST-DLR (2001b) SCIAMACHY operations concept II. Timelines: Generation, planning & execution rules and reference timelines (PO-TN-DLR-SH-0001/2). Technical Document, DLR

SOST-DLR (2003) SCIAMACHY operations concept III. Instrument states and onboard tables (PFM) – (PO-TN-DLR-SH-0001/3). Technical Document, DLR

Chapter 5
Calibration and Monitoring

R. Snel, G. Lichtenberg, S. Noël, M. Krijger, S. Slijkhuis, and K. Bramstedt

Abstract Spaceborne spectral measurements over long time periods require calibration and monitoring of the instrument as a crucial prerequisite for successful retrieval of atmospheric parameters. Calibration applies a sequence of steps to the measurement data while monitoring assesses the optical performance thus permitting degradation corrections. The parameters characterising the instrument were obtained in a sequence of on-ground calibration runs under different environmental conditions. They are stored as *Key Data* and serve as input when deriving calibrated spectra. Relevant calibration steps include the memory effect and non-linearity, wavelength calibration and both spectral and spatial stray light corrections. Since SCIAMACHY is sensitive to the polarisation state of the incoming light, polarisation needs to be thoroughly taken into account. The final step performs the radiometric calibration. Once in orbit, the optical performance monitoring establishes information concerning the channel and wavelength dependent degradation. From the combination of the results for the different light paths it is even possible to learn how individual optical components degrade.

Keywords Calibration • Key data • Calibration equation • Polarisation correction • Optical performance monitoring

Spaceborne spectral measurements require the translation of measured signals into physical quantities and need to maintain this process with high precision over long time periods. Therefore calibration and monitoring of the instrument is a crucial prerequisite for any successful retrieval of atmospheric geophysical parameters. Calibration of the instrument has to be valid at any time during the mission. Calibration measurements, which cannot be performed in flight, need to be obtained before launch on ground. Once in orbit the instrument is expected to change. It needs to be calibrated in flight where possible, and monitored where in-flight calibration is not feasible.

The goal of the calibration is to convert electronic signals of detectors (Binary Units – BU) into physical units (e.g. $W/m^2/nm$). This is achieved by applying a complex sequence of individual calibration steps to measurement data. A detailed description of each step can be found in Slijkhuis (2000a) and Lichtenberg et al. (2006). Monitoring aims at giving an as precise as possible characterisation of the optical performance in order to correct for degradation effects throughout the instrument's lifetime. Therefore, monitoring serves as a general prerequisite for continuous high data product quality. Monitoring related to the optical performance of SCIAMACHY is to a large degree linked to the instrument calibration and characterisation status. It establishes in-flight information which permits proper application of on-ground calibration corrections and modelling of the in-orbit environment.

5.1 On-ground Calibration Philosophy

Establishing a valid and accurate on-ground calibration strategy facilitates calibrating the data once the instrument is in orbit. This includes environmental considerations. The experience gained from GOME flying on-board the ERS-2 satellite, where various air-vacuum effects led to calibration problems, showed that spectrometers should ideally be calibrated under thermal vacuum (TV) conditions. In the case of SCIAMACHY a range of incidence angles on the

R. Snel (✉) and M. Krijger
SRON, Netherlands Institute for Space Research, Sorbonnelaan 2, 3584 CA Utrecht, The Netherlands
e-mail: r.snel@sron.nl

G. Lichtenberg and S. Slijkhuis
Remote Sensing Technology Institute, German Aerospace Center (DLR-IMF), Oberpfaffenhofen, 82234 Wessling, Germany

S. Noël and K. Bramstedt
Institute of Environmental Physics/Institute of Remote Sensing (IUP-IFE), University of Bremen, Otto-Hahn-Allee 1, 28359 Bremen, Germany

M. Gottwald and H. Bovensmann (eds.), *SCIAMACHY - Exploring the Changing Earth's Atmosphere*, DOI 10.1007/978-90-481-9896-2_5, © Springer Science+Business Media B.V. 2011

mirror(s) and mirror-diffuser combinations had to be covered in the calibration, requiring rotation of the instrument. The available vacuum chamber hardware did not allow rotation of the instrument itself, and only provided views in a limited angular range centred on nadir and limb in the flight direction. Therefore, a combination of thermal vacuum and ambient measurements was used. Radiometric sensitivity and polarisation sensitivity of the instrument were measured under TV conditions for *one* reference angle and all necessary instrument modes (limb, nadir and irradiance). The TV on-ground calibration was performed during several campaigns using the OPTEC facility (see section "The Making of SCIAMACHY" of Chapter 3). SCIAMACHY had been placed inside the vacuum chamber with the thermal hardware being replaced by a system based on liquid nitrogen and heaters to reach and maintain the correct temperature. Optical windows in the tank allowed the light from external optical stimuli to enter. In the Key Data (see below) the effect of the optical window had been compensated for. The OPTEC facility was first used for requirement verification tests, i.e. a check to see if the instrument met its requirements. Later, calibration measurements were performed in OPTEC. Since optimising the instrument required major hardware changes, several OPTEC campaigns had to be scheduled and executed. They ran from summer 1997 till spring 2000.

In order to be able to calibrate all incidence angles on the mirrors (or diffusers), component level measurements of all possible mirror combinations and the mirror/ESM diffuser combination were made under ambient conditions. These ambient measurements were performed for a set of angles – including the reference angle measured under TV conditions – and a set of selected wavelengths. From such ambient measurements the scan angle correction was calculated. The reference angle measurement was used to transfer the results from the ambient measurement to the TV conditions. Measurements included unpolarised and linearly polarised light. The combination of TV and ambient measurements gave the instrument response for all incidence angles at Begin-of-Life (BOL) of the instrument. Implicit assumptions for the combination of the TV and ambient measurements had been that the polarisation dependence of the mirrors and diffusers are the same in air and in vacuum and that there is no temperature dependence. Both assumptions were reasonable for SCIAMACHY because of using uncoated mirrors. Critical points in the transfer of ambient and TV measurements comprised the geometry (incidence angles on the mirrors or diffusers), the illumination conditions and the detector used for the component measurements. Obviously, errors in the geometry would have led to incorrect angle dependence for the calibration quantity to be measured. Light levels during instrument measurements and during component measurements were certainly different. While the footprint of the light source on the component could be matched to the footprint during the instrument measurements, it was impossible to recreate the exact illumination conditions. This could have introduced systematic errors into the calibration. Finally, care had to be taken that the detector used during the measurements under ambient conditions did not introduce artefacts since it differed from the detectors used on-board SCIAMACHY. The ambient calibration was executed between December 1997 and April 1998 in a dedicated set-up, the ARCF (Absolute Radiometric Calibration Facility). In order to allow a rotation of the mirror (s) or the mirror/diffuser combination to any required position, they were placed on a special optical bench. Having two mirrors or a mirror plus a diffuser on the optical bench permitted direct measurement of the combined response of both optical elements and calibration of all SCIAMACHY instrument modes at the appropriate angles. A monochromator and polarisers were used to obtain the response for different wavelengths and polarisations. All measurements took place in a class 100 cleanroom with a controlled temperature of 20°C and 50% air humidity.

In order to minimise potential errors from the measurements performed under ambient conditions, only ratios of measurements were used for the calibration where possible. The individual calibration parameters derived from the on-ground measurements were combined into a set of data files, the *Key Data* files. These Key Data are applied by the data processor to derive calibrated spectra.

5.2 The General Calibration Equation

SCIAMACHY is a scanning instrument measuring only the spatially integrated light of one ground pixel at a time. Tracing the light from the Earth through the telescope, slit, spectrometer, detector, and on-board processing can be described by

$$\vec{I}_{tel}(\lambda, \alpha, \beta) = \mathbf{P_{tel}} * \vec{I}(\lambda, lat, long) \tag{5.1}$$

$\vec{I}$ is the polarised intensity at wavelength λ arriving at the telescope from the geolocation given by *lat/long*. $\vec{I}$ is written in Stokes notation as a vector. $\mathbf{P_{tel}}$ is the point spread function of the telescope which may affect the polarisation state of the light, represented here as a matrix. Optical distortion due to the telescope is implicitly included in $\mathbf{P_{tel}}$ which thus may depend on the coordinates α and β. They are defined in the focal plane of the telescope at the entrance slit to the spectrometer part of the instrument. $\vec{I}_{tel}$, the intensity after the telescope, results from a convolution of $\mathbf{P_{tel}}$ and $\vec{I}$ denoted by the operator '*'.

After having passed through the telescope, the light crosses the entrance slit where the intensity masked out by the slit is removed:

$$\vec{I}_{slit}(\lambda, \alpha, \beta) = \vec{I}_{tel}(\lambda, \alpha, \beta) \cdot F(\alpha, \beta) \quad (5.2)$$

where F is a function of the focal plane coordinates α and β and is zero where the light is blocked and unity where the light passes unhindered through the slit.

The optics after the slit project the image of the slit on the detector and add wavelength dispersion to obtain a spectrum. For instruments with an integrating Instantaneous Field of View (IFoV), like SCIAMACHY, the detector can be considered as one-dimensional:

$$I_{det}(x) = \int_{detector} \mathbf{P_0} * \vec{I}_{slit}(\lambda, \alpha, \beta) + B \quad (5.3)$$

where $\mathbf{P_o}$ is the wavelength and polarisation dependent point spread function of the optics, including the dispersion introduced by the spectrometer. The coordinate x is the detector pixel index. Integration is performed over each pixel individually. Possible self-emission of the instrument, i.e. thermal background radiation, is covered by B which may be position dependent. Note that polarisation information is lost once the signal has been projected on the detector.

As soon as the signal is integrated over the detector pixels, it can be treated electronically. Up to this point the signal has been considered purely linear in *intensity*, but the detector material and electronics may introduce non-linearity and hysteresis:

$$S_{det} = E(I_{det}, t_{exp}, I_{det}(t)) \quad (5.4)$$

where E is the function describing the transfer of I_{det} into the digitally sampled detector signal S_{det}, taking into account the exposure time t_{exp}. The function E may depend on the history (memory effect – see below) of I_{det} represented here as $I_{det}(t)$. Note that in case of SCIAMACHY, on-board processing can modify the digitally sampled signal S_{det} only in the form of co-adding multiple readouts of the detector. Combining all of the equations above, while

- Considering a dedicated correction for polarisation sensitivity (see below)
- Describing the effects of the point spread function of the optics $\mathbf{P_o}$ on the recorded spectrum as an additive component
- Introducing explicit temperature dependence of the detector quantum efficiency

(5.4) can be worked out to

$$S_{det} = I(\lambda) \cdot \Gamma_{inst}(\lambda) \cdot QE(T_{det}, \lambda) + S_{stray} + DC + S_{elec} \quad (5.5)$$

where Γ_{inst} is the total transmission of the instrument, QE the detector temperature dependent quantum efficiency, S_{stray} the stray light, DC the total dark signal and S_{elec} electronic effects such as non-linearity. This equation must be solved for every detector pixel. In order to obtain the spectrum as a function of wavelength λ for each pixel, the wavelength has to be determined and the equation has to be inverted to calculate the intensity I. Generally, the transmission of the instrument is dependent on the polarisation state of the incoming light.

5.3 Detector Corrections

Several corrections related to the electronics of the detectors and the detectors themselves, i.e. the terms S_{elec} and DC in (5.5), have to be applied (see Fig. 5.1). The UV-VIS-NIR channels 1–5 and the SWIR channels 6–8 must be treated separately during the calibration due to their different detector material and readout electronics. Signals are described in terms of Binary Units (BU). The Analogue-to-Digital Converter (ADC) of SCIAMACHY digitises the signal of the detector with 16 bit resolution, meaning that detector signals, also referred to as 'fillings', range from 0 to 65535 BU.

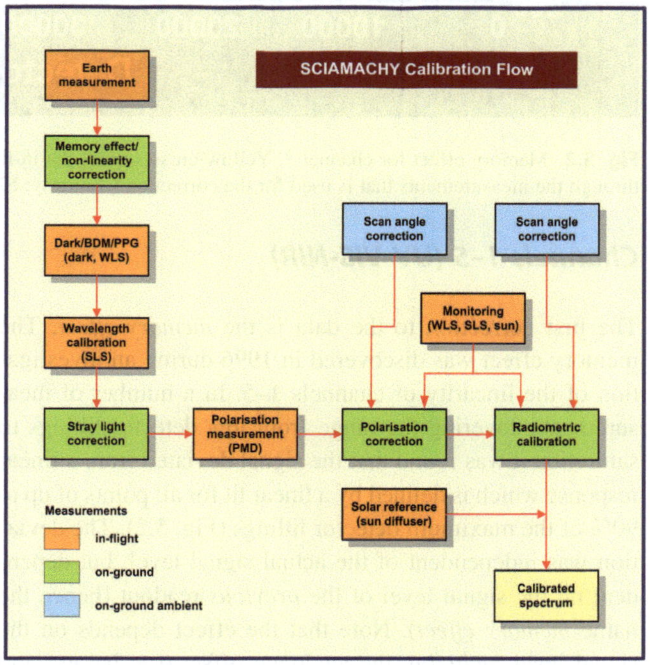

Fig. 5.1 Calibration concept for SCIAMACHY. The final calibrated Earth radiance spectra are obtained by applying several calibration steps to the measured Earthshine signals. They include in-flight calibration measurements (*red*), on-ground measurements performed under thermal vacuum conditions (*green*) and component measurements from on-ground ambient tests (*blue*). The optical performance monitoring (*red*) provides additional corrections (Courtesy: SRON).

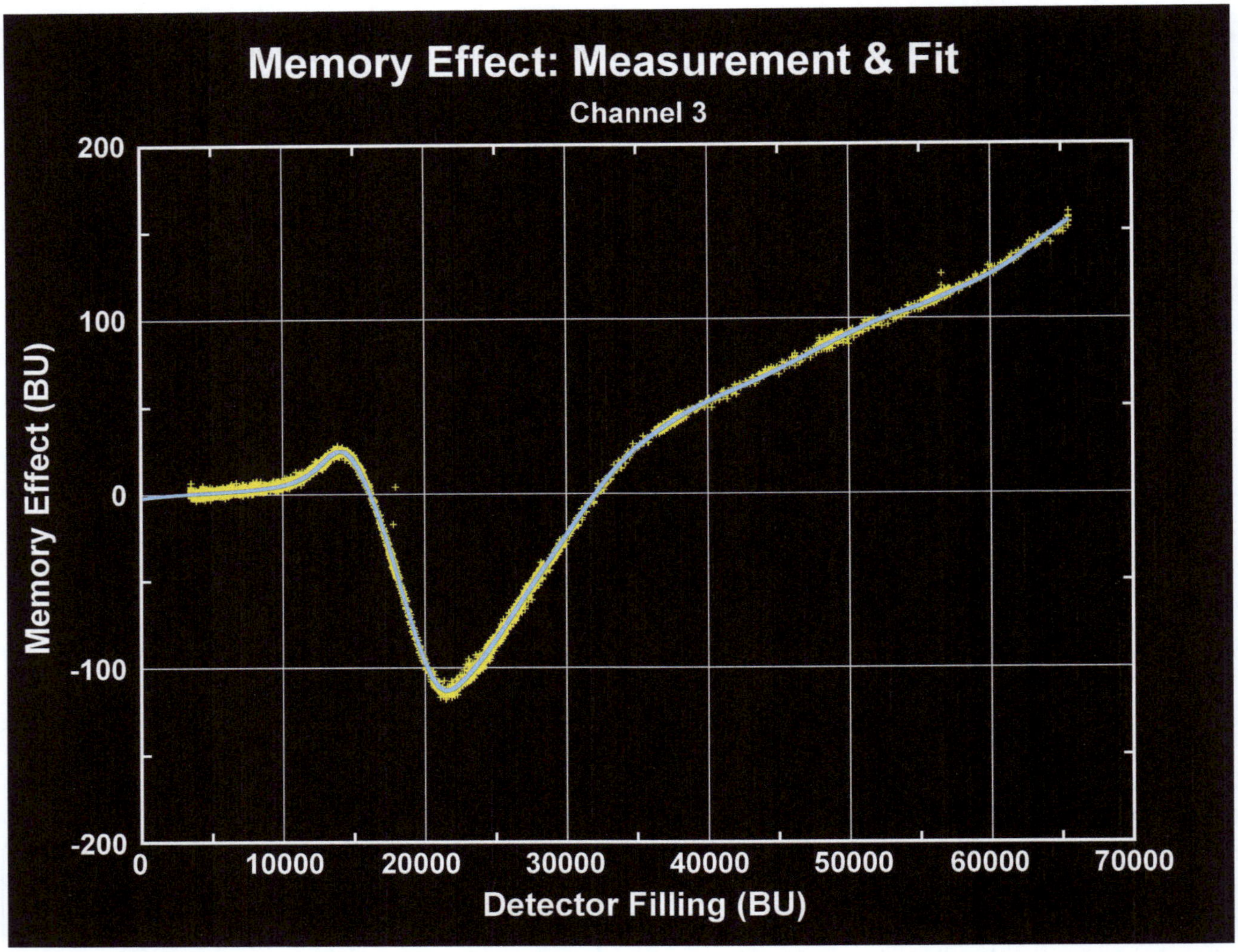

Fig. 5.2 Memory effect for channel 3. Yellow crosses mark the in-flight measurement of the memory effect. The blue solid line is a spline fit through the measurements that is used for the correction (Courtesy: SRON).

Channels 1–5 (UV-VIS-NIR)

The first correction to the data is the *memory effect*. The memory effect was discovered in 1996 during an investigation of the linearity of channels 1–5. In a number of measurements covering the range from low detector fillings to saturation it was found that the signal deviated from a linear response which is defined by a linear fit for all points of up to 90% of the maximum detector fillings (Fig. 5.2). The deviation was independent of the actual signal level, but dependent on the signal level of the *previous* readout (hence the name *memory effect*). Note that the effect depends on the signal level *including* the analogue offset (see below) and dark current. Thus it has to be applied before any other correction. In order to characterise the memory effect, White Light Source (WLS) measurements followed by several dark measurements were performed on-ground and in-flight. The difference between the first dark measurement after the WLS measurement and subsequent dark measurements gives a correction value as a function of detector filling. This value needs to be subtracted from the data to correct for the memory effect which is assumed to be the same for all pixels. The total correction for a single readout ranges from −0.61% to 0.21% of the detector filling of the previous readout with a maximum effect at fillings around 19000–21000 BU, depending on the channel. More information can be found in Lichtenberg (2003).

The second detector correction to be applied is the *dark signal* correction. The dark signal is measured in every orbit during eclipse using five different states. In channels 1–5 the dark signal consists of two components: the analogue offset (AO) and the leakage current (LC). The analogue offset is independent of exposure time, it is a fixed signal added to the measured signal to avoid negative signals. The leakage current is caused by thermally created electron-hole pairs. The total dark signal for channels 1–5 is

$$DC_{ch15} = f_{coadd} \cdot AO + f_{coadd} \cdot t_{PET} \cdot LC \quad (5.6)$$

where f_{coadd} and t_{PET} are the co-adding factor of the cluster and the pixel exposure time, respectively. Note that the analogue offset is only multiplied with the co-adding factor since it is independent of the integration time and is added to the signal for each detector readout. Linear fitting to dark measurements with different integration times yields the in-flight dark signal correction. The dark signal in the UV-VIS-NIR channels is dominated by the analogue offset while the leakage current amounts to only 0.04–0.5 BU/s and has roughly doubled since launch.

The final detector related corrections concern the *Pixel-to-Pixel Gain* (PPG) and the *detector etalon*. Generally the pixels in each channel do not show the same response to incoming light. This effect is caused by the electronics and the detector and is thus associated with the individual pixels but not with the wavelength. For channels 1–5, PPG is however small and its correction has turned out to be very stable. The PPG correction used for the UV-VIS-NIR channels had been characterised on-ground. It was derived by first smoothing a WLS measurement, assuming the spectrum is flat. Then the original spectrum was divided by the smoothed measurement, leaving only the high frequency variations that were caused by the different pixel gains in the result.

The etalon correction takes care of in-flight changes of the detector etalon, a spectral structure caused by interference between the boundaries of the protecting SiO layer and the light detecting Si layer. The etalon structures have changed after launch compared to on-ground measurements, perhaps due to contamination on the top layer. The structures also vary with time, particularly after decontaminations. For SCIAMACHY the detector etalon correction is far smaller than for GOME. It can be derived from in-flight WLS measurements.

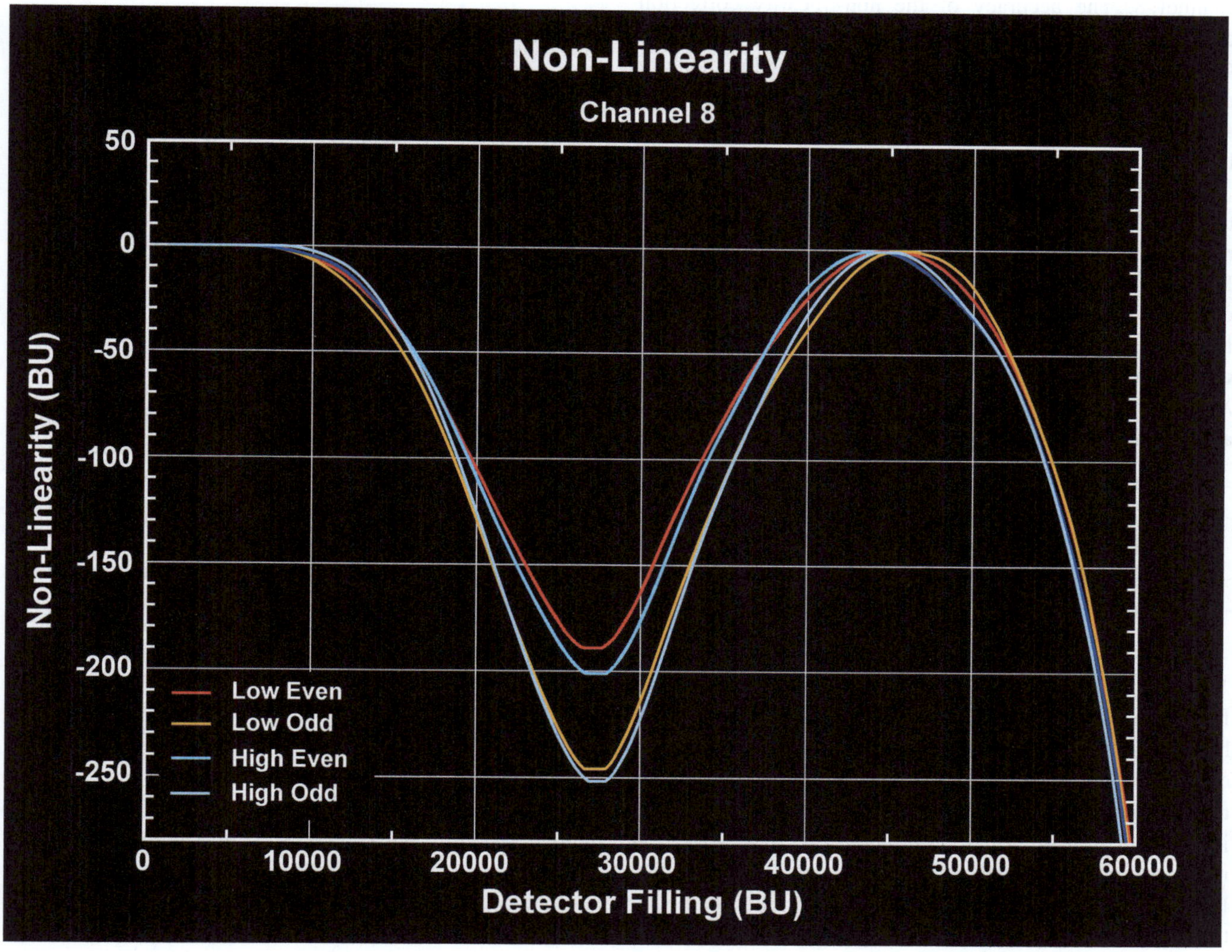

Fig. 5.3 Non-linearity in channel 8 for different pixel regions (indicated by colours). 'Low' pixels are those with pixel numbers below 512. The pixel numbering starts at 0 (Courtesy: SRON).

Channels 6–8 (SWIR)

The SWIR channels do not suffer from the memory effect. However, these channels display a significant non-linearity, i.e. a deviation in the detector response from a (chosen) linear curve. The non-linearity has been measured during the on-ground calibration campaign and a correction algorithm was defined. The maximum value of the non-linearity is around 250 BU which can be significant for weak absorbers such as CO. A separate non-linearity correction for the channels 6, 6+, 7 and 8 has been derived. Within these channels the non-linearity differs for odd and even pixels (starting pixel numbering with 0) because of the different multiplexers used for these two groups of pixels. Additionally, there is a clear difference in the non-linearity between pixel numbers higher and lower than pixel number 511. This leads to 14 correction curves, four per channel with the exception of channel 6+, which covers only pixels 794–1024. Figure 5.3 shows the non-linearity curves derived for channel 8. The accuracy of the non-linearity correction corresponds to 5–21 BU for detector fillings from 10000–40000 BU, depending on the channel. As for the memory effect correction, the non-linearity has to be corrected before any other correction is applied. More details about the non-linearity can be found in Kleipool (2003).

In addition to the non-linearity, Channels 6+, 7 and 8 contain a significant number of unusable pixels due to the lattice mismatch between the light detecting InGaAs layer and the InP substrate. These channels are doted with a higher amount of Indium (see Chapter 3). It changes the lattice constant of the light detecting layer so that it no longer matches the lattice constant of the substrate on which the detecting layer is grown. Any degraded pixels are called 'bad' or 'dead' pixels. There are various effects making these pixels unusable:

- Disconnected pixels preventing any signal readout
- *Random Telegraph* (RT) pixels which spontaneously and unpredictably jump between two levels of dark current leading to different detected signals for the same intensity
- Other effects including excessive noise or too high leakage current that saturates the detector

All these effects were measured on-ground and a Bad and Dead Pixel Mask (BDPM) was created. Pixels of the BDPM have to be ignored in any retrieval. As a result of radiation damage to the detectors in orbit, previously sound pixels may become flagged as 'bad' or 'dead'. A dynamic BDPM is determined in-flight based on monitoring and calibration measurements, and updated as pixels change their status. In order to be less affected by noise on the measurements, the dynamic BDPM is smoothed in time.

After the application of the non-linearity and the BDPM, the dark signal has to be corrected. The dark signal correction in channels 7 and 8 is complicated by the presence of a large thermal background BG_{th} and the unforeseen growth of an ice layer on the detector (see section "Thermal Performance" of Chapter 6). The ice layer slowly changes the detector temperature and attenuates the signal on the detector, including the thermal background. The dark signal in these channels becomes

$$DC_{ch68} = f_{coadd} \cdot (AO + t_{PET} \cdot LC + t_{PET} \cdot \Gamma_{ice} \cdot QE(T_{det}, \lambda) \cdot BG_{th}(\phi)) \tag{5.7}$$

where Γ_{ice} is the transmission coefficient that changes due to the ice layer and QE is the quantum efficiency for the detector.

For channels 6+ and 8 the quantum efficiency changes with the detector temperature T_{det}, whereas the first part of channels 6 and 7 shows no significant temperature dependence. The thermal background is caused by the thermal radiation of the instrument and is the dominant part of the dark signal (about 4000 BU/s) in channel 8. It depends on the orbit phase φ because the temperature gradients in the instrument are not completely stable but vary over one orbit due to the changing angle of solar irradiation. The variation of the dark signal over the orbit can reach up to 60 BU/s which has significant impact on the retrievals of trace gases. In flight the orbital variation is measured once a month during a dedicated calibration orbit in which only dark signal measurements are performed by looking to deep space at a tangent height of 250 km in limb mode. The variation of the transmission makes the dark signal correction time dependent meaning that for channels 7 and 8 a dark signal correction, calculated from measurements in the same orbit, must be used.

The PPG variations in the SWIR channels are larger than in channels 1–5. Differences, although rather stable, of a few percent can be observed. They are monitored on-board but their application is currently not required in data processing (see section "Operational Level 0-1b Processing" of Chapter 8). Due to the design of the detectors no detector etalon effect is present in channels 6–8.

5.4 Wavelength Calibration

In-flight spectral calibration of SCIAMACHY data uses the internal Spectral Line Source (SLS) measurements with the exception of channels 7 and 8 (see below). For selected lines the Falk algorithm (Falk 1984) determines the pixel positions. These are then fitted to theoretical line positions provided with the calibration data. From the polynomial coefficients of the fit the wavelength for each pixel can be calculated. Measurements of solar Fraunhofer lines serve as a quality check. In channels 7 and 8 a calibration with the internal SLS lamp is impossible because in these channels not enough useful lines are available to calculate the wavelength calibration with sufficient accuracy. In channel 8 this is caused by bad pixels interfering with the determination of the line position. Channel 7 only contains two strong doublet lines preventing an accurate determination of line positions over the whole channel. In both channels data from on-ground gas cell absorption measurements establish the wavelength calibration.

An additional effect discovered during the on-ground calibration is the *blocking shift*: During the spectral calibration on-ground, the internal SLS and an external SLS were used. A comparison of the measurements done with the two lamps revealed a wavelength shift of up to 0.07 nm. The reason is a partial blocking of the light path during internal SLS measurements. The blocking shift was characterised and is part of the calibration data. Verification of the spectral calibration in flight has proven that SCIAMACHY is spectrally very stable.

5.5 Stray Light

There are two types of stray light, S_{stray} in (5.5), the spectral stray light and the spatial stray light. It is characterised as a fraction of the total measured intensity for a given pixel.

Spectral stray light is light of a certain wavelength which is scattered to a detector pixel 'belonging' to a different wavelength. It can lead to distortions in the shape of the spectrum. This type of stray light may be caused by a reflection in the instrument after the dispersion of the light beam, or by periodic errors in the spacing of the ruled grooves in a diffraction grating. The source of spectral stray light can be within the same channel, referred to as *intra-channel* stray light, or it can scale with the intensity in a different channel, referred to as *inter-channel* stray light.

Spatial stray light is light entering the telescope from outside the IFoV. It is dispersed just like light from the observation target. Depending on the source of the stray light, the spatial stray light component can add an additional offset to the spectrum and/or distort the spectrum, if the primary source of the stray light has spectral characteristics that differ significantly from the observed target.

Spectral Stray Light

Spectral stray light was characterised on-ground using measurements employing a monochromator. A monochromator produces light in a narrow, predefined spectral band. The centre wavelength of the spectral band can be adjusted. In the derivation of the stray light fractions from monochromator measurements it is assumed that any signal in detector pixels outside this spectral band is caused by stray light. During the on-ground calibration the spectral stray light was measured by changing the central wavelength of the monochromator spectral band, thus covering the whole wavelength range of SCIAMACHY. Dividing the integrated light of the monochromator peak(s) by the light detected outside the peak yielded the stray light fraction. The resulting data is part of the calibration data set and is used to correct the spectral stray light in flight.

In a full matrix approach, the spectral stray light determination would measure the stray light contribution from each individual pixel to all other pixels separately. In practice, however, this is not always possible. In the case of SCIAMACHY a 8192×8192 matrix would be needed making the calculation of stray light too slow. Thus, initially the spectral stray light for SCIAMACHY was separated into three types: uniform stray light, ghost stray light and channel 1 stray light. Ghost stray light is caused by a more or less focused reflection of one part of a spectrum to another part of the spectrum. It can distort the shape of the 'true' spectrum, because it does not add signal to all pixels. During the on-ground measurements many tens of ghost signals were detected in channels 3–8, of which the 20 strongest ones were characterised. The total sum of ghost stray light in a channel is at maximum 1% of the incoming intensity. As it turned out, the split into uniform and ghost stray light did not provide a sufficient correction of the stray light, so the uniform stray light correction was expanded to include a reduced matrix correction.

The reduced stray light matrix describes any of the remaining stray light as a matrix multiplication of an input spectrum and a stray light matrix. A measured spectrum is resampled to lower resolution (1022 pixels instead of 8192), and yields a stray light spectrum of 2048 pixels after matrix multiplication. This stray light spectrum is interpolated to the full 8192 pixel grid and subtracted from the input spectrum.

For channel 1 the situation is less favourable with respect to stray light levels. The on-ground measurements revealed that the spectral stray light in channel 1 can reach levels of up to 10% of the incoming signal for a typical input spectrum. It is also highly wavelength dependent. The main reason for the larger stray light fraction in channel 1 is the high dynamic range of the spectra in this channel, with the lowest signal 3 orders of magnitude smaller than the highest signal. Therefore a dedicated method for channel 1, already formulated before launch, combines the correction of uniform and ghost stray light in a modified matrix approach taking into account polarisation. In order to avoid signal-to-noise problems during the spectral stray light measurements, ten wavelength bands were defined separately for x- and y-polarised light leading to a total of 20 bands (see polarisation section below). Characterising intra channel stray light required for both polarisation directions nine bands in channel 1 and one band in channels 2–5. The channel 1 detector material is not sensitive for light with wavelengths longer than 1,000 nm so the SWIR channels did not need to be considered. For each band the stray light contribution to all detector pixels was calculated leading to a $10 \times 1{,}024$ matrix for both polarisation directions. The stray light fraction in channel 1 ranges from less than 1% to as much as 10%. The correction has an accuracy of around 25% and reduces the stray light by an order of magnitude leaving at most 1% stray light in the spectrum after correction.

Spatial Stray Light

Shortly after ENVISAT emerges from eclipse and passes the North Pole, the Sun shines directly into the limb port. In this orbit region spatial stray light cannot be avoided, i.e. the particular effect was foreseen and the data are flagged accordingly. In order to minimise spatial stray light, the ASM is rotated such that the edge of the mirror/diffuser plate points into flight direction, with the diffuser looking to the instrument side during all measurements using the ESM only.

However both on-ground performance measurements and in-flight limb measurements indicated that there was still a small fraction of spatial stray light present. Dedicated in-flight measurements confirmed the performance measurements and indicate periodic structures in the optics before the slit, resulting in a small fraction of the light being dispersed as by a grating. This has no significant impact for nadir measurements, but limb measurements with a dynamic range of several orders of magnitude over a few degrees do suffer from the spatial stray light. At the moment no corrections for spatial stray light exist (see also section "Optical Performance" of Chapter 6).

5.6 Polarisation

SCIAMACHY is – as all grating spectrometers without a polarisation scrambler – sensitive to the polarisation of the incoming light, i.e. the response will not only depend on the intensity but also on the polarisation of the light. Thus polarisation correction is required. It uses the *Mueller matrix* approach (see e.g. Azzam and Bashara 1977; Coulson 1988). Measurements of polarised light can be expressed by a Mueller matrix **M** and a Stokes vector. Since the detectors only yield a single measurement value per pixel, they can be regarded as polarisation insensitive detectors, and any polarisation sensitivity can be included in the Mueller matrix of the optical components between the incoming light and the detector. Thus, only the top row of the end-to-end Mueller matrix of the instrument is relevant, and can be regarded as a polarisation sensitivity vector:

$$S_{det} = \mathbf{M} \cdot \vec{I} \tag{5.8}$$

SCIAMACHY has multiple viewing geometries, selected by appropriate configuration of the scan mirrors and diffusers. Behind the scanner, the Optical Bench Module (OBM) is identical for limb and nadir measurements. The end-to-end Mueller matrix of the instrument can be split up into a Mueller matrix for the scanner, which is configuration dependent, and a fixed polarisation sensitivity vector of the OBM. This yields

$$S_{det} = \left(M_1^{OBM}\ M_2^{OBM}\ M_3^{OBM}\ M_4^{OBM}\right) \cdot \begin{pmatrix} M_{11}^{sc} & M_{12}^{sc} & M_{13}^{sc} & M_{14}^{sc} \\ M_{21}^{sc} & M_{22}^{sc} & M_{23}^{sc} & M_{24}^{sc} \\ M_{31}^{sc} & M_{32}^{sc} & M_{33}^{sc} & M_{34}^{sc} \\ M_{41}^{sc} & M_{42}^{sc} & M_{43}^{sc} & M_{44}^{sc} \end{pmatrix} \cdot \begin{pmatrix} I \\ Q \\ U \\ V \end{pmatrix} \tag{5.9}$$

On the left hand side a scalar describes the light as detected by the instrument, and on the right hand side we have the polarisation sensitivity vector $\vec{M}^{OBM}$ of the OBM, and the Mueller matrix $\mathbf{M}^{sc}$ defining the response of the scanner to the incoming light represented by another Stokes vector. The first element of this Stokes vector, I, denotes the total intensity of the light. Q is a measure for the polarisation along the x- or y-axis of a chosen reference frame and can be described as $Q = I_x - I_y$. U is a measure for the polarisation along the 45° direction and is defined as $U = I_{45}–I_{-45}$. Finally V is the circular polarisation component of the incoming light, which is negligible for atmospheric light. Note that the total intensity can be written as $I = I_x + I_y$ or $I = I_{45} + I_{-45}$. Often Q and U are normalised to the total intensity I. We will denote normalised fractions with q and u.

All $\mathbf{M}^{sc}$ Mueller matrix elements depend on wavelength and on the incidence angle of the light on the scan mirror(s) or diffuser(s). In the calibration, ambient measurements on component level and instrument TV measurements have to be combined meaning that the actual instrument matrix has to be calculated by a multiplication of the matrix for the scanner (combination) and the OBM. Note that though the end-to-end circular polarisation sensitivity of the instrument is irrelevant when V is zero, the scanner may – through its Mueller matrix – cause a circular polarisation component of the light, which requires the circular polarisation sensitivity of the OBM to be known. Although originally not foreseen for on-ground calibration, an update of the calibration concept introduced the circular polarisation sensitivity into the calibration equations. Defining the relative polarisation sensitivity vector of the OBM as

$$\vec{\mu}^{OBM} = \left(1,\ \mu_2^{OBM},\ \mu_3^{OBM},\ \mu_4^{OBM}\right) = \left(1,\ \frac{M_2^{OBM}}{M_1^{OBM}},\ \frac{M_3^{OBM}}{M_1^{OBM}},\ \frac{M_4^{OBM}}{M_1^{OBM}}\right) \tag{5.10}$$

the response of the instrument to polarised light becomes

$$S_{det} = M_1^{OBM} \cdot \vec{\mu}^{OBM} \cdot \mathbf{M}^{sc} \cdot \vec{I} \tag{5.11}$$

Combination of the polarisation response of the scanner and the OBM using

$$\vec{\mu}^{instr} = \vec{\mu}^{OBM} \cdot \mathbf{M}^{sc} \tag{5.12}$$

and ensuring normalisation of $\vec{\mu}^{instr}$ such that the first element $\vec{\mu}^{instr}$ remains unity, yields the equation describing the instrument response as expressed in total intensity and fractional linear polarisation:

$$S_{det} = M_1^{instr} \cdot I \cdot (1 + \mu_2^{instr} \cdot q + \mu_3^{instr} \cdot u) \tag{5.13}$$

where M_1^{instr} is the radiometric sensitivity of the instrument – with implicit dependence on wavelength for the main science channels, and determined as well for all seven PMD channels – given by

$$M_1^{instr} = \left|\vec{\mu}^{OBM} \cdot \mathbf{M}^{sc}\right|_1 \tag{5.14}$$

$\vec{\mu}^{OBM}$ is provided in the calibration Key Data as determined from on-ground TV measurements. $\mathbf{M}^{SC}$ can be calculated with a scanner model using as input the actual viewing geometry of the instrument, the complex index of refraction of the mirror material as determined during on-ground ambient calibration and optionally any contamination built up on the scan mirrors since launch.

The term in brackets of (5.13) is the inverse of the polarisation correction factor, c_{pol}. It depends on the polarisation sensitivity of the instrument and the polarisation of the incoming light q and u, assuming a zero circular component (Fig. 5.4). The problem of correcting the response of the instrument for polarisation can thus be divided into two parts: (1) determining the polarisation sensitivity of the instrument and (2) determining the polarisation of the incoming light during the science measurements in flight. SCIAMACHY exhibits a different sensitivity to different polarisation states of the light. Thus, in the on-ground calibration, measurements with fully linearly polarised light at a range of polarisation angles and with the instrument under thermal vacuum conditions were used to derive the OBM polarisation sensitivities μ_2^{OBM}, μ_3^{OBM}, and μ_4^{OBM}.

The Mueller matrix for the scanner was determined from ambient measurements with fully linearly polarised light with different polarisation directions. Care was taken to ensure that the intensity in all these measurements was the same. With the definitions given above the polarisation correction factor in terms of on-ground measurements is expressed as

$$c_{pol} = (1 + \mu_2^{instr} \cdot q + \mu_3^{instr} \cdot u)^{-1} \tag{5.15}$$

The second step, the determination of the polarisation of the incoming light is done by determining the ratio of the signal in the PMD channels – which is fully polarised due to the Brewster reflection at the pre-disperser prism (see section "Optical Assembly" of Chapter 3) – and the corresponding signal in the science channel for each individual measurement. During calibration this ratio was determined for different directions of linearly polarised light. The comparison of the in-flight ratio with the calibration data gives seven polarisation values for the whole spectrum, one for each PMD channel. Polarisation values q are calculated from PMD A-F needing the corresponding value of u. The ratio u/q, which depends only on the polarisation angle, is assumed to be constant, such that

$$u_{meas} = q_{meas} \cdot \left(\frac{u}{q}\right)_{const} \tag{5.16}$$

In the original calibration concept, for UV-VIS wavelengths below 600 nm the polarisation angle from single scattering theory was planned to be used (see below), whereas for higher wavelengths the ratio u/q from PMD D and PMD 45°, both centred around 850 nm, had to be taken. The values of q and u here are derived by iteration, until the u needed to calculate q from PMD D and the q needed to calculate u from PMD 45°, match. In-flight it was noted that PMD 45° delivers signals which are systematically 10–15% higher than expected, even for unpolarised sources such as the Sun. As there are indications that this PMD suffers from stray light, it remains currently unused. Instead, u/q is taken from single scattering theory for the complete wavelength range. From POLDER satellite measurements of u/q this appears to be a sufficiently accurate assumption (Schutgens et al. 2004). Note that for small values of q, this ratio becomes very large, thereby amplifying small measurement errors on q_{meas} into large measurement errors on u_{meas}. However, since the instrument is much more sensitive to q than to u, this has little impact on the radiometric calibration, even though u is notably unreliable. In limb, the calculated polarisation displayed an unexpected drift with increasing tangent height, which is probably due to increasing significance of the spatial stray light contribution as the limb intensity

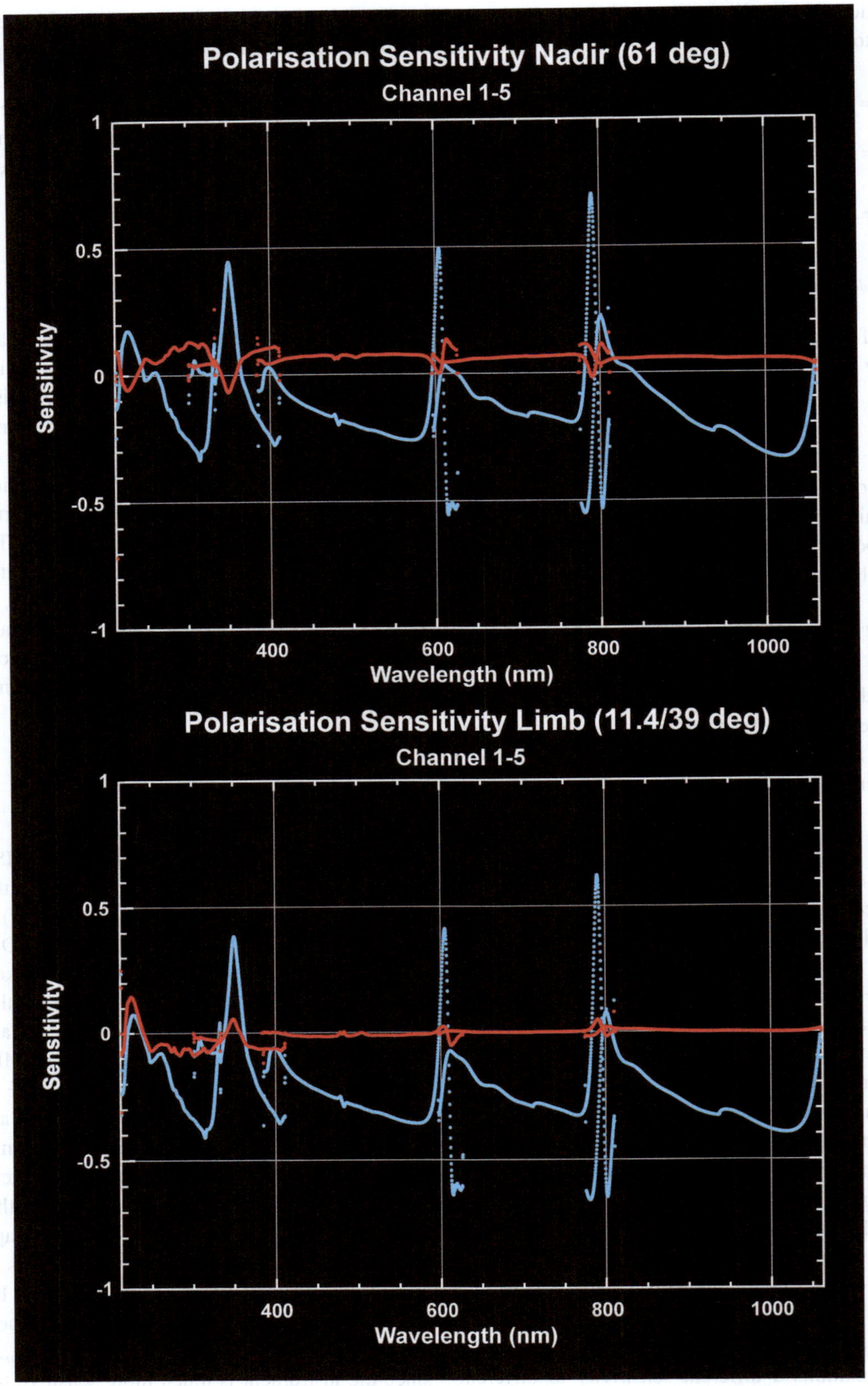

Fig. 5.4 q (*blue*) and u (*red*) sensitivity from (5.13) for nadir (elevation angle of 61°, *top*) and for limb (elevation angle of 11.4° and azimuth angle of 39°, *bottom*) for channels 1–5. Note that these sensitivities are multiplied with the polarisation fractions to get c_{pol} and the correction will thus be smaller than displayed for lower polarisation (Courtesy: SRON).

decreases. Therefore the PMD measurements are only used up to 30 km. Radiative transfer calculations show that above this height the depolarisation remains constant (Mc Linden et al. 2002), such that for higher limb tangent heights we are able to scale the measured polarisation at 30 km with a value obtained from single scattering theory.

The PMD channels cover only the instrument channels 2–8. For channel 1 the backscattered radiation is dominated by single scattering as can be inferred from radiative transfer calculations. Similarly to the GOME instrument, a theoretical value based on single scattering geometry is used here (Slijkhuis 2000b; Tanzi 1999; Tilstra et al. 2003). The transition region from single scattering to multiple scattering and/or ground reflection in the region between approximately 300–325 nm requires special attention. For GOME, a parameterisation of the degree of polarisation was derived as a function of wavelength (see Balzer et al. 1996), known as the 'general distribution function' (GDF) for polarisation. The GDF is characterised by the single scattering value plus three parameters. These parameters are currently obtained using a simplified version of the algorithm from Schutgens and Stammes (2002) where the dependence on scene albedo and ozone content is neglected. More polarisation information may be derived from the channel overlaps of channels 1–6 (five polarisation points) where the different polarisation sensitivities of each channel lead to two independent measurements for the two variables q and u. However, due to calibration inconsistencies, these polarisation points are currently not reliable.

The polarisation values q and u on the level 1b products are specified in an 'atmospheric' coordinate frame which is different from the coordinate frame used for the on-ground calibration and Key Data specification. The 'atmospheric' coordinate frame is related to the geometry of the scattering of light in the atmosphere. The choice has been to define q as parallel to the local meridian plane – the plane through satellite, zenith, and centre-of-FoV (where its Z axis points in the travel direction of light, i.e. towards the instrument). This plane is depicted in Fig. 5.5 for nadir viewing geometry. For limb viewing geometry, the polarisation plane in the figure is rotated 90° towards the line-of-sight as the line-of-sight is approximately in the flight direction. All derived polarisation values and instrument sensitivities are converted to this 'atmospheric' coordinate frame.

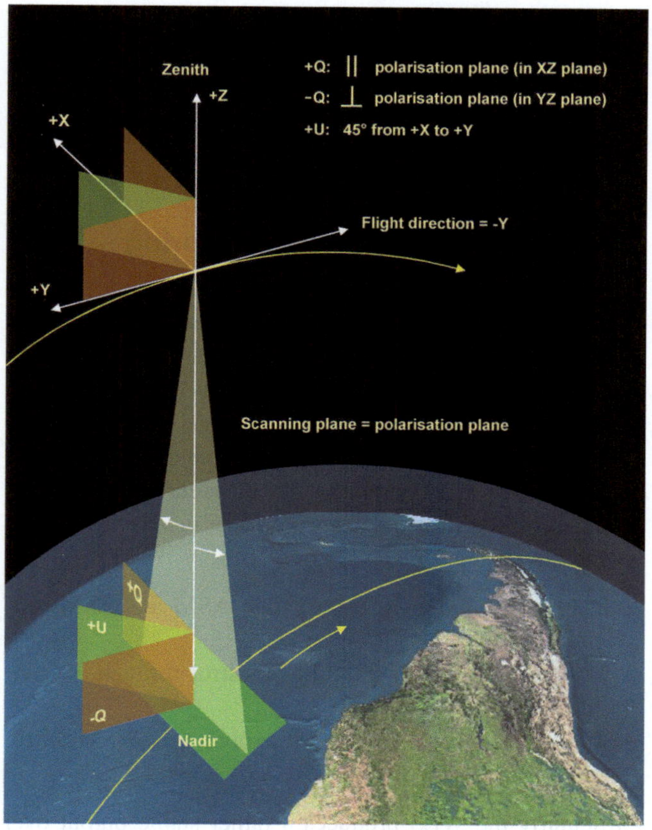

Fig. 5.5 Definition of the nadir coordinate frame used in the output of the data processor for polarisation values q = Q/I, u = U/I (Courtesy: DLR-IMF).

5.7 Radiometric Calibration

The final step in the calibration of the data is the radiometric calibration. The retrieval of trace gases generally uses the reflectance, the ratio of Earth radiance and solar irradiance. The solar irradiance is measured with on-board diffusers in flight. From (5.13) and (5.15) the reflectance can be written as:

$$R = \frac{\pi}{\mu_0} \cdot \frac{I_{Earth}}{I_{sun}} = \frac{\pi}{\mu_0} \cdot \frac{M_1^{sun}}{M_1^{N,L}} \cdot \frac{S_{det}^{Earth} \cdot c_{pol}}{S_{det}^{sun}} \tag{5.17}$$

I_{Earth}, I_{sun} and S_{det}^{Earth} are the Earth or Sun intensity and the measured signal, μ_0 is the cosine of the solar zenith angle, $M_1^{N,L}$ the instrument radiometric response for limb (L) or nadir (N) and M_1^{sun} the instrument radiometric response for Sun diffuser measurements, as calculated from the Mueller matrix for the scanner and the polarisation sensitivity of the OBM. All matrix elements in (5.17) are detector dependent. For a proper calibration the instrument responses have to be determined as a function of wavelength λ and incidence angle α. As mentioned in section "On-ground Calibration Philosophy" the radiometric response was measured on instrument level under TV conditions while the mirror and the mirror/diffuser combination were measured under ambient conditions. By combining ambient and TV measurements it is therefore possible to derive the angular dependence of the instrument throughput, i.e. the Mueller matrix elements.

Together with the measured signals S_{det} the reflectance R can be determined from (5.17). Finally those solar measurements where the neutral density filter is in the light path require correcting. It is done by dividing the recorded signal with the throughput of the NDF.

5.8 Optical Performance Monitoring

One of the main long-term monitoring activities to be performed over the mission's lifetime is to trace the degradation of optical components (Noël et al. 2003). It applies regular trend analyses to measurement data obtained with the internal WLS and of observations of the unobscured Sun above the atmosphere. In order to monitor the different SCIAMACHY light paths, solar measurements are taken in various viewing geometries (Fig. 5.6):

- In limb/occultation geometry (via ASM and ESM mirrors)
- In nadir geometry (via the ESM mirror through the subsolar port), and
- Via the *calibration light path* involving the ASM mirror and the ESM diffuser

Particularly the WLS produces a rather stable output over time – except for some degradation in the UV – which makes it well suited for throughput monitoring.

In the early phase of the mission the optical performance of the SCIAMACHY instrument was monitored based on the analysis of level 0 data which had been corrected for dead/bad pixels, dark current (fixed value from August 2002), scan angle dependencies, quantum efficiency changes and the seasonally varying distance to the Sun. Meanwhile the light path monitoring relies on fully calibrated data, yielding so called *m-factors* (Bramstedt 2008). The m-factor is defined as the ratio between a measured spectrum of a constant light source – usually the Sun – at a certain time to a spectrum obtained for the same optical path at a reference time. M-factors therefore describe how the individual light paths degrade, i.e. they provide the information required for correcting it. For the science channels m-factors are multiplicative when an absolute radiometric calibration is performed. This is different to the PMD channels where the m-factors impact the polarisation correction in a non-linear fashion.

M-factors are generated for each of the SCIAMACHY light paths (nadir, limb and the calibration light path), taking into account times of reduced instrument performance, e.g. switch-offs or decontamination periods. Moreover, prominent solar features, i.e. Fraunhofer lines used in the derivation of solar data products like solar activity monitoring via the Mg II index are masked out. For channels 1–6 the m-factors are further spectrally smoothed using a 9 pixel triangular filter to avoid the introduction of additional spectral features in the degradation corrected spectra. The m-factors are operationally produced on a daily grid involving, if necessary, interpolations. They can be fed into operational data processing ensuring that the measured signals are fully matched to the performance of SCIAMACHY. Particularly crucial in this context is the calibration of the solar and WLS data used in the m-factor generation, which needs to be consistent with the operational data processing. Therefore, the m-factors are specific for one level 1 product version and thus possibly need to be recalculated after each update of the operational 0–1b data processing software.

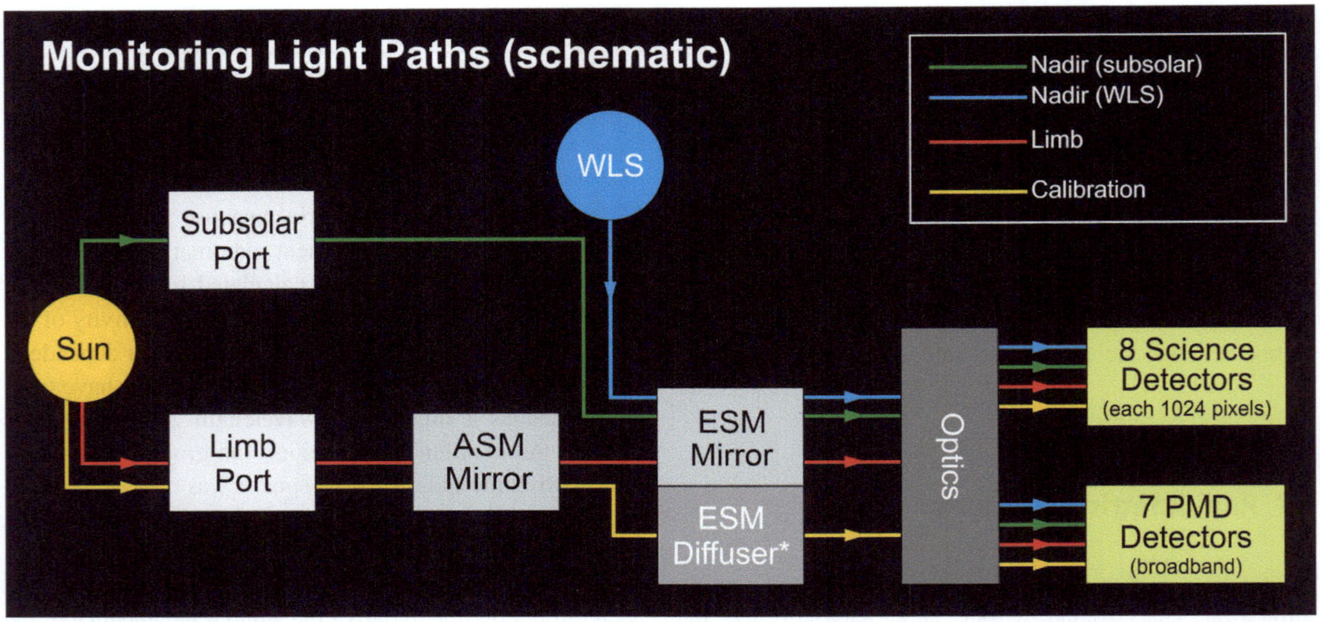

Fig. 5.6 Schematic view of SCIAMACHY light paths used in performance monitoring (Courtesy: IUP-IFE, University of Bremen).

From the combination of the results for the different light paths it is possible to derive information about the degradation of individual optical components such as mirrors and diffusers (Fig. 5.6). The degradation of the ASM mirror, for example, may be determined from the ratio of the limb to the nadir light path degradation. To determine the ESM mirror degradation it is necessary to combine the limb light path results with dedicated measurements involving the extra mirror which is located inside the instrument, only rarely used and thus assumed not to degrade. Finally, the degradation of the ESM diffuser can be computed from the combination of the nadir, limb, and calibration light path. A comparison of the limb and nadir light path monitoring results indicates that the major degrading element in the SCIAMACHY optical train seems to be the ESM mirror.

References

Azzam RMA, Bashara NM (1977) Ellipsometry and polarized light. Elsevier Science Publishers, Amsterdam

Balzer W, Aberle B, Loyola D, Spurr R (1996) GOME Level 0 to 1b Algorithm Description (ER-TN-DLR-GO-0022). Technical Document, DLR

Bramstedt K (2008) Calculation of SCIAMACHY M-Factors (IFE-SCIA-TN-2007-01-CalcMFactor). Technical Document, IUP-IFE, University of Bremen. Available at http://www.iup.uni-bremen.de/sciamachy/mfactors/mfactor-TN-1-0.pdf

Coulson KL (1988) Polarisation and intensity of light in the atmosphere. A. Deepak Publishing, Hampton, VA

Falk WR (1984) Data reduction from experimental histograms. Nucl Instrum Methods Phys Res 220:473–478

Kleipool Q (2003) SCIAMACHY: Recalculation of OPTEC 5 non-linearity (SRON-SCIA-PhE-RP-013). Technical Document, SRON. Available at http://www.sron.nl/index.php?option=com_content&task=view&id=509&Itemid=787&lang=en

Lichtenberg G (2003) SCIAMACHY channel 1-5 memory effect I: Key data implementation and in-flight measurements (SRON-PhE-RP-11). Technical Document, SRON. Available at http://www.sron.nl/index.php?option=com_content&task=view&id=509&Itemid=787&lang=en

Lichtenberg G, Kleipool Q, Krijger JM, van Soest G, van Hees R, Tilstra LG, Acarreta JR, Aben I, Ahlers B, Bovensmann H, Chance K, Gloudemans AMS, Hoogeveen RWM, Jongma R, Noël S, Piters A, Schrijver H, Schrijvers C, Sioris CE, Skupin J, Slijkhuis S, Stammes P, Wuttke M (2006) SCIAMACHY level 1 data: Calibration concept and in-flight calibration. Atmos Chem Phys 6:5347–5367

McLinden CA, McConnell JC, Griffioen E, McElroy CT (2002) A vector radiative-transfer model for the Odin/OSIRIS project. Can J Phys 80:375–393

Noël S, Bovensmann H, Skupin J, Wuttke MW, Burrows JP, Gottwald M, Krieg E (2003) The SCIAMACHY calibration/monitoring concept and first results. Adv Space Res 32:2123–2128

Schutgens NAJ, Stammes P (2002) Parametrisation of Earth's polarisation spectrum from 290 to 330 nm. J Quant Spectr Rad Transfer 75:239–255

Schutgens NAJ, Tilstra LG, Stammes P, Bréon F-M (2004) On the relationship between Stokes parameters Q and U of atmospheric ultraviolet/visible/near-infrared radiation. J Geophys Res 109: D09205. doi:10.1029/2003JD004081

Slijkhuis S (2000a) ENVISAT-1 SCIAMACHY level 0 to 1c processing: Algorithm theoretical basis document. Technical Document, DLR

Slijkhuis S (2000b) Calculation of polarisation from Rayleigh single scattering (ENV-TN-DLR-SCIA-0043). Technical Document, DLR

Tanzi CP (1999) Considerations on Stokes parameters derived from memo. The seventh point polarisation algorithm by P. Stammes. Internal Report, SRON

Tilstra LG, Schutgens NAJ, Stammes P (2003) Analytical calculation of Stokes parameters Q and U of atmospheric radiation (WR-2003-01). Scientific Report, KNMI

Chapter 6
SCIAMACHY In-Orbit Operations and Performance

M. Gottwald, K. Bramstedt, R. Snel, M. Krijger, G. Lichtenberg, S. Slijkhuis, C. von Savigny, S. Noël, and E. Krieg

Abstract Since the launch in early 2002 SCIAMACHY has successfully operated in low-Earth orbit for more than 8 years. For the first several months a challenging Commissioning Phase programme was executed. It successively brought SCIAMACHY into full operation mode and verified the instrument's functional capabilities. In early August 2002 quasi-routine measurements executing nominal mission scenarios could start. In January 2003 the routine operations phase commenced. Since then SCIAMACHY is kept under strict configuration control. Because of the harsh space environment the instrument is subject to degradation, both optically and thermally. The optical performance is described by the throughput which is a measure for how optical components in a light path age with time. It also includes characterisation of optical imperfections such as scan angle dependence, channel 7 light leak and spatial stray light. Illustrating the thermal performance includes decontaminations, used to tackle the ice layers in channels 7 and 8 and configuration of the thermal control systems to respond to degradation. Finally the improvement of the line-of-sight performance by determination of mispointing angles achieved the best possible pointing knowledge. This was especially needed for the retrieval of accurate limb data products. The current excellent status of SCIAMACHY is a prerequisite for successfully accomplishing the intended ENVISAT mission extension until 2013.

Keywords Commissioning Phase • Routine operations • Optical performance • Thermal performance • Line-of-sight performance • Mission extension

On 1 March 2002 at 1:07 UTC, SCIAMACHY was lifted into space from Kourou as part of the ENVISAT mission (Fig. 6.1). The highly precise injection into a sun-synchronous orbit saved fuel so that from an orbit maintenance point of view an extension of the mission lifetime beyond the specified 5 years became possible. At about 02:53:51 UTC, ENVISAT crossed for the first time the Earth's equator on the night side corresponding to the start of absolute orbit no. 1. Since then, until the end of 2009, more than 40000 orbits have been executed.

SCIAMACHY's initial operational programme reflects the requirements of the overall ENVISAT mission and the specific needs of the instrument. It was the goal to reach routine operations as soon as possible but also to perform a thorough in-orbit functional check-out and verification of the instrument. SCIAMACHY mission phases consisted of the launch and early operation phase (LEOP), the switch-on and data acquisition phase (SODAP), the main validation phase with quasi-routine operations and finally the routine operations phase. SODAP and main validation phase formed the Commissioning Phase. A time schedule is presented in Table 6.1.

6.1 Commissioning Phase

Establishing the instrument activities in the Commissioning Phase, particularly SODAP, required assembling a plan including engineering and specific measurement tasks. This plan had to provide a continuous, conflict-free schedule at instrument as well as on ENVISAT level which finally permitted the declaration that SCIAMACHY was ready for routine operations (Table 6.1). The approach was to start with separate planning of engineering and measurement

M. Gottwald (✉), G. Lichtenberg, S. Slijkhuis, and E. Krieg
Remote Sensing Technology Institute, German Aerospace Center (DLR-IMF), Oberpfaffenhofen, 82234 Wessling, Germany
e-mail: manfred.gottwald@dlr.de

K. Bramstedt, C. von Savigny, and S. Noël
Institute of Environmental Physics/Institute of Remote Sensing (IUP-IFE), University of Bremen, Otto-Hahn-Allee 1, 28359 Bremen, Germany

R. Snel and M. Krijger
SRON, Netherlands Institute for Space Research, Sorbonnelaan 2, 3584 CA Utrecht, The Netherlands

M. Gottwald and H. Bovensmann (eds.), *SCIAMACHY - Exploring the Changing Earth's Atmosphere*,
DOI 10.1007/978-90-481-9896-2_6, © Springer Science+Business Media B.V. 2011

Fig. 6.1 ENVISAT launch (Photo: ESA).

tasks, to integrate both to obtain a complete SCIAMACHY flow and to insert this flow into the overall ENVISAT SODAP plan.

Engineering Tasks

Instrument operations on command and control level are described in the Instrument Operation Manual (IOM). The IOM provides the ENVISAT Flight Operation Control Center (FOCC) with all information necessary to properly operate and maintain the instrument. During SODAP the instrument capabilities, which may later be used on a routine basis, had to be functionally tested and verified both in nominal and non-nominal situations. In addition, engineering settings for dedicated subsystems had to be derived. The engineering SODAP tasks comprised:

- Mode transitions with associated parameters
- Thermal operations including decontamination
- Flight procedures
- Routine monitoring
- Processor patch and dump

These were supplemented by dedicated operations that occurred only once during SODAP. Major activities included the release of the azimuth and elevation aperture cover mechanisms and the opening of the SCIAMACHY Radiant Cooler (SRC) door.

Table 6.1 SCIAMACHY activities from launch to end of Commissioning Phase

Phase	Instrument activity	Date	Orbit
LEOP	OFF-Leo mode	1/7 March 2002	
SODAP	First switch-on	11 March 2002	147
	First MPS driven operations	17 March 2002	238
	First decontamination	18 March 2002	253
	AZACM cover released	3 April 2002	477
	SRC released	15 April 2002	653
	First lunar measurement	22 April 2002	753
	Final ATC/TC settings loaded	10 June 2002	1454
	ELACM cover released	20 June 2002	1594
	β states loaded	17 July 2002	1982
	Timelines with β states loaded	18 July 2002	1990
	End SODAP (remaining SODAP measurements inserted as *Δ SODAP* in validation phase)	2 August 2002	2204
Validation	Start validation	2 August 2002	2204
	End Δ SODAP measurements	14 December 2002	4127
	Final flight states loaded	15 December 2002	4143
	Timelines with final flight states loaded	16 December 2002	4151
	First non-nominal decontamination started	19 December 2002	4204
Routine Operations	Nominal measurement programme	since 6 January 2003	4457

Many of the engineering functions had been tested during phase C/D on-ground. Thermal operations, however, were for the first time executed under in-orbit conditions. Therefore emphasis was put on a thorough verification of the thermal subsystems Active Thermal Control (ATC), Thermal Control (TC) and Radiant Cooler. The goal was to characterise the subsystems well enough in order to be able to routinely select, after the end of SODAP, the correct parameter settings suitable to maintain Optical Bench Module (OBM) and detector temperatures within specified limits.

Measurement Tasks

The flexibility of the instrument required verification of many different functionalities and characterisation of a large set of parameters. In order to ensure a thorough testing of all instrument capabilities, a call for proposals was issued to all parties involved in the instrument development. All proposals underwent a review process and finally in total more than 110 proposals were accepted. Some of them requested verification that states defined for routine operations worked as expected but for the majority new states had to be specified. Each new state corresponded to a new instrument on-board configuration that had to be commanded via the upload of measurement parameter CTI (Configurable Transfer Item) tables. Individual states, new ones and those already existing, were assembled to generate specific timelines. Execution of the states was triggered via the start of such SODAP specific timelines scheduled via the ENVISAT Mission Planning System (MPS).

Since some commissioning objectives required particular instrument configurations – e.g. aperture covers released, a certain thermal status – or needed as a precondition for the output of other objectives, it was impossible to generate the full SCIAMACHY SODAP measurement plan by simply concatenating all the objectives. Furthermore, all lunar measurements could only be performed in the short monthly visibility periods. Therefore, engineering and measurement tasks were integrated at the level of individual objectives thus ensuring that all instrument prerequisites were fulfilled and the overall ENVISAT SODAP schedule adhered to. The result was the SCIAMACHY On-Board Operation Plan (SCOOP), an Excel based database of the complete SODAP period. It split this period into engineering and measurement windows. In an engineering window, SCIAMACHY operation was procedure driven while in a measurement window, the instrument operations were timeline driven and controlled by the ENVISAT MPS. The measurement windows were arranged so that on entry and exit the instrument was in a well defined configuration and all timelines and states scheduled therein executed just those measurements compliant with this configuration.

The SODAP Sequence

At the time of launch the SODAP specific planning information – SCOOP, states and timelines – was available and ready for activation. The ENVISAT SODAP plan had scheduled SCIAMACHY's first switch-on for orbit 147 (11 March), 11 days after launch. Six days later (17 March), the instrument was controlled for the first time by the ENVISAT MPS when the first timeline was executed which ran successfully the *Full Functional Test*. Because all aperture covers were still closed, no external light was collected and only light from the internal calibration sources was used. The sequence of engineering and measurement activities continued until 3 April. That day the first aperture cover, the azimuth aperture cover mechanism (AZACM), was released and the light path via the limb port opened, permitting limb and occultation observations. The long time delay between launch and the first appendage release was required to avoid possible contamination of the instrument due to outgassing from the platform. Another important milestone was reached on 15 April with the opening of the SRC. This event started the passive cooling of the detectors to their nominal temperatures, i.e. from this release on, detectors could be operated under in-flight thermal conditions. Furthermore thermal tests were now possible, aiming to find the final settings for the instrument. A particular challenge for mission planning and scanner control occurred at the end of April with the first lunar measurement window which was successfully passed. In June SODAP had progressed so far that the final flight settings for the ATC and TC could be uploaded. OBM and detectors were now under continuous thermal control with modifications only being triggered by seasonal effects or the status of the SRC. On 20 June, the third and final cover, the elevation aperture cover mechanism (ELACM), was removed from the light paths. It permitted light to enter SCIAMACHY via the nadir port. From then on the engineering and measurement programme focused on finalising SODAP with the goal to begin the validation part of the Commissioning Phase in early August. With the set of *β states*, originating from the evaluation of earlier measurements, and with the associated timelines a configuration was specified and uploaded mid July which already came very close to the envisaged final flight definitions. SODAP ended on 2 August, but leaving a few measurements still to be done. This was mainly due to the occurrence of anomalies which made it necessary to rearrange the tight SODAP plan. Also some of the objectives required seasonally dependent observing conditions, available only in the second half of the year.

At the end of SODAP, after a period of 130 days, SCIAMACHY had successfully executed more than 21200 macrocommands (MCMD). Almost 5500 timelines were started which triggered more than 78000 individual states. This impressive operational record required the upload of 5700 parameter tables and 560 timelines. Each parameter table was equivalent to reconfiguring one functionality or characteristic of the instrument. SODAP proved that the operational concept and the flight operations ground segment interfaces were well developed to handle the complex mission.

Although SODAP did not show major instrument malfunctions, the original plans had to be modified due to several anomalies. A rather persistent recurring anomaly hampered SCIAMACHY between May and July 2002. Subsequent detailed failure analysis had led to the conclusion that a bug in the Instrument Control Unit (ICU) can temporarily block its interface to the ENVISAT Payload Management Computer causing a transfer to a safe instrument mode lower than *Measurement*. The error is known as *MCMD CCA check error*. In October 2002 a software patch was uploaded to correct for this ICU bug. It did not fully cure the problem – complete repair would have required a more extended patch – but since then the average rate of the check error was reduced to a very low level of about 1/year. Single Event Upsets (SEU) were another type of anomaly triggering operational interrupts. The instrument can suffer from a SEU when high-energy particles, most likely protons, hit electronic components and switch their status information. The particle flux increases in low-Earth orbits during phases of high solar activity in general and when crossing the South Atlantic Anomaly (SAA) of the Earth's magnetic field in particular. Obviously SEU were not only spoiling the Commissioning Phase activities but have turned out to be the most frequent reason for unexpected instrument anomalies in the routine operations phase.

The Validation Sequence

From an instrument point of view, during the validation phase SCIAMACHY was operated in a quasi-routine fashion. Only some inserted *Δ SODAP* measurements interrupted the nominal measurement plan. During this part of the mission, the extensive support of the ground segment was gradually reduced. SCIAMACHY now executed a continuous measurement programme which reflected the mission scenarios for routine operations. Validation scientists were provided with planning information – exact overpass times for nadir and limb observations – permitting synchronisation of ground-based, airborne and balloon-borne campaigns with SCIAMACHY operations. Details are described in Chapter 9.

The end of the validation phase corresponded to the end of the Commissioning Phase. Therefore the definition and upload of the final flight states and timelines was the ultimate goal to be accomplished. It occurred mid December 2002. After another decontamination the instrument was prepared and ready for the start of the routine operations phase.

6.2 Routine Operations Phase

In January 2003, SCIAMACHY commissioning had ended and transfer to the routine operations phase was initiated. In spite of the few discrepancies described above, the instrument entered this phase with excellent performance. The start of routine operations also meant transfer of operational responsibilities on the instrument provider side from EADS Astrium to the SCIAMACHY Operations Support Team (SOST).

Routine operations are characterised by their continuity. Since spaceborne remote sensing of the atmosphere requires measurement data to be acquired under stable conditions, both in terms of instrument configuration and performance, SCIAMACHY has the goal to maintain the baseline measurement programme as long as possible. The majority of measurement sequences consist of a solar occultation each orbit, matching limb/nadir observations on the dayside and specific eclipse observations, including calibration and monitoring in the eclipse phase. Swath width in nadir and limb states is set to wide, i.e. 960 km, with an upper limit for the limb horizontal scans close to the top of the atmosphere at 100 km. This is supplemented by calibration and monitoring activities on a daily, weekly and monthly timescale. For a typical orbital mission scenario 92% of the orbital period is covered by measurements. The remaining 8% are idle gaps required for potential command and control activities or are caused by the fact that the smallest possible time slice in a timeline is the duration of a state. Therefore the continuous seasonal changes of solar and lunar constellations cannot always be perfectly matched and cause gaps up to the duration of a state.

By the end of 2009, SCIAMACHY had already executed more than 7 years of scientific measurements, where initially only 5 years were planned. For most of the routine operations phase, a high level of instrument availability could be ensured (Fig. 6.2). The continuous improvement of calibration and monitoring (Chapter 5) permits deriving spectra with high relative (Fig. 6.3) and absolute (Fig. 6.4) accuracy.

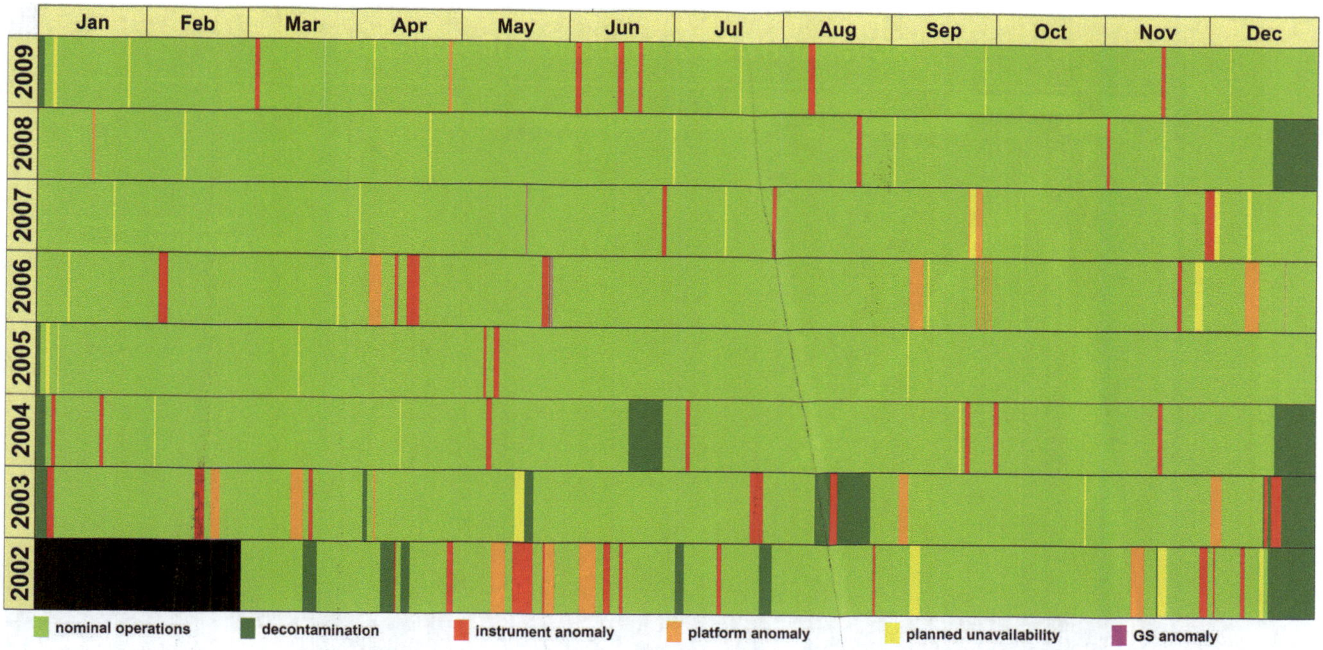

Fig. 6.2 Operations summary for the commissioning and routine operations phase. Colour coding indicates instrument availability (Courtesy: DLR-IMF).

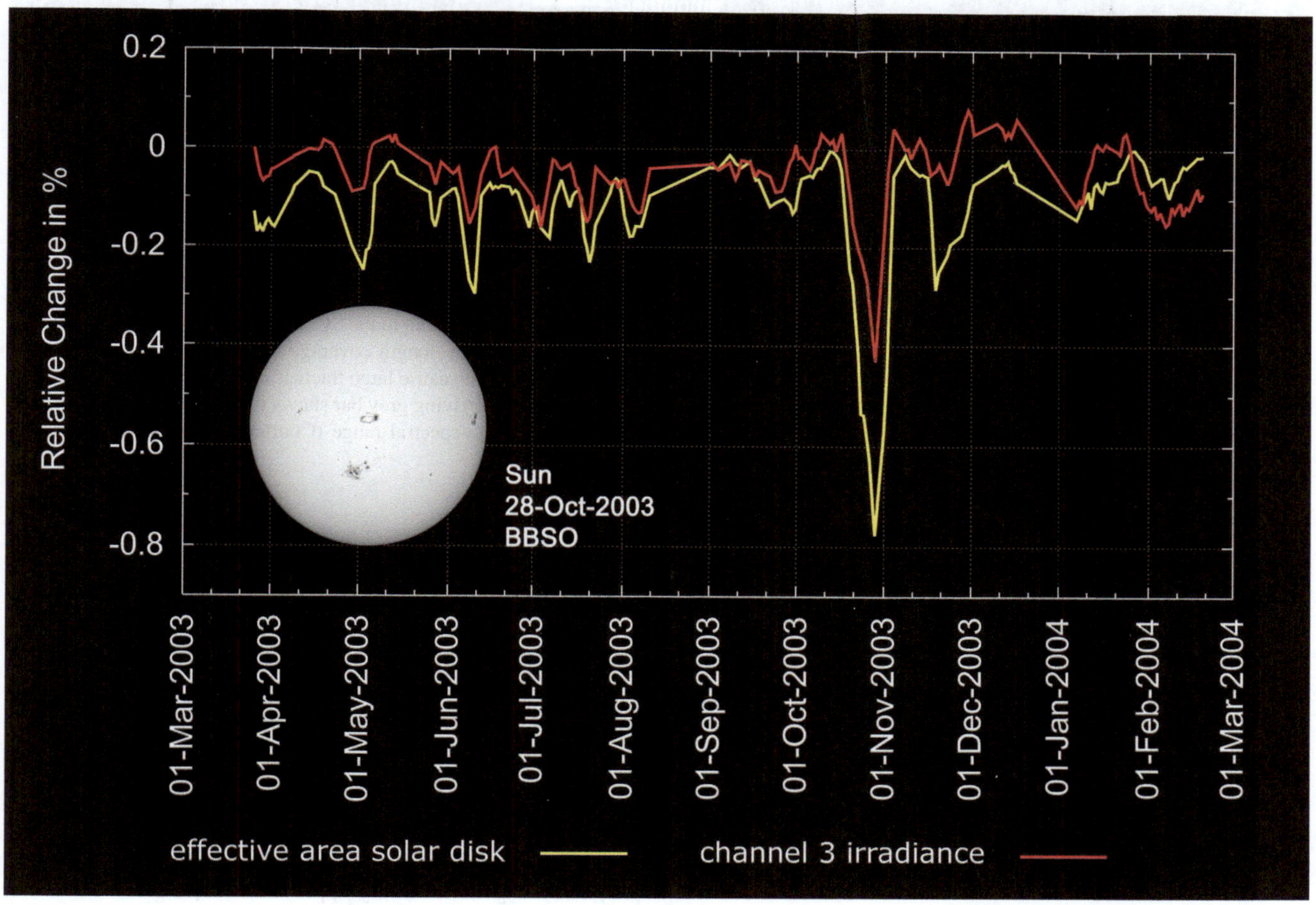

Fig. 6.3 Measured irradiance in channel 3 (*red*) and effective solardisk area (*yellow*) between March 2003 and March 2004. The large sunspots at the end of October caused a significantly reduced irradiance. The inset shows the active solar disk in white light on 28 October 2003 (Courtesy: IUP-IFE, University of Bremen; Sun photo: Big Bear Solar Observatory).

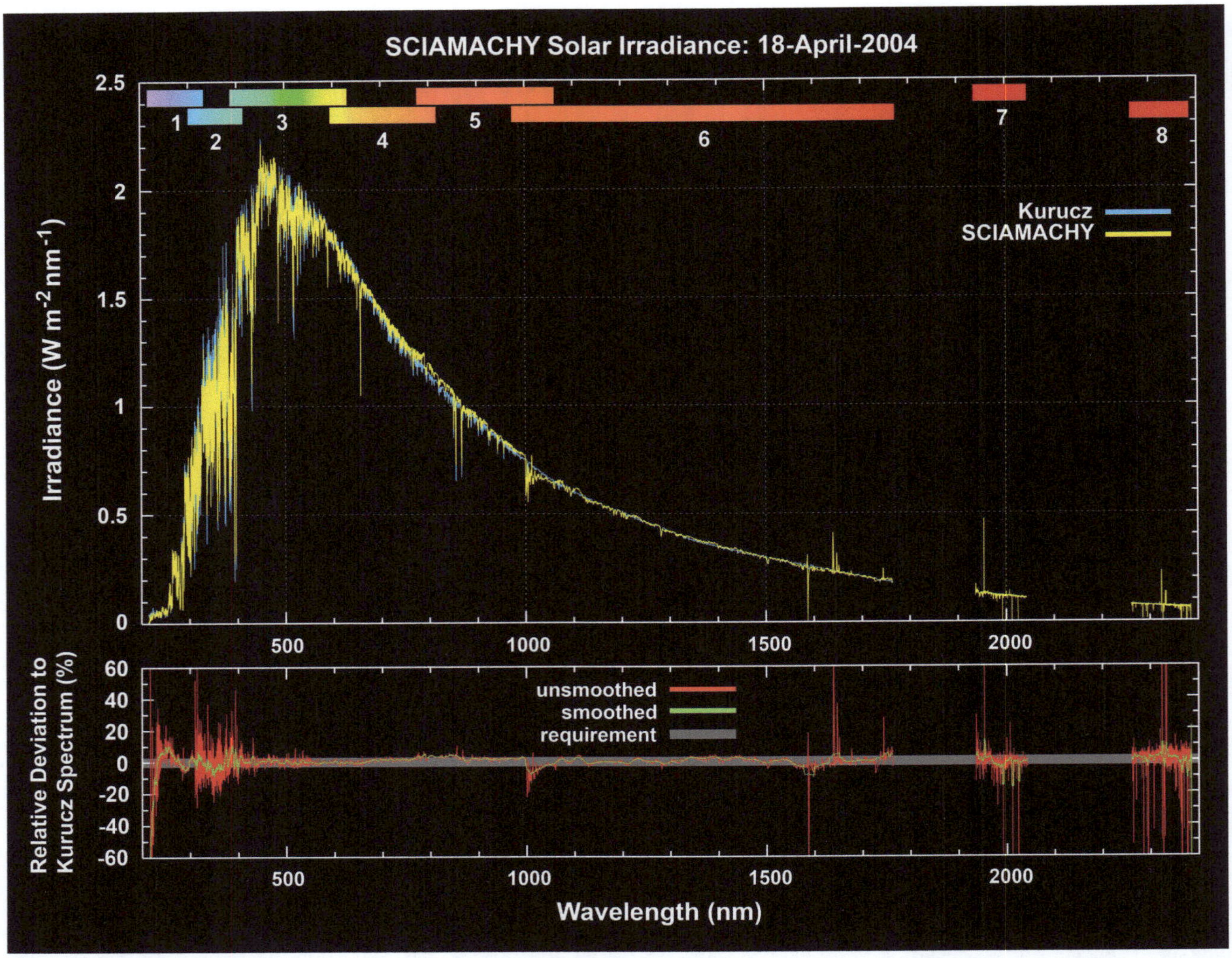

Fig. 6.4 Calibrated SCIAMACHY solar irradiance spectrum measured 18 April 2004, in comparison to the Kurucz solar reference spectrum. SCIAMACHY data have been normalised to 1 AU solar distance. Kurucz data have been interpolated to the SCIAMACHY wavelength grid. *Top*: SCIAMACHY (*yellow*) and Kurucz (*blue*) irradiance as function of wavelength with the wavelength coverage of all eight channels. *Bottom*: Relative deviation between the two data sets. Small deviations in the wavelength calibration can cause large fluctuations in the ratio of the spectra. Therefore, the deviation is shown as unsmoothed (*red*) and smoothed (*green*) curve. The underlying grey bar shows the ±3% requirement for the absolute radiometric calibration of SCIAMACHY which is fulfilled over almost the whole spectral range (Courtesy: IUP-IFE, University of Bremen).

Together with the progress in the development of retrieval algorithms (Chapter 7), data processing (Chapter 8) and validation (Chapter 9), this forms the basis for finally obtaining excellent scientific results (Chapter 10).

Operation Change Requests and Final Flight Configuration

From January 2003 on, SCIAMACHY followed the specified mission scenarios. Deviations were only possible when a strict configuration controlled procedure, the Operation Change Request (OCR), was pursued and approved by project management. The OCR process was introduced early in 2003 because of the high number of configurable parameters describing the instrument status. Any modification of mission scenarios, states or timelines has to be requested via an OCR. The OCR procedure includes technical analyses by SOST of the proposed change and possible implementation options. Upon recommendation by SOST, and endorsement by the SCIAMACHY Science Advisory Group (SSAG) ensuring that no other scientific requirements are violated, project management finally approves or rejects the OCR.

An OCR may ask for a temporary modification of operations, or result in permanent changes. In the latter case the final flight configuration, either for states or timelines, has to

Table 6.2 Changes in final flight configurations from 2003–2009

Instrument activity	Date	Orbit
Modification of PET in Moon states	10 March 2003	5358
Modification of dark current sequence in eclipse and monthly calibration timelines	4 April 2003	5712
Modification of TCFoV in nadir states	8 April 2003	5771
Modification of tangent height in limb dark current measurement	26 May 2003	6456
Modification of *WLS over diffuser* sequence in monthly calibration timelines	13 July 2003	7151
Modification of altitude range in *limb mesosphere* state	21 July 2003	7265
Modification of PET in dark current and NDFM monitoring states	21 July 2003	7276
Modification of limb altitude range and new timeline set for improved limb/nadir matching	15 October 2003	8489
New timeline set for improved limb/nadir matching in early orbit phase	22 May 2004	11638
Modification of nadir states and new timeline set for increased signal-to-noise at high latitudes	6 September 2004	13172
New eclipse timelines with extended *limb mesosphere* coverage	6 September 2004	13172
Increase of sub-solar pointing rate and reduction of sub-solar fast sweep scanning rate	1 October 2006	23978
Reduction of sub-solar observation rate to 1 every third day	1 October 2006	23978
New limb mesosphere-thermosphere state to be executed 30 orbits/month	3 November 2008	34922

be altered. Such a modification is permanently implemented in the flight operation ground segment command database and must be made operational at each ICU re-initialisation. Temporary changes are reset to the appropriate initial final flight settings. Since introducing the OCR process, more than 40 requests to change the instrument configuration have been issued. All of them were successfully implemented and executed. On average this was equivalent to modify the measurement setup every 2 months. Some of the OCR requested rather trivial changes. Others were rather demanding such that SCIAMACHY's operational capabilities had to be used at their limits. The final flight versions as of December 2002 were modified several times (Table 6.2). However the overall state and timeline definitions remained rather stable throughout the routine operations phase so far. Mainly minor adjustments to state parameters or state sequences in timelines were required, but no major restructuring of their concepts.

6.3 Optical Performance

Optical Throughput

The regular monitoring of the optical light paths has generated a continuous record of optical throughput measurements, both averaged in each channel and spectrally resolved. The overall throughput behaviour is best illustrated by the channel averages. Although the results for the various paths differ slightly, all results provide a consistent view of the optical performance (Figs. 6.5 and 6.6).

- UV: Even with the observed degradation of 60% (channel 1) and 35% (channel 2) in 7 years, SCIAMACHY still maintains a high throughput at the UV wavelengths. The GOME mission with similar detectors suffered from a faster and larger decrease in sensitivity at short wavelengths.
- VIS-NIR-SWIR: For detectors 3–6, annual variations are detected on a sub-percent level. The excellent stability permits correlation of the overall throughput with large scale features due to seasonal variations. Over short timescales even solar activity sensed via the effective sunspot area is reflected in the data.
- SWIR: Throughput analysis in channels 7 and 8 provides a means for monitoring the status of the ice layers. The record of decontaminations has led to the conclusion that throughput measurements do not necessarily reflect the water content on the detectors but only in the optical path. As was learnt by the sequence of decontaminations between December 2003 and January 2005 and December/January 2008/2009, water might condense either on a second cold trap mimicking a water-free detector or on optical surfaces. In the first case the throughput remains at high levels after having achieved values close to 100% immediately after the decontamination while in the second case it can be drastically reduced. While detectors 7 and 8 display common trends after each decontamination, a more detailed look reveals differences not fully understood. The light path in detector 7 can become virtually ice-free but sensitivity in detector 8 is always reduced by about 20–30% within 3–4 months.

An example for a spectrally resolved, i.e. wavelength dependent instrument degradation is depicted in Fig. 6.7. It displays

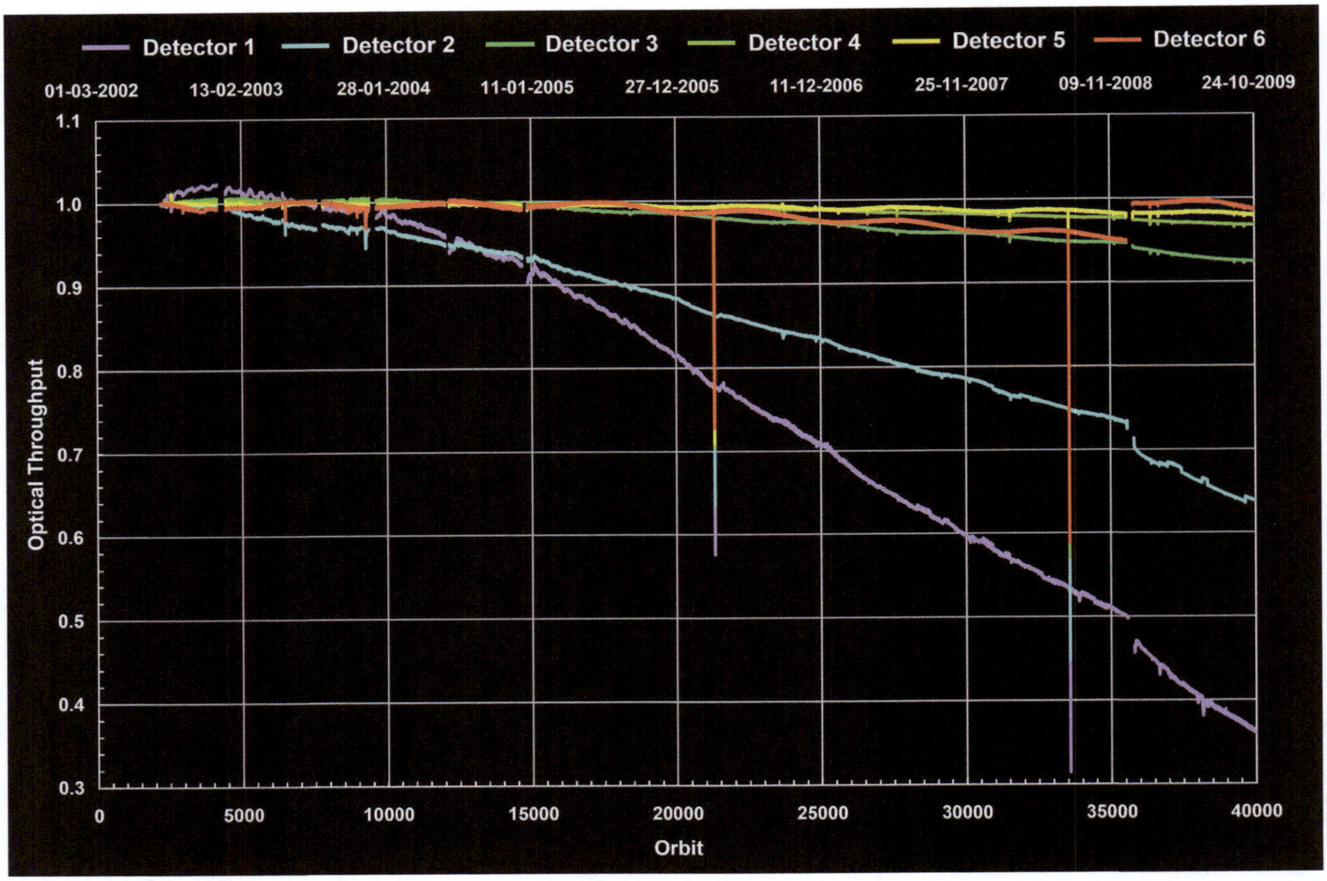

Fig. 6.5 Seven years of optical throughput for channels 1–6 starting with quasi-routine measurements on 2 August 2002. Decontamination intervals and periods after instrument saving are omitted. The information has been derived from the operational light path monitoring using the Sun via the ASM and ESM mirror. The two strong temporal reductions just before orbit 22000 and 34000 are caused by total solar eclipses (Courtesy: IUP-IFE, University of Bremen).

for channel 2 the relative variation of the nadir throughput – light enters the spectrometer via the ESM mirror only – as a function of time and wavelength, derived from m-factors (see section "Optical Performance Monitoring" of Chapter 5). Grey bars denote times of reduced instrument performance, e.g. switch-offs or decontamination periods. The nadir light path time series shown in Fig. 6.7 is derived from a combination of WLS data and two different types of sub-solar measurements (fast sweep scan and pointing). As in the case of the channel averaged throughput, the measured signals are referenced to 2 August 2002 at about orbit 2200. It is obvious from Fig. 6.7 that the degradation is wavelength dependent. Degradation in channel 2 peaks around 345 nm and in the overlap regions between the channels. Additionally, a spectrally broadband increase of degradation towards lower wavelengths can be identified. However, the channel 4 monitoring data presented in Fig. 6.8 show the excellent absolute radiometric stability of SCIAMACHY. Here the degradation does not exceed about 2%, except for the channel overlaps. A similar trend can be observed in the other VIS channels. The optical throughput as displayed in Figs. 6.5–6.8 is the result of the performance of the complete optical path. It includes detector specific ageing effects. Over the past 7 years the detectors for channels 1–5 showed very stable behaviour. Only the detector material of channel 2 exhibits several spots where the pixel sensitivity degrades with the exact mechanism being unknown. A higher degree of pixel degradation was detected in the SWIR channels 6+, 7 and 8 (Fig. 6.9). This is not unexpected because they are more demanding in terms of stable operating conditions. Particularly high-energy particles in the radiation environment, as often found in the South Atlantic Anomaly, can severely impact the status of individual pixels. Affected pixel will suddenly obtain a significantly higher dark current and corresponding higher dark noise, in many cases resulting in effective loss of the pixel. No reliable mechanism for permanent recovery of lost pixel exists, though the decontamination periods have resulted in temporary recovery of affected pixels. At launch approx. 20% of the pixels

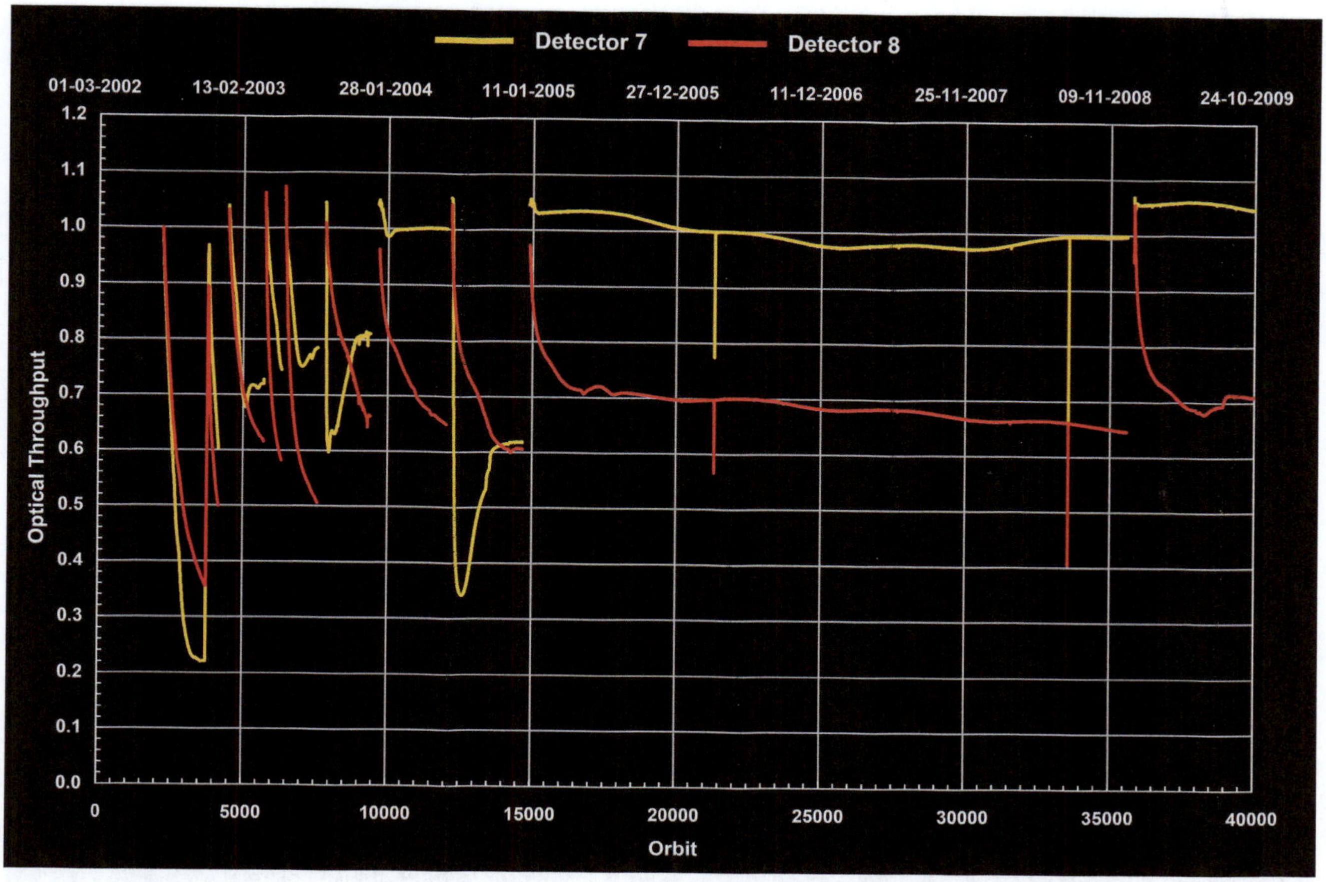

Fig. 6.6 Optical throughput for the SWIR channels 7 and 8. Similarly to Fig. 6.15, a growing ice layer and its decontamination induced evaporation determines the shape of the curve (Courtesy: IUP-IFE, University of Bremen).

in channels 6+, 7 and 8 were unusable. Since then this fraction has increased with a rate of 6% pixel loss per year.

Scan Angle Dependence

The measured signal depends on the viewing geometry, i.e. on the incidence angles on the mirrors or diffusers (see Chapter 5). This is not only because the reflectivity of the mirrors/diffusers changes when viewed under a different scan angle, but also the polarisation sensitivities. The default calibration corrects for this scan angle dependence under the implicit assumption that it has not changed since launch. While several science products indicate increasing scan angle dependence, no method for directly monitoring this degradation effect exists. Experience with GOME on ERS-2 shows that the effect is likely due to a contamination layer, built up on the scan mirrors since launch. This introduces a scan angle dependent degradation which varies with wavelength. Intrinsic scan angle dependence is expected, based on the Fresnel equations describing the polarised reflection of light off dielectric media (e.g. a mirror), including additionally added dielectric layers (e.g. a layer of built-up contamination). When the properties of the dielectrics or the layers change, as in contamination built up on a mirror surface, not only the unpolarised but also the polarised and scan angle dependent reflectance changes. These effects are all coupled through the Fresnel equations. If sufficient information is available from, e.g. component data and in-flight monitoring at multiple wavelengths, polarisation conditions, or multiple angles, the properties of the changed mirror or coating can be derived and corrected for. This was successfully done for the GOME scan mirror for the period 1995–2001, where the contaminant thickness with time and contaminant optical properties with wavelength could be derived from solar and lunar monitoring data. Due to the superior monitoring capabilities of SCIAMACHY, it is expected that deriving such information for the SCIAMACHY scan mirrors can be achieved as well. If the SCIAMACHY ESM mirror is behaving similarly as the GOME scan mirror, a scan angle dependent degradation of

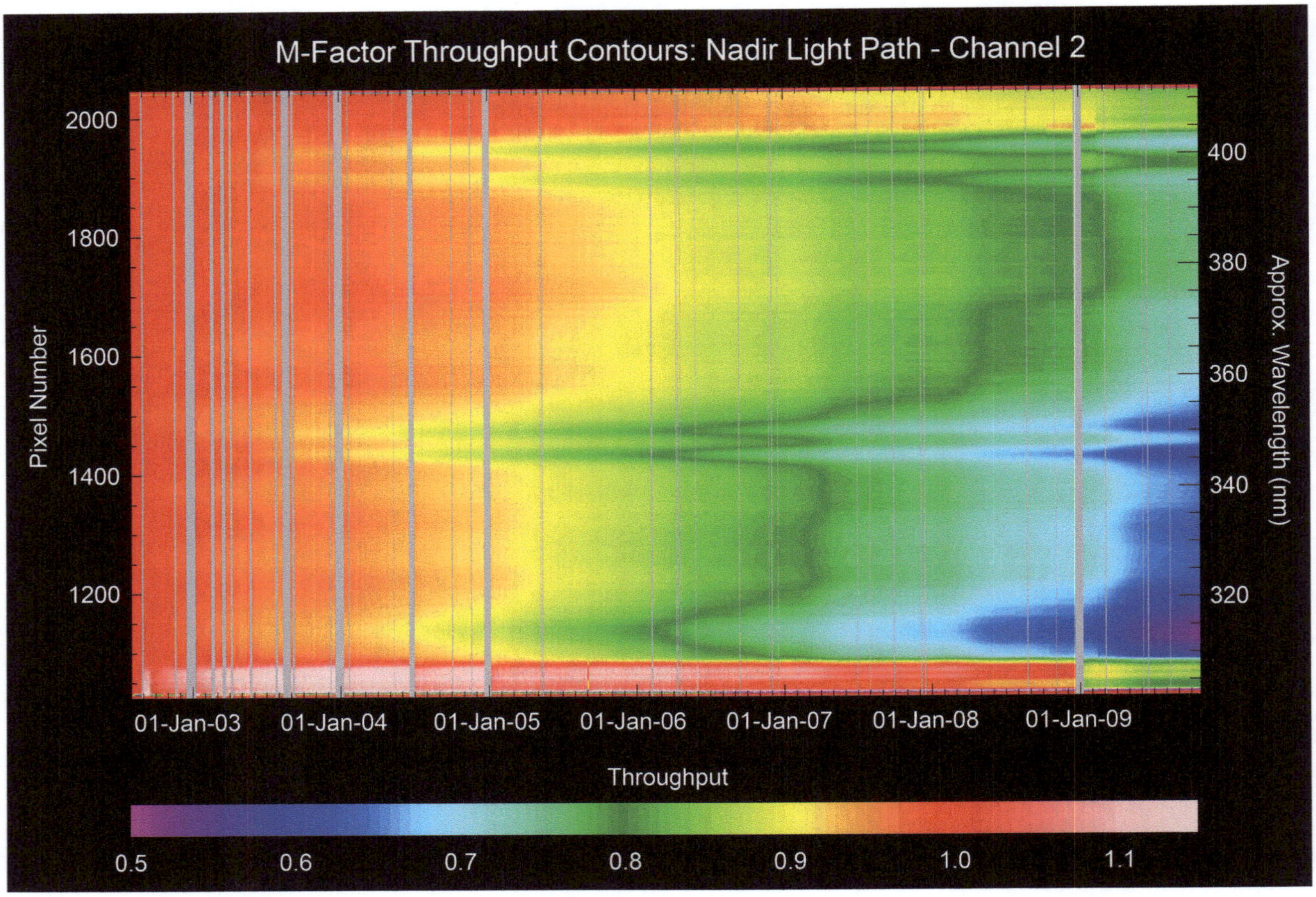

Fig. 6.7 Variation of the throughput of the SCIAMACHY nadir light path in channel 2, determined from m-factors. The contours plot shows the spectrally resolved throughput as a function of time. Grey bars mask out times of reduced instrument performance, i.e. switch-offs or decontamination periods (Courtesy: IUP-IFE, University of Bremen).

the order of a few percent is estimated (Snel and Tilstra 2008). It occurs mostly in the UV (channels 1 and 2) for extreme scan angles, compared to central viewing, due to accumulated contamination over the in-orbit mission lifetime of more than 7 years.

Light Leak in Channel 7

After launch it was discovered that channel 7 shows a spurious signal in limb dark measurements that is much higher than the spatial stray light found in the other channels. This signal has no spectral signature and is a broadband feature, excluding that it is caused by light passing through the optics of the instrument. The explanation requires a tiny hole – a *light leak* – in the channel 7 detector module, where light can enter the instrument and illuminate the detector without being dispersed first. Investigations to characterise the light leak were done using limb dark current measurements at 250 km altitude. At this tangent height and after correction of the data for non-linearity, dark signal and spectral stray light, the residual signal should be caused by the light leak alone, because no light from the Earth is expected. No contribution from the spatial stray light mentioned below has to be considered since this is only a fraction of the light leak signal and can be neglected here.

Figure 6.10 displays the mean residuals after correction of all analysed data, i.e. the signal caused by the light leak. This signal is spectrally smooth, systematically varying as a function of orbit phase. However, variations over one month can be quite large (50 BU/s), nearly comparable to the size of the light leak signal. Therefore, the light leak is not only a function of orbit phase but more likely a function of viewing geometry combined with the presence of regions with high albedo, e.g. caused by clouds. Correction of the light leak signal is required for the generation of high quality calibrated spectra in channel 7 and subsequent trace gas retrieval.

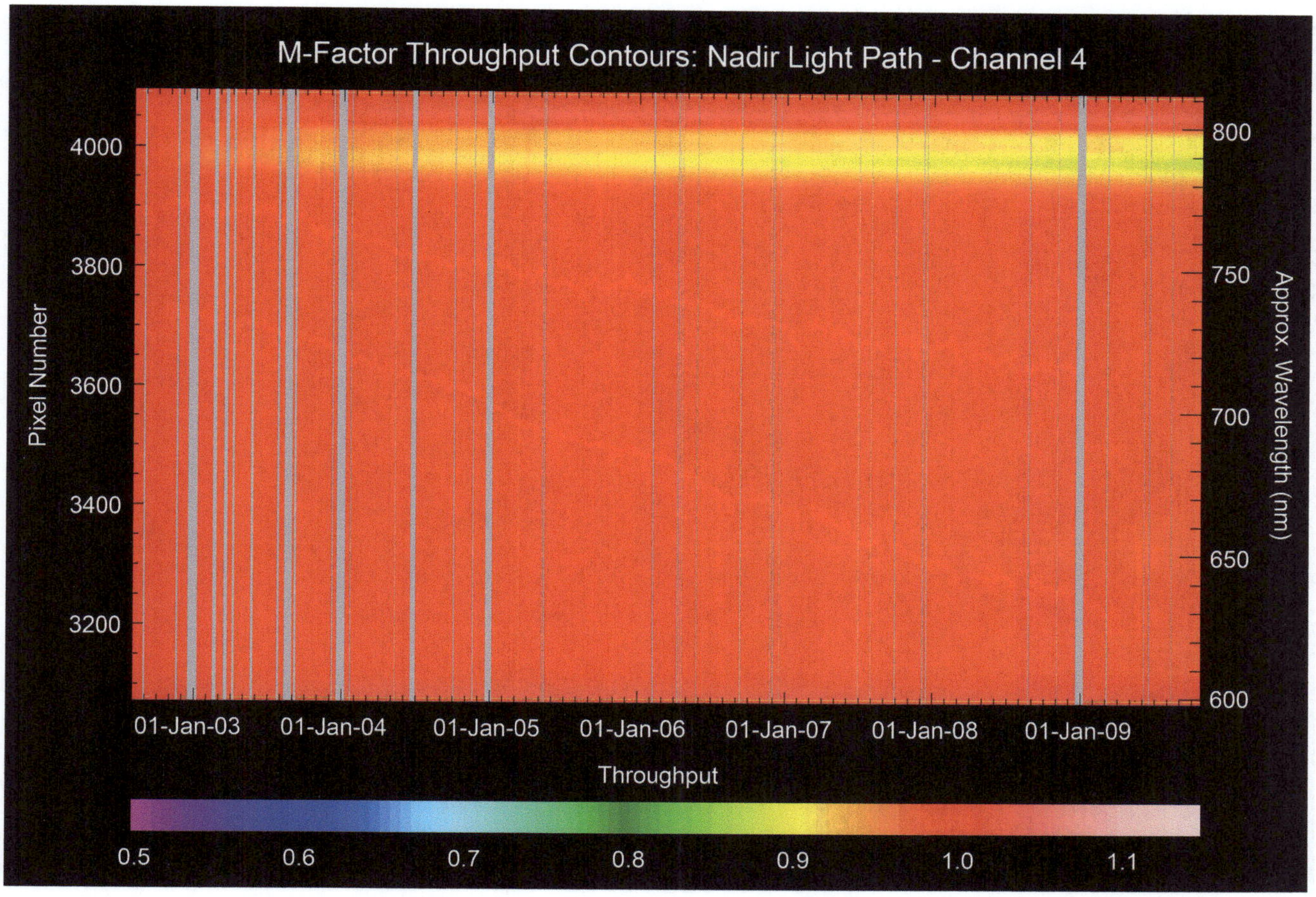

Fig. 6.8 Same as Fig. 6.7, but for SCIAMACHY channel 4 (Courtesy: IUP-IFE, University of Bremen).

Spatial Stray Light in Limb Measurements

During analysis of limb data it was discovered that these measurements suffer from spatial stray light. This was first noticed in measurements taken at 150 km tangent height where no atmospheric light should be present. The measurements at 150 km were originally intended to determine the orbital variation of the dark signal in channel 8 serving as an optional dark correction for limb measurements. Investigations (van Soest 2005) show that the additional signal was not caused by a light leak because spectral structures such as air glow emissions and atmospheric absorptions are visible in the measurements. This means that the signal is spectrally dispersed and thus goes through the optics of the instrument. Comparison with MERIS data revealed that the stray light does not correlate with the intensity of the scene at the subsatellite point ruling out the possibility that light leaking through the nadir port is subsequently directed into the telescope. Measurements of limb scans at high altitude and at a lower tangent altitude of 10 km result in a good correlation confirming that the stray light is caused by light entering the instrument through the slit from regions outside the IFoV. The stray light impact is highest in channels 2–4 and is very low in channels 1, 5 and 6. In order to avoid corrupted limb dark current measurements, the final flight instrument configuration has been adjusted early during the routine operations phase by raising the tangent altitude from 150 to 250 km. At this height the spatial stray light is reduced by an order of magnitude to 5–10 BU/s, thus allowing an estimation of the orbital dark variation.

Further investigations into the spatial stray light in 2007 revealed that a significant fraction of the stray light most likely originates from residual grooves in the diamond-turned telescope mirror (Fig. 6.11). The regular grooves act as a grating before the entrance slit and cause a wavelength dependent shifted scene to be recorded with the intended scene. The shift is below 1° at UV wavelengths and can be seen to extend to about 3° at 1000 nm and longer.

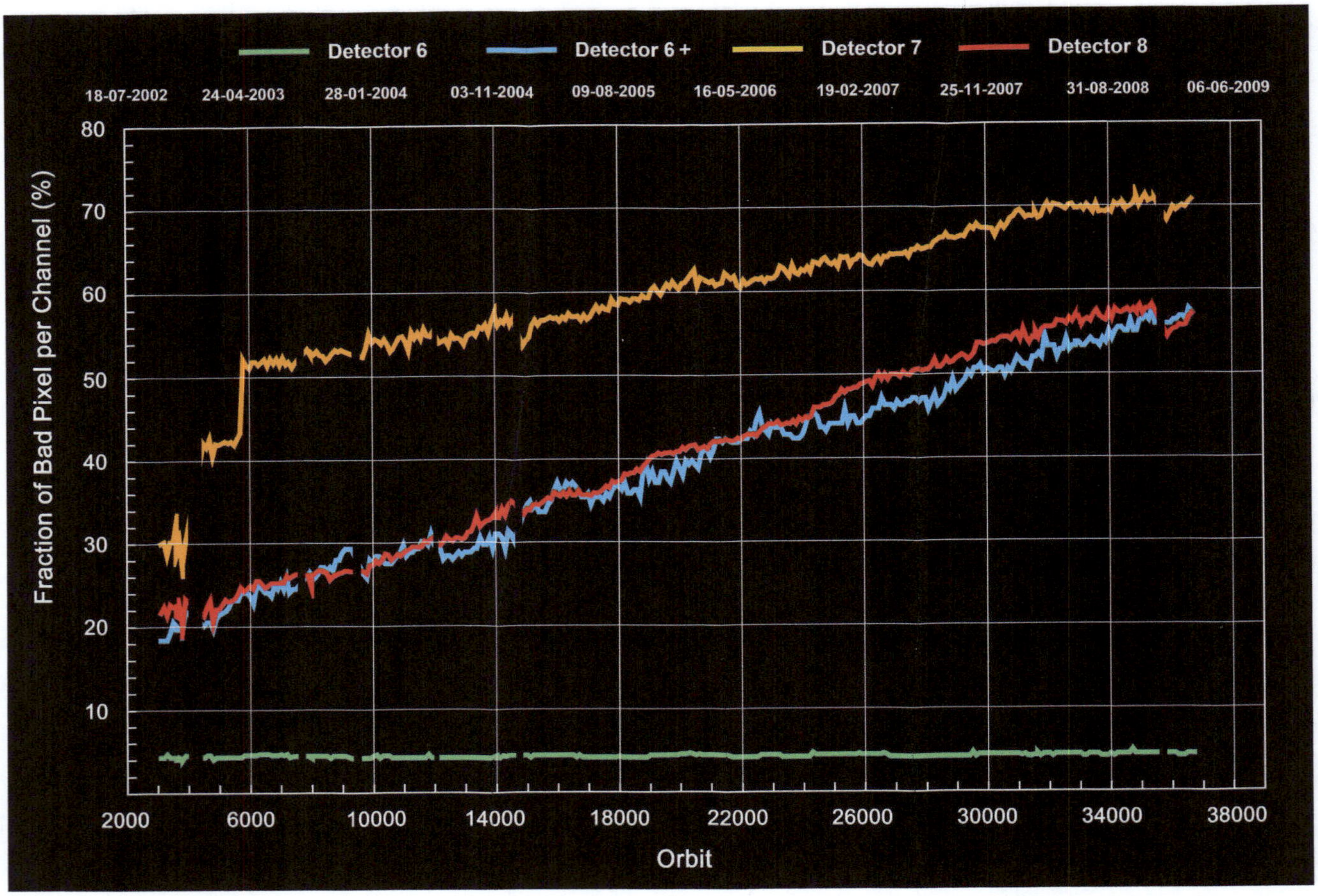

Fig. 6.9 Degradation of the pixel quality in the SWIR channels. The number of bad pixels is continuously increasing. For comparison, the first half of channel 6 is illustrated which consists of a material being more robust in the space environment and is thus much less subject to degradation (Courtesy: SRON).

6.4 Thermal Performance

Ice on Detectors 7 and 8

Already during SODAP it became obvious that the infrared channels 7 and 8 began to show a significant loss of radiance response in the weeks after the SRC had been opened. Investigations indicated that an ice layer growing on top of the cylindrical lens covering the detectors was responsible for this. It affected only channels 7 and 8 because these are the detectors operated at lowest temperatures. A likely source of the contaminant water is the carbon-fibre-reinforced plastic structure of ENVISAT. The water contained in the compound started to outgas once the platform was in orbit and condensed on the cold surfaces in channels 7 and 8. Obviously the venting holes in the multilayer insulation covering SCIAMACHY could not efficiently support the outgassing of the instrument. Over a period of only a few months the ice layer reduced the throughput in channels 7 and 8 by almost 80% (see Fig. 6.6). Methods to stop accumulation of ice were limited and only the application of decontamination means was finally selected to become the operational countermeasure.

Decontamination

Decontamination capabilities are among the requirements for thermal operations. Detailed pre-launch analysis had shown that the efficiency of the RRU on the SRC to dissipate energy from the detectors to open space might decrease with time due to contamination of volatile molecules on the RRU surface. Cooling via the RRU usually yields detector temperatures below the lower limit. Therefore trim heaters counterbalance this effect by additional heating. When contamination decreases RRU efficiency, detectors become less

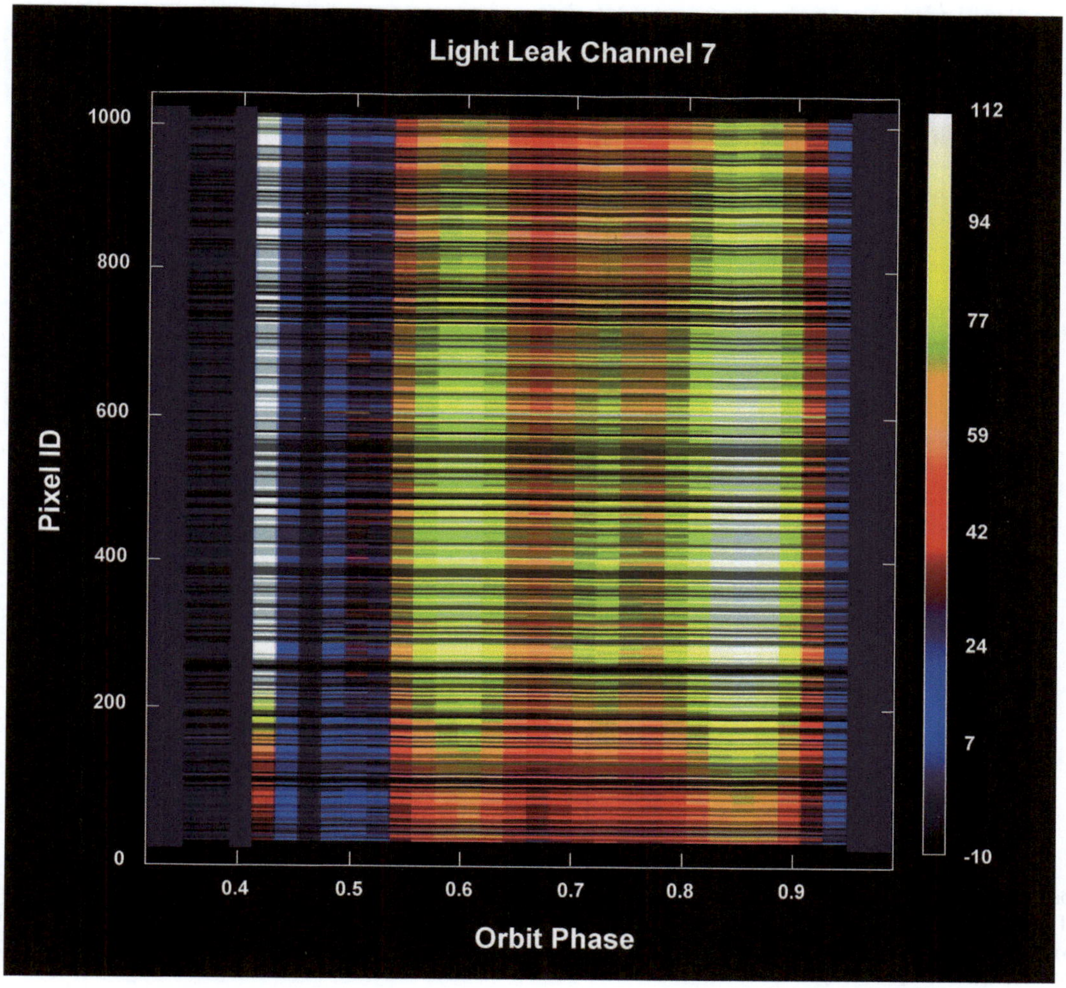

Fig. 6.10 Mean light leak signal for all pixels in channel 7 as a function of orbit phase. The x-axis shows the orbit phase (eclipse corresponds to phase 0.95–0.40, sunrise occurs at phase 0.41), the y-axis the pixel number in channel 7. The colours indicate the size of the light leak signal in BU/s (Courtesy: DLR-IMF/SRON).

cold and thus TC heater power (which is used to raise temperatures to keep detectors within limits) approaches zero. To re-establish the initial RRU efficiency, a decontamination mode had been originally foreseen with the goal of removing any contaminants from the RRU surface by heating up the SRC for a few days. During this decontamination procedure any measurements would have been stopped. The SRC decontamination heaters would have been turned on for the warm-up phase while ATC and TC heaters would have remained at their current operational levels. Such a SRC decontamination would either have been required when one of the TC heaters would have reached a power of 0 W or, as originally required, at least twice per year.

Because of the necessity to heat up the detectors as much as possible to effectively get rid of the ice layers on channels 7 and 8, this decontamination procedure was redefined in the Commissioning Phase to form a Non-Nominal Decontamination (NNDEC) to be used during routine operations. During a NNDEC not only the SRC decontamination heaters provide energy to the optical subsystem but also ATC and TC heaters are switched to their maximum power. Measurements continue throughout warm-up and cool-down, contrary to what had been defined for the original decontamination procedure. In the warm-up phase of NNDEC, channels 7 and 8 reach temperatures of 267 K and the OBM approaches a temperature of −3° C. Also the duration of the warm-up phase was extended to 15 days. This method no longer creates a long data gap since data analysis still permits retrieval of – somewhat degraded – information from the UV-VIS channels even at elevated temperatures.

Active Thermal Control

To keep the OBM and detectors within the specified temperature limits is one of the key requirements of instrument

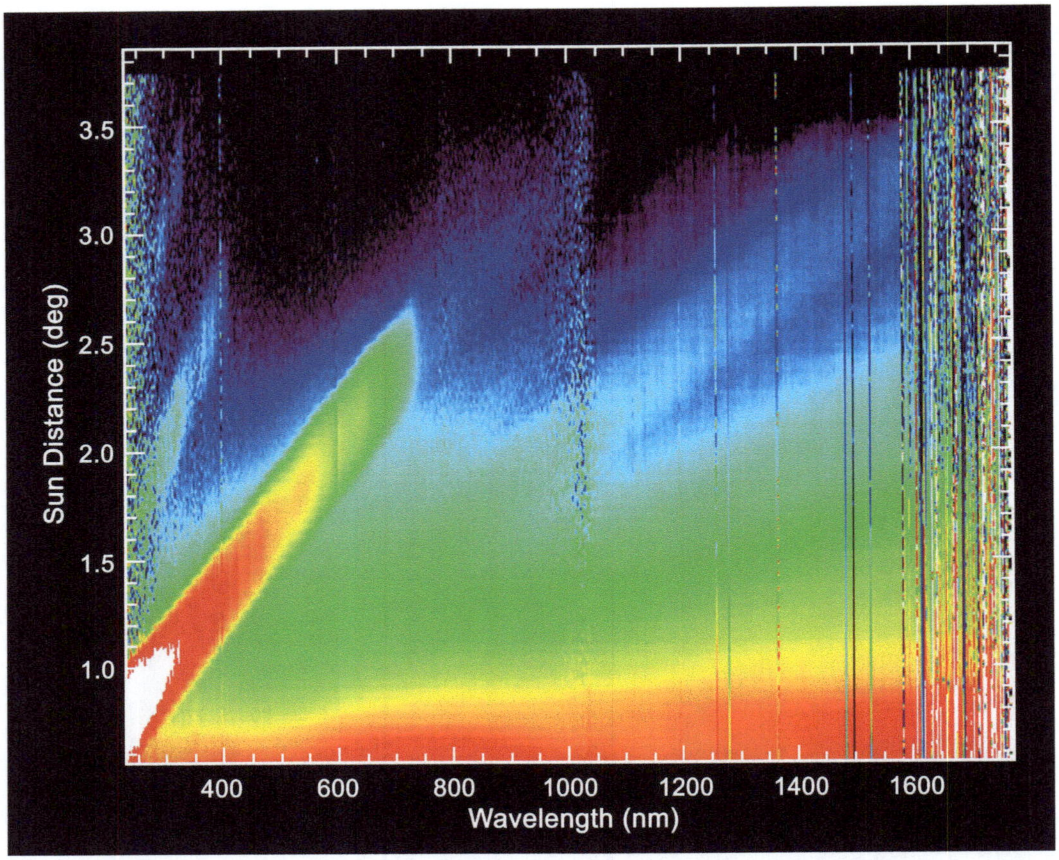

Fig. 6.11 Spatial stray light in channels 1–6 as a function of the LoS distance from the Sun. Close to the Sun stray light contributions are strongest (*red areas*). The tilted bands of different strengths are caused by diffraction from residual grooves in the telescope mirror. The measurements shown here were collected in close vicinity of the Sun. By varying the distance between Sun and IFoV characterising the wavelength dependent diffracted signal was possible (Courtesy: DLR-IMF).

performance. Several of the parameters determining the precision of calibrated spectra, e.g. dark current or wavelength calibration, are highly temperature sensitive. Therefore, with the start of routine operations, monitoring of the thermal status of the OBM and detectors was one of the prime tasks. Each week, average OBM and detector temperatures per orbit, together with ATC heater power consumption, have to be monitored, following the procedures outlined in the IOM. The OBM temperature, since controlled via the ATC, is kept autonomously within limits. Only when an ATC heater power approaches its lower limit, new setpoints have to be defined.

In the first years of operations, the ATC settings as of June 2002 provided a very stable average orbital OBM temperature of −17.90°C (Fig. 6.12). No adjustments were necessary for several years into the mission. The ATC heater power history displayed the expected seasonal modulation and a decrease – most pronounced in the ATC-Nadir heater – indicated signs of degradation (Fig. 6.13). This was, however, much smaller than predicted. During the months of the annual seasonal minimum, i.e. November and December, the orbital mean ATC-Nadir power slowly approached the specified lower limit. In October 2008 modified ATC settings had to be commanded to ensure safe ATC operations and a stable OBM temperature for the coming years of the extended mission. During this readjustment the ATC-Nadir power was increased and the ATC-Limb power decreased resulting in almost no effect on the OBM temperature.

Thermal Control

Contrary to the stable ATC behaviour, the detector temperature control requires occasional manual adjustments due to environmental variations. Whenever the weekly monitoring indicates that one of the detectors will violate the limit in the near future, FOCC is requested to command new trim heater power settings to bring the particular channel temperature well back in range. Since channels 4 and 5 have the highest temperature sensitivity, most of the TC adjustments were caused by these detectors. Early 2003 the temperatures

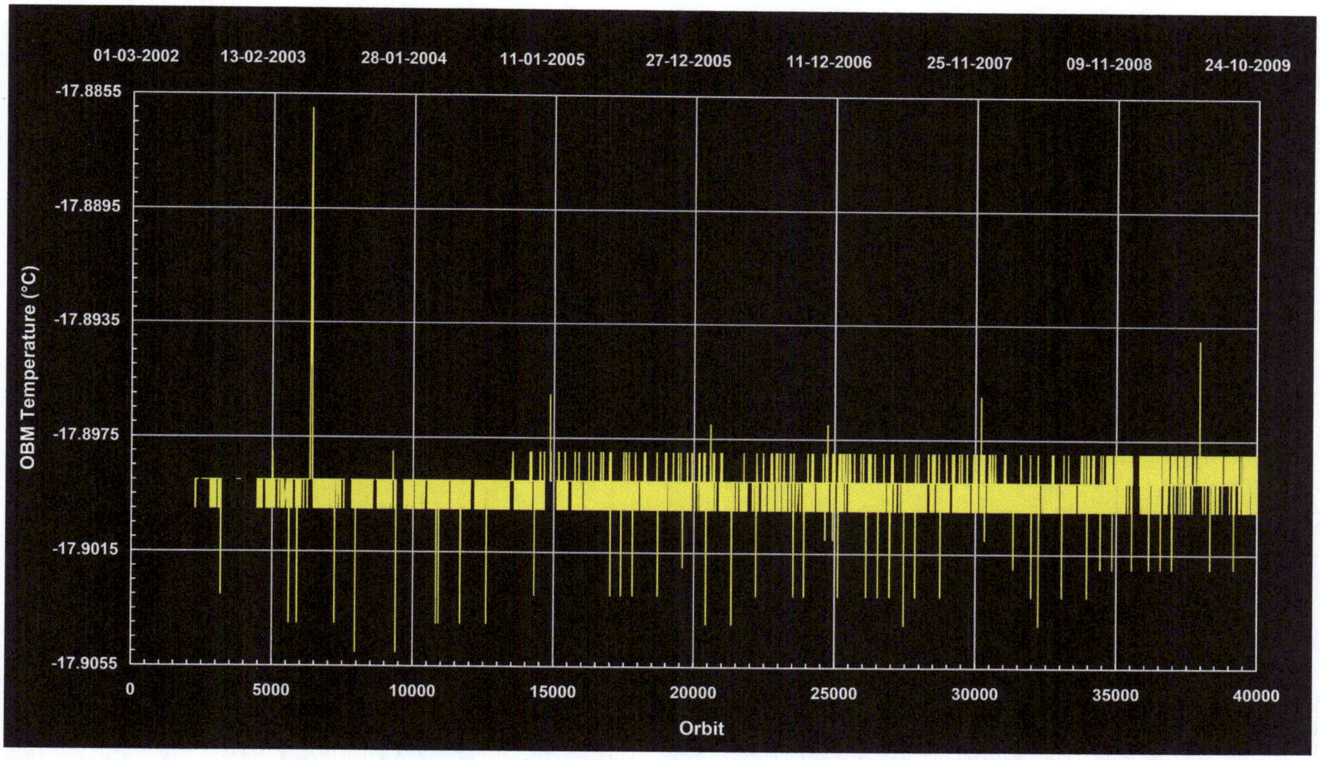

Fig. 6.12 Mean OBM temperature per orbit. On orbit average, the ATC system keeps the OBM within 0.001°C of the selected temperature (Courtesy: DLR-IMF).

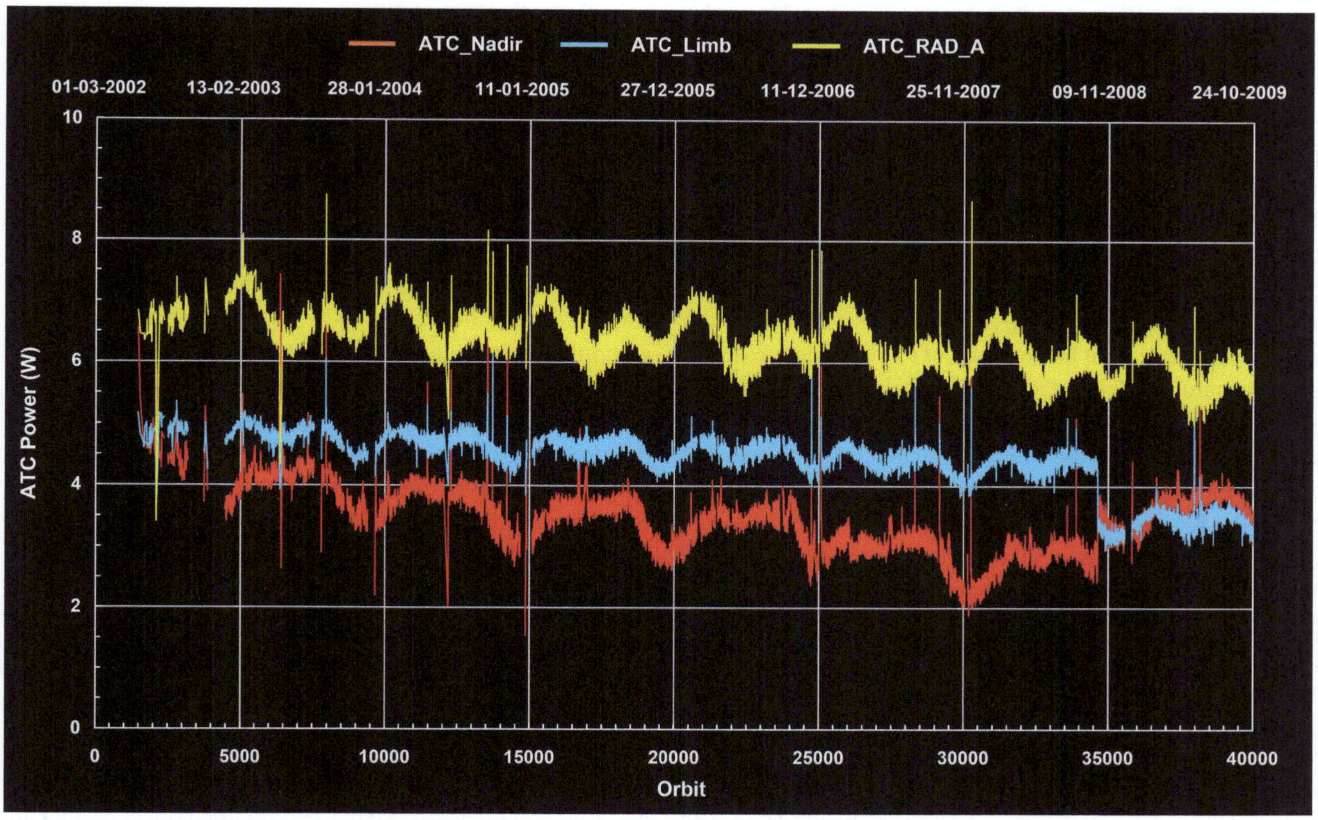

Fig. 6.13 Power consumption of the three heaters (Nadir, Limb and RAD A) of the ATC system. The curve follows a seasonal variation with an expected long-term decrease most prominent for the ATC-Nadir heater. The glitch in ATC-Nadir and ATC-Limb powers in October 2008 is caused by the ATC readjustment (Courtesy: DLR-IMF).

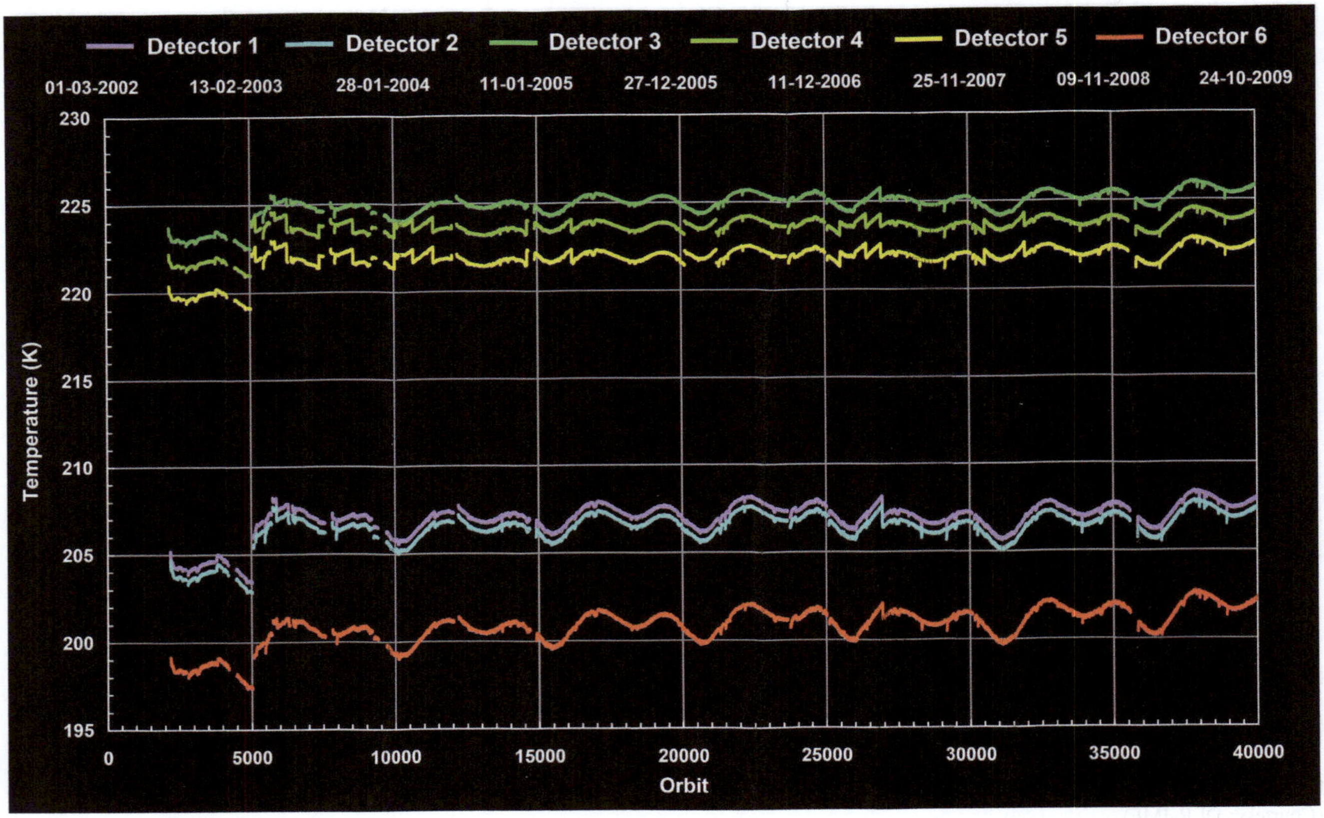

Fig. 6.14 Mean detector temperatures per orbit for channels 1–6. Decontamination intervals with elevated temperatures and periods after instrument saving with too low temperatures are omitted. The resulting curves display the seasonal variation. Steps are due to TC adjustments, mainly affecting channels 4 and 5. The increase around orbit 5000 corresponds to an overall TC correction to establish new temperature ranges (Courtesy: DLR-IMF).

selected by the TC settings for June 2002 were considered not to be optimum. Therefore in a number of TC adjustments during February 2003, the detector temperatures were brought into new ranges (see Table 3.4). Since then, control of the seasonal thermal variations requires only very few adjustments each year (Fig. 6.14). With the calibration improving over time it had turned out to even tolerate detector temperatures a few 0.1°C above or below their limits. For a period of 5–6 orbits after the adjustment, quality of the measurement data of the modified channel can be reduced because the temperature gradient exceeds the specified limit of 0.1–0.3 K/orbit for channels 5–8.

Decontamination is a vital part of thermal operations. Monitoring of the TC heater power led to the conclusion that RRU efficiency does not degrade with the rate estimated before launch. It was therefore decided to abandon the original requirement of two decontaminations per year and initiate a NNDEC only when the ice induced throughput reduction of channels 7 and 8 had reached unacceptable levels. Early during the routine phase decontaminations occurred more frequently since experience had to be gained about the most appropriate duration of the warm-up phase. Then, with a frequency of twice per year, NNDEC were executed, one in summer and one in winter. However these NNDEC achieved only a temporary removal of the ice layers. Due to accidental non-nominal events, either in the warm-up or cool-down, almost none of the executed NNDECs was identical. Improvement of the throughput for channel 7 after decontamination showed erratic behaviour without a clear explanation. Only when the cool-down phase of the NNDEC was modified by introducing a well specified transfer to STANDBY mode with subsequent 'recovery', the throughput in channels 7 and 8 remained at high values for long periods. Hypothesis is that a second cold trap exists which may be activated by forcing low temperatures in the cool-down. At the beginning of the cool-down water is still uncondensed and can deposit on the second cold trap thus leaving the optical path of detectors 7 and 8 to a large extent void of ice. The new procedure was implemented operationally for the decontamination in winter 2004/2005. It removed almost all the ice from the light path of channel 7 and kept the reduction in channel 8 to less than 30%. After operating channels 7 and 8 without another decontamination for 4 years, the same NNDEC procedure was applied in winter 2008/2009 and delivered a similar recovery of the SWIR throughput.

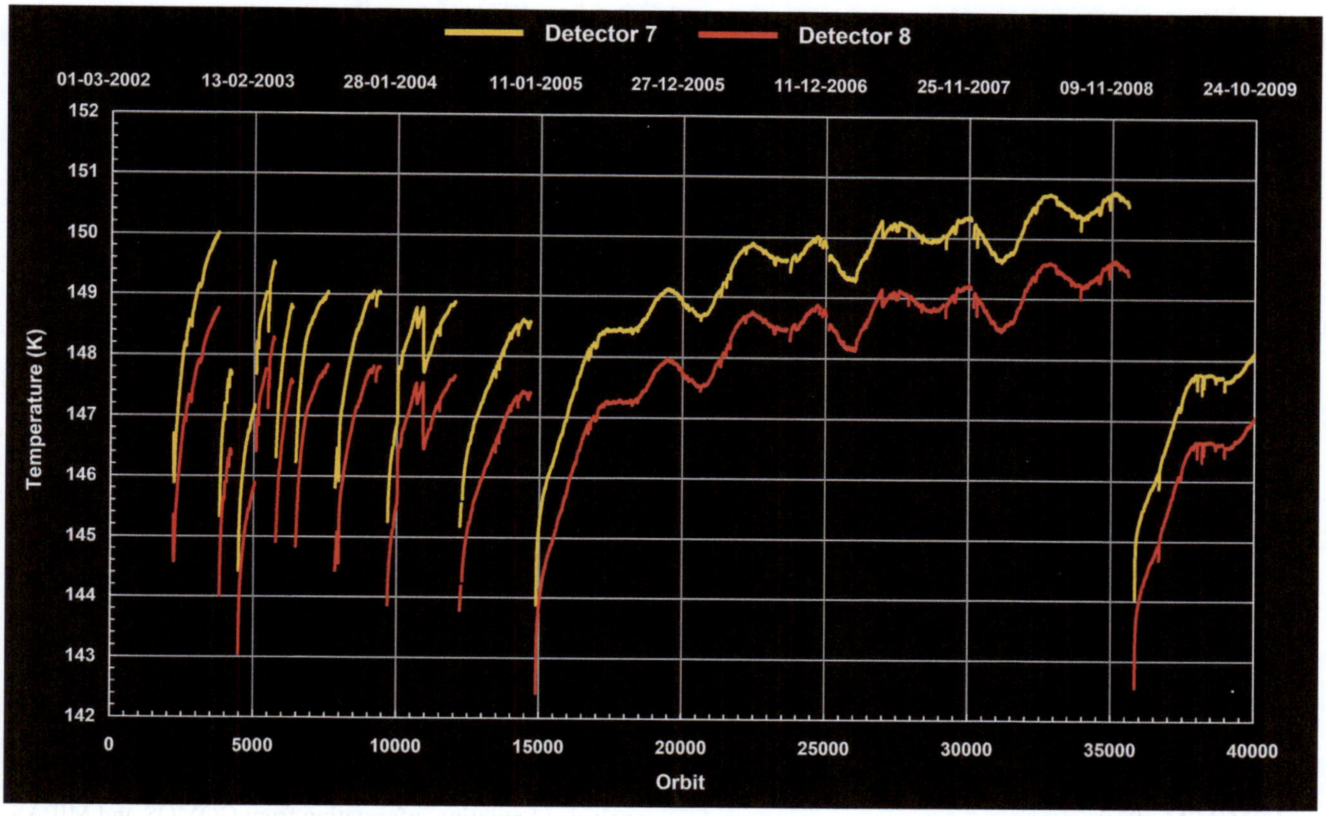

Fig. 6.15 Mean detector temperatures per orbit for the SWIR channels 7 and 8. Displayed data are as in Fig. 6.14. The graphs are the result of growing ice (increase) and occasional decontaminations (drop to lower values). Between orbit 10600 and 10900 two TC adjustments had a significant impact on the SWIR detector temperatures (Courtesy: DLR-IMF).

The temperature behaviour of the infrared channels 7 and 8 is largely driven by the ice conditions. Ice also covers the gold plated aluminium structures of the detector suspension leading to an increased infrared absorption and thus radiatively heated detectors. This results in a slow but steady rising temperature. Immediately after a decontamination ice is removed and temperatures are at the selected cold level from where they start to increase – caused by the growth of the ice layer – until the next decontamination is started or an equilibrium with a stable ice layer is reached (Fig. 6.15).

6.5 Line-of-Sight Performance

Scanners

Because of their design, both the ESM and the ASM were not considered to be critical mechanical subsystems. When the end of the nominal specified mission lifetime was reached in March 2007, however, the scanners became subject of regular monitoring. This ensured early detection of potential degradation of the scanner performance and development of appropriate countermeasures. The scanner current parameters, which are downlinked as part of the Housekeeping (HK) telemetry stream, have been considered suitable for the purpose of scanner monitoring. Since state 65 (*scanner maintenance and ADC calibration*) is scheduled each orbit, the acquired HK telemetry provides good statistics even though the measurement phase of state 65 only lasts about 45 s. With a rate of 1/16 Hz about 2–3 telemetry readings are obtained at each state execution. Due to the design of state 65 a distinct pattern of the ASM and ESM clockwise, counterclockwise and mean current is expected. As long as this pattern does not change, no degradation of the scanner performance is deduced. Over the past 7 years the ASM current pattern remained rather stable. At the end of 2007 the ESM mean current began to deviate from its previous curve, reached a plateau value and started to decrease again. This behaviour is not fully understood and therefore subject to further investigations. Because continuous ESM operations occur in both nadir and limb states usage of this scanner is much more frequent than that of the ASM. Therefore degradation effects might emerge earlier. However the pointing performance of both scanners remained stable and seems unrelated to the observed scanner currents effect.

Extra Mispointing

Since the Commissioning Phase it is well known that SCIAMACHY's LoS exhibits small inconsistencies. These include in the SO&C window an elevation jump of about −0.04° when acquiring the Sun with the Sun Follower in elevation above the atmosphere after following the solar trajectory with a predicted elevation rate in the ESM control. In the sub-solar window an elevation jump of about −0.02° occurs, again when the Sun Follower captures the Sun in elevation above the atmosphere. Before launch the misalignment of the instrument LoS in all three axes had been measured. The corresponding parameters are given in Table 6.3. They are stored on-board as part of the engineering parameters and compensated in the scanner control via the *optical zero correction*. Thus the observed pointing inconsistencies could not be attributed to this known misalignment. Their origin was instead modelled by an extra mispointing which is not necessarily only reflecting an additional misalignment of the instrument axes but could be associated with all components impacting the complete measurement planning and execution chain, as there are

- Extra instrument misalignment (pitch, roll, yaw)
- ESM/ASM offset
- Extra platform attitude mispointing (pitch, roll, yaw)
- FOCC planning and scheduling s/w
- Scanner control s/w
- Sun Follower control loop

Best agreement with the pointing inconsistencies in solar observations was achieved when assuming extra mispointings as listed in Table 6.3. The values in pitch and roll turned out to be so large that a significant impact had to be expected when retrieving altitudes from limb measurements (Gottwald et al. 2007).

Table 6.3 SCIAMACHY instrument misalignment measured before launch and modelled extra mispointing.

Axis	Instrument misalignment (°)	Extra mispointing (°)
Pitch	+0.000630	−0.026 ± 0.003
Roll	+0.001662	−0.020 ± 0.001
Yaw	−0.227464	+0.009 ± 0.008

Tangent Height

SCIAMACHY's limb mode is a powerful technique to sense the atmosphere with global coverage and high vertical resolution. A necessity for obtaining useful measurements is the accuracy of reconstructing altitudes in general and tangent heights in particular from elevation and azimuth angles (von Savigny et al. 2005). Because of the large distances involved in limb geometry – about 3200 km from the instrument to the Earth horizon – even small pointing uncertainties translate into large tangent height errors. For SCIAMACHY on ENVISAT, a pitch pointing error of 1 arcmin shifts retrieved profiles by 1 km.

Early in the mission, it was recognised that the operationally generated tangent height information differs by as much as 3 km from what was expected. The reference tangent heights were derived from the TRUE (Tangent Height Retrieval by UV-B Exploitation) method (Kaiser et al. 2004), which uses the so-called 'knee' as a spectral signature for retrieving limb pointing information. The TRUE results are considered reliable as long as the technique is restricted to tropical latitudes, e.g. between 20°N and 20°S, where stratospheric and lower mesospheric ozone variability is small. The offset between the operational tangent heights and those derived via the 'knee' method showed a strong seasonal variation with a mean amplitude of 0.8 km and a constant bias of 0.5 km. The sinusoidal seasonal modulation was superimposed on a linear trend with a gradient of about 0.4 km/year.

SCIAMACHY's findings were compliant with a detailed analysis of pointing information from GOMOS and MIPAS. ENVISAT's on-board processing of state vector parameters uplinked from ground was identified to be the source of the observed pointing inaccuracy. In a corrective action the s/w algorithm for deriving such parameters on-ground was upgraded and finally implemented in December 2003 around orbit 9300 (Bargellini et al. 2006). This resulted in a reduction of the tangent height jumps observed around the times of the daily updates of the on-board state vector and of the seasonal variation of the tangent height offset. When using residual ENVISAT platform pointing information in pitch, roll and yaw the mean amplitude of the offset amounted now to about 0.2 km with the linear gradient reduced by a factor of 3. However the bias increased to 1.5 km and remained practically stable over a year (Fig. 6.16).

Exploiting the full capabilities of limb data obtained by SCIAMACHY required additional improvement in altitude retrievals. It was indeed achieved via the extra mispointing model. Adding the correction angles from Table 6.3 in the calculation of geolocations reduced the tangent height bias to almost 0 km as obvious in Fig. 6.16 with an uncertainty of 100 m. Having obtained such a high accuracy in altitude determination permitted investigations on even smaller residual effects. A seasonal trend with an amplitude of 150 m became obvious. In addition indications for a tangent height trend with a rate of a few 10 m/year exist. Further investigations are ongoing to prove their validity, understand their origin and correct for it.

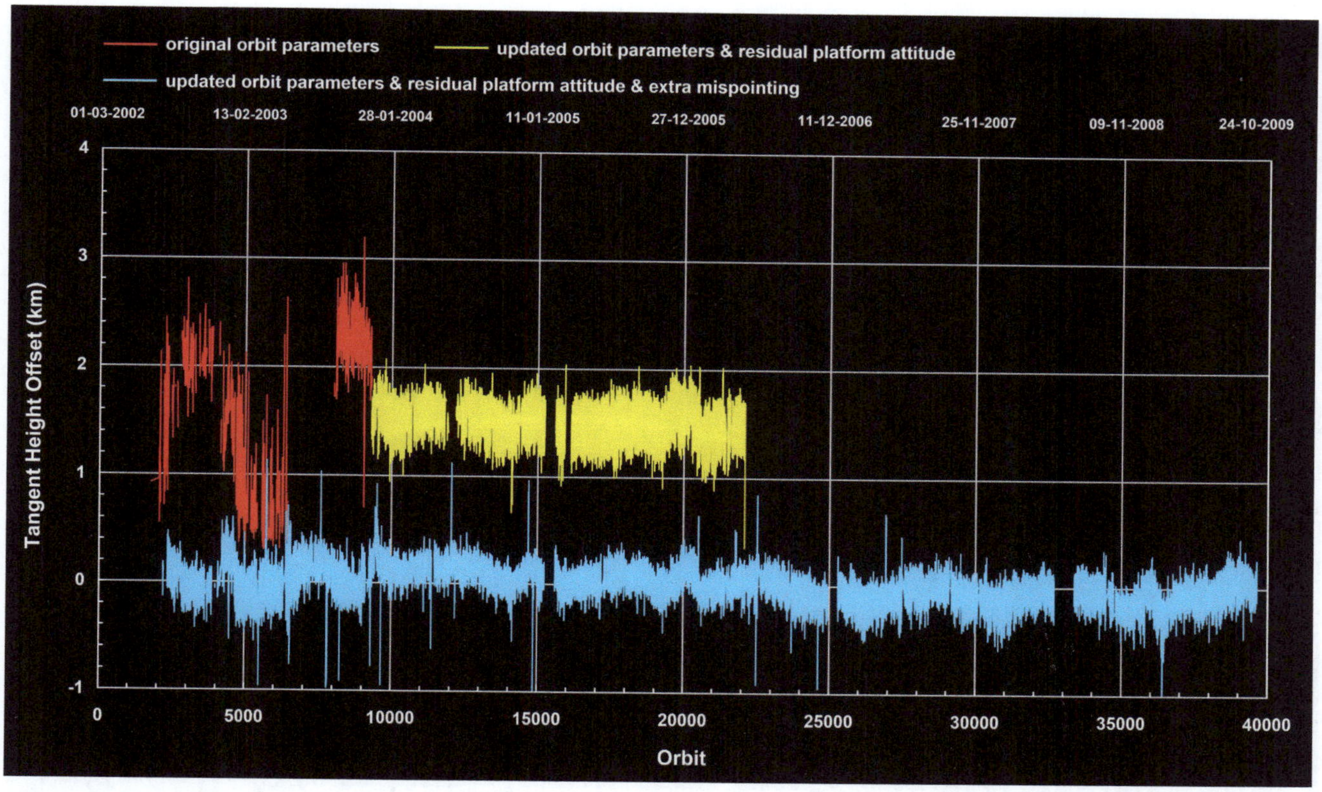

Fig. 6.16 Tangent height offsets as determined from operational data products and TRUE retrieved profile information. Prior to December 2003 the bias had a strong harmonic variation (*red*). After the update of the on-board propagator model and including residual platform mispointing the variation was reduced but a constant offset persisted (*yellow*). When the extra mispointing is introduced in geolocation retrieval this offset vanishes (*blue*) (Courtesy: IUP-IFE, University of Bremen).

6.6 Mission Extension

Since its launch in March 2002 the ENVISAT mission proved to be very successful, concerning both the platform and the instrument payload complement. When the end of the specified nominal mission lifetime was reached in early 2007, it was decided to continue with platform and instrument operations for another 3.5 years until 2010. In this phase of the mission extension the high mission performance could be maintained. After platform and sensors had convincingly demonstrated that even after 2010 very valuable information about the Earth system can be acquired, ESA and the participating national agencies approved another mission extension period lasting from the second half of 2010 to the end of 2013, at least.

Orbit Modification

One of the major mission limiting factors is available fuel (see Chapter 2). With the requirement to deposit ENVISAT at the end of the mission in an orbit with different altitude to reduce the risk of collisions with other Earth Observation satellites, sufficient hydrazine must be kept in the fuel tanks for the final orbit lowering manoeuvre which consumes for an altitude change of 1 km about 2 kg of fuel. It has been decided that the disposal orbit shall be commanded when the first phase of the mission extension ends. Thus the second phase will be operated in an orbit different from the nominal ENVISAT orbit. All the fuel remaining after the altitude manoeuvre, estimated to be slightly more than 20 kg, will then be available for orbit control. If the fuel demanding inclination maintenance is discarded, altitude manoeuvres will be possible until 2013 (Frerick et al. 2007).

Assuming that the altitude is decreased by 17.4 km, the modified ENVISAT orbit in the second mission extension phase will be characterised by a repeat cycle of 30 days, i.e. 431 orbits. Its orbital period of 6014 s is shorter than the nominal one of about 6036 s. No longer readjusting the inclination is equivalent to a slightly drifting inclination at a rate of 0.046°/year. This affects the Mean Local Solar Time (MLST), which will also leave the specified value of 22:00 ± 5 min after a while. The maximum excursion occurs 1.5 years after the orbit modification when a value of 22:09 is reached. Then the MLST approaches 22:00 again and will start to drift to earlier values with the end of 2013 (Fig. 6.17).

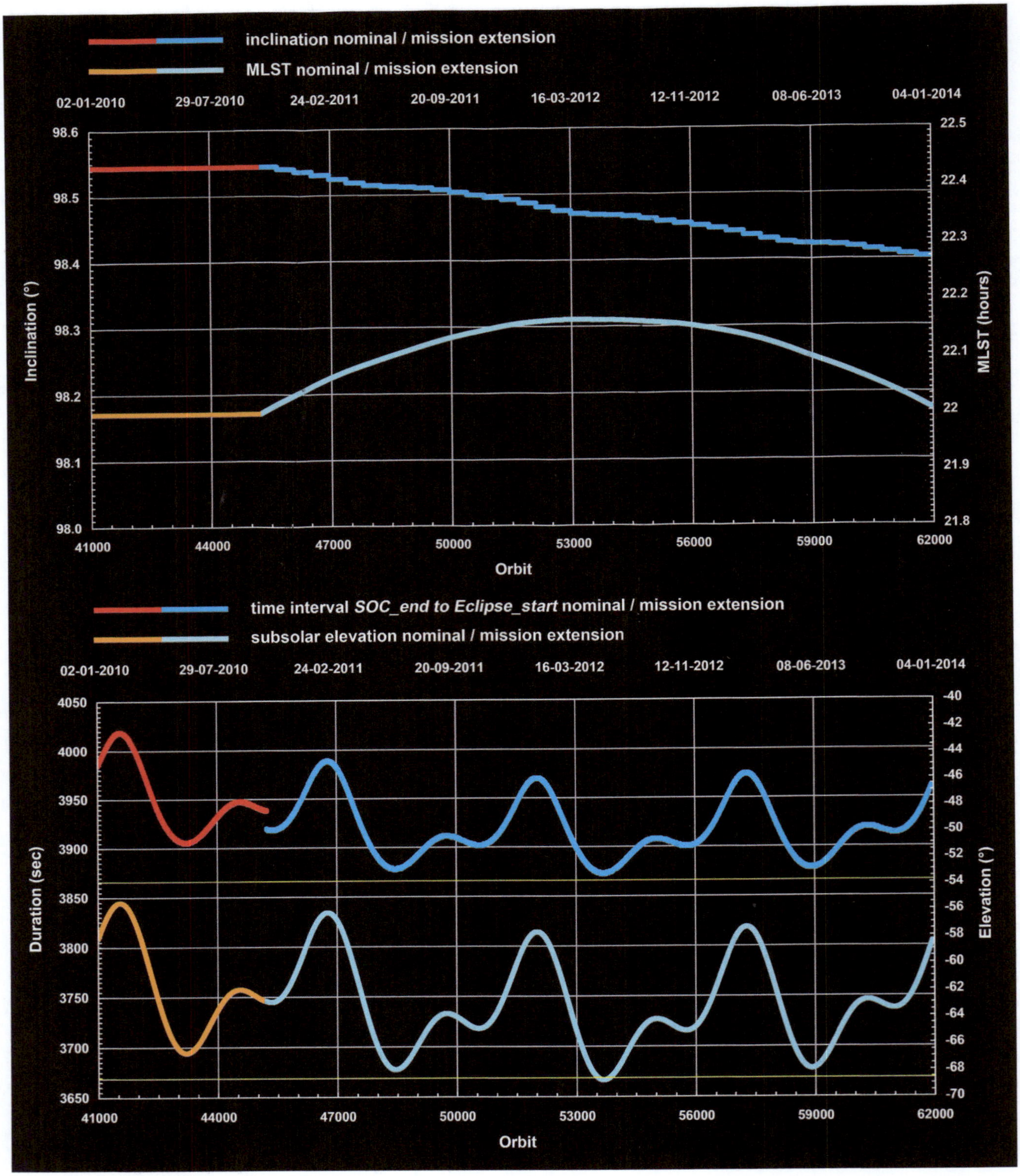

Fig. 6.17 Inclination (*left axis*) and MLST (*right axis*) drift of the ENVISAT orbit during the mission extension phase until 2013 (*top*). The bottom panel shows an example how the orbit interval from end of the SO&C window to the start of the eclipse (*left axis*) and the solar elevation at sub-solar condition (*right axis*) changes when the orbit altitude is lowered. The yellow lines indicate the lower and upper limit of the NCWM window (Courtesy: DLR-IMF).

Impact on Instrument Performance

Operating SCIAMACHY in the 2010–2013 timeframe means to more than double the instrument's in-orbit lifetime. This will partially occur in a different orbit. Therefore impacts on the instrument performance are expected from

- Orbit: SCIAMACHY as an instrument with multi-viewing capabilities is strongly dependent on the status of the LoS during measurements. Because of the modified orbit the visibility of terrestrial and celestial targets along the orbit changes.
- Time: With evolving time the subsystems of the instrument degrade. Having extended the lifetime well beyond the specified time span stronger degradation than predicted for nominal End-of-Life (EOL) must be expected.

Degradation in the extended mission is treated in two ways. One is via reconfiguring the affected subsystems as long as sufficient margins in configurable parameters are available. The other concerns improvements in calibration and monitoring which helps to use measurement data obtained in a degraded environment for retrievals.

The flexibility of SCIAMACHY operations control permits to compensate for the modified orbit to a large degree. It requires a reconfiguration of the instrument by updating certain engineering and measurement parameters as well as states and timelines (Gottwald et al. 2009). Usually the TCFoV for each viewing geometry is so wide that the Sun and the Moon are still visible even when the MLST approaches its extreme value. The only exception occurs when the Sun hits the upper edge of the sub-solar window for a short time period making sub-solar measurements impossible (Fig. 6.17). Thus the mission scenarios need not be changed and the scientific requirements can be fulfilled even until 2013.

SCIAMACHY's life limited items (LLI) require particular attention in extended mission operations. They shall only execute the specified activations defined by individual in-flight LLI budgets from on-ground tests and analyses. Safe operations aim at not exceeding these budgets over the mission lifetime, including extension phases. The usage of LLIs is therefore monitored regularly and compared with the corresponding in-flight budget. Assuming that the mission scenarios remain stable, the LLI executions can be extrapolated until the end of 2013. For none of the LLI except the Nadir Calibration Window Mechanism (NCWM) will the in-flight budget be reached in more than 11 years of operations. Only the accumulated NCWM activations will slightly violate the allocated limit. However this is uncritical since the rate of sub-solar measurements can be adjusted accordingly any time during the mission.

References

Bargellini P, Garcia Matatoros MA, Ventimiglia L, Suen D (2006) ENVISAT attitude and orbit control in-orbit performance: An operational view. Proceedings of 6th international conference on guidance, navigation and control systems, Loutraki, Greece, ESA SP-606

Frerick J, Duesmann B, Canela M (2007). 2010 and beyond – The ENVISAT mission extension. Proceedings of ENVISAT Symposium 2007, Montreux, Switzerland, ESA-SP-636

Gottwald M, Krieg E, Slijkhuis S, von Savigny C, Noël S, Bovensmann H, Bramstedt K (2007) Determination of SCIAMACHY LOS misalignements. Proceedings of ENVISAT symposium 2007, Montreux, Switzerland, ESA-SP-636

Gottwald M, Krieg E, Lichtenberg G, Noël S, Bramstedt K, BovensmannH, Snel R (2009) SCIAMACHY instrument status – from 2010 to 2013. Proceedings of ESA atmospheric science conference, Barcelona, Spain, ESA-SP-676

Kaiser JW, von Savigny C, Eichmann K-U, Noël S, Bovensmann H, Burrows JP (2004) Satellite pointing retrieval from atmospheric limb scattering of solar UV-B radiation. Can J Phys 82:1041–1052

Snel R, Tilstra G (2008) SCIAMACHY Scan Angle Degradation (SRON-SCIA-PhE-RP-23). Technical Document, SRON

van Soest G (2005) Investigation of SCIAMACHY limb stray light (SRON-EOS-RP-05-006). Technical Document, SRON. Available at http://www.sron.nl/~SCIA_CAL/SCIACALdocuments.php/

von Savigny C, Kaiser JW, Bovensmann H, Burrows JP, McDermid IS, Leblanc T (2005) Spatial and temporal characterization of SCIAMACHY limb pointing errors during the first three years of the mission. Atmos Chem Phys 5:2593–2602

Chapter 7
From Radiation Fields to Atmospheric Concentrations – Retrieval of Geophysical Parameters

H. Bovensmann, A. Doicu, P. Stammes, M. Van Roozendael, C. von Savigny, M. Penning de Vries, S. Beirle, T. Wagner, K. Chance, M. Buchwitz, A. Kokhanovsky, A. Richter, A.V. Rozanov, and V.V. Rozanov

Abstract Satellite-based atmospheric remote sensing aims at deriving the properties of trace gases, aerosols and clouds, as well as surface parameters from the measured top-of-atmosphere spectral radiance and reflectance. This requires, besides high quality spectra, an accurate modelling of the radiative transfer of solar radiation through the atmosphere to the sensor (*forward model*) and methods to derive the constituent properties from the measured top-of-atmosphere spectra (*inversion methods*). Many trace gases have structured absorption spectra in the UV-VIS spectral range serving as the starting point for determining their abundance by applying Differential Optical Absorption Spectroscopy (DOAS) or similar methods. In the UV-VIS-NIR and SWIR spectral regions the solar radiation is strongly scattered by clouds and aerosols. Therefore the presence of clouds and aerosol particles and their properties can also be inferred from the outgoing radiance measured by space-based instruments. Contrary to the forward model, the inversion methods allow to derive characteristics of the atmospheric state based on the measured quantities. A common product of the inversion of satellite measurements in limb, nadir or occultation geometry are total columns or height-resolved profiles of trace gas concentrations and aerosol parameters. Retrieving trace gas amounts in the troposphere constitutes a specific challenge. SCIAMACHY's unique limb/nadir matching capability provides access to tropospheric columns by combining total columns obtained from nadir geometry with simultaneously measured stratospheric columns obtained from limb geometry.

Keywords Atmospheric composition retrieval • Absorption • Scattering • Radiative transfer • Differential Optical Absorption Spectroscopy (DOAS) • Inversion theory

Satellite remote sensing of the atmosphere with spectrometers such as SCIAMACHY gives the top-of-atmosphere (TOA) spectral radiance. It remains a challenging task to quantitatively derive the atmospheric composition – trace gases, aerosols, clouds – from these measurements. The main source of radiation for passive remote sounding of the atmosphere by SCIAMACHY in the UV-SWIR regions is the Sun. The absorption and scattering characteristics of the Earth's atmosphere can be determined by comparing the radiance reflected from the Earth's surface and scattered by the atmosphere to the sensor with the unaffected extraterrestrial solar irradiance entering at the top of the atmosphere. A quantitative analysis requires:

- Earth and solar radiance spectra of high spectral and radiometric quality
- An accurate modelling of the transfer of the solar light through the atmosphere to the sensor (*radiative transfer model* or *forward model*)
- Techniques to relate the measured TOA spectra to the constituent properties (usually referred to as *inversion methods*)

The retrieval of information on atmospheric trace gases relies on the knowledge of the absorption, emission and

H. Bovensmann (✉), C. von Savigny, M. Buchwitz, A. Kokhanovsky, A. Richter, A.V. Rozanov, and V.V. Rozanov
Institute of Environmental Physics/Institute of Remote Sensing (IUP-IFE), University of Bremen, Otto-Hahn-Allee 1, 28359 Bremen, Germany
e-mail: heinrich.bovensmann@iup.physik.uni-breman.de

A. Doicu
Remote Sensing Technology Institute, German Aerospace Center (DLR-IMF), Oberpfaffenhofen, 82234 Wessling, Germany

P. Stammes
Royal Netherlands Meteorological Institute (KNMI), Wilhelminalaan 10, 3732 GK De Bilt, The Netherlands

M. Van Roozendael
Belgian Institute for Space Aeronomie (BIRA-IASB), 3 Avenue Circulaire, 1180 Brussels, Belgium

M. Penning de Vries, S. Beirle, and T. Wagner
Max Planck Institute for Chemistry, Johann-Joachim-Becher-Weg 27, 55128 Mainz, Germany

K. Chance
Harvard-Smithsonian Center for Astrophysics (SAO), 60 Garden Street, 02138 Cambridge, MA, USA

DOI 10.1007/978-90-481-9896-2_7,

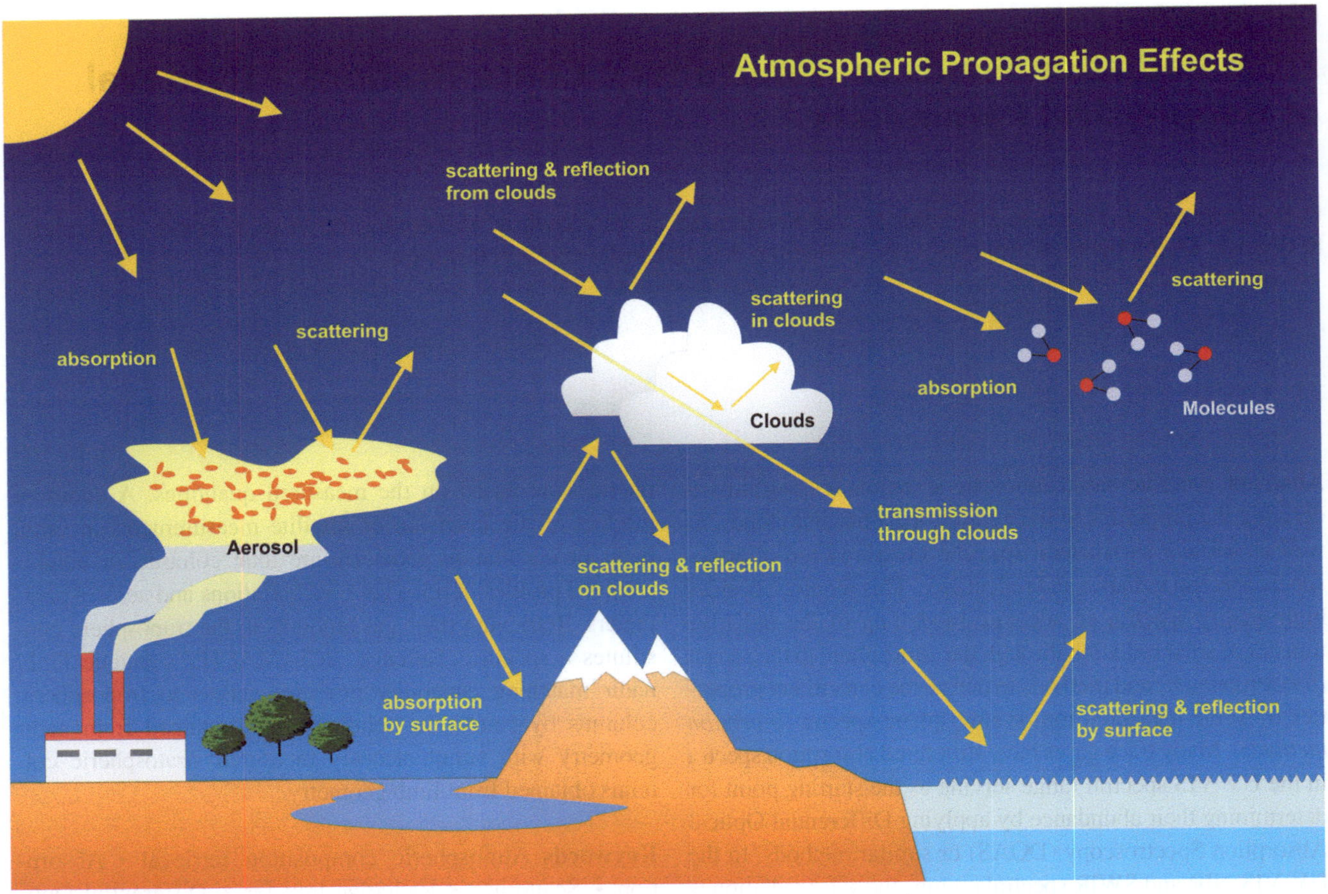

Fig. 7.1 Scheme of the relevant interactions of solar light with the Earth's atmosphere and surface (Courtesy: DLR-IMF).

scattering of electromagnetic radiation in the atmosphere. In the UV-VIS-NIR and SWIR spectral ranges, the radiative transfer through the atmosphere is affected by (see Fig. 7.1):

- Scattering by air molecules (Rayleigh and Raman scattering)
- Scattering and absorption by aerosol and cloud particles (Mie scattering)
- Absorption and emission by trace gases
- Refraction due to the density gradient in the atmosphere
- Surface reflection and absorption

These are the processes which must be quantitatively taken into account when retrieving atmospheric geophysical parameters from SCIAMACHY measurements.

Trace gases usually exhibit characteristic fingerprint spectra in emission or absorption, originating from:

- Rotational transitions: primarily observed in the far infrared and microwave spectral regions
- Vibrational–rotational transitions: can be measured in the thermal and short-wave infrared
- Electronic transitions: mainly detected in the UV-VIS and NIR spectral regions

SCIAMACHY with its wide spectral coverage from the UV to the SWIR detects trace gases mainly via their electronic and vibrational–rotational transition spectra.

7.1 Radiative Transfer in the Earth's Atmosphere

Important tools to simulate changes in the solar radiation due to atmospheric scattering and absorption processes are radiative transfer models (RTM), also referred to as *forward models*. They provide the synthetic radiances, as they would be measured by the sensor, for a specified state of the atmosphere. These models are an important part of any retrieval process. Considering the entire atmosphere, one usually talks about the radiation field, which describes the angular and spatial distribution of the radiation in the atmosphere.

The main characteristic of the radiation field in the atmosphere is the radiance I (also often referred to as 'intensity'), which is defined as the flux of energy in a given direction per unit time per unit wavelength range per unit solid angle per unit area perpendicular to a given direction (Liou 2002). All interactions between the radiation and the atmosphere are classified by the sign of the change of the radiation intensity as a result of the interaction. Processes which reduce the intensity in the direction under consideration by absorption as well as scattering processes from the original direction into other directions are commonly named *extinction*. Processes which increase the intensity in the direction under consideration for example by scattering into the beam from other directions, by thermal or other emission processes within the volume, are referred to as *source function*.

In the description of radiative transfer presented here we neglect the polarisation state of light for reasons of simplicity. However, we note that polarisation is important for SCIAMACHY for two reasons, namely in order to

- Accurately simulate radiances in the UV and VIS
- Account for the polarisation sensitivity of the instrument when determining the true radiance (see discussion in Chapter 5)

The general form of the radiative transfer equation describes all processes affecting the radiation field as a result of its interaction with a medium, taking energy conservation into account. It has the form

$$\frac{dI}{ds} = -\alpha\,(I - J) \tag{7.1}$$

where I is radiance (or intensity) in a given direction, s the light path, α is the extinction coefficient describing the fraction of the energy which is removed from the original beam by absorption and scattering and J is the source function which describes the increase of the radiance I into the original direction due to scattering and/or emission.

If the amount of light travelling in a certain direction through the atmosphere can only be increased due to the scattering processes – as is the case for the spectral range covered by SCIAMACHY when neglecting emissions – the source function depends on the intensity falling on the elementary volume from all directions

$$J = \frac{\omega}{4\pi}\int p(\gamma)\, I\, d\Omega \tag{7.2}$$

with γ being the scattering angle, i.e. the angle between the directions of the incident and scattered radiation, and ω is the single scattering albedo representing the probability that a photon, which interacts with a volume element, will be scattered rather than being absorbed. The term $p(\gamma)d\Omega/4\pi$ denotes the probability that the radiation is scattered into a solid angle $d\Omega$ about a direction forming an angle γ with the direction of the incident radiation. The quantity $p(\gamma)$ is called the phase function.

The total radiation field can be split into two components

$$I = I_{dir} + I_{dif} \tag{7.3}$$

with I_{dir} being the direct radiation, which is never scattered in the atmosphere or reflected from the Earth's surface, and the diffuse radiation I_{dif}, which is scattered or reflected at least once. Since there is no relevant process in the atmosphere which increases the intensity of the direct solar radiation, the radiative transfer equation for the direct radiation leads to the homogeneous differential equation

$$\frac{dI_{dir}}{ds} = -\alpha\, I_{dir} \tag{7.4}$$

with a solution described by Lambert–Beer's law

$$I_{dir} = I_0\,\exp\left(-\int \alpha(s)\, ds\right) \tag{7.5}$$

where I_0 being the incident radiance at the top of the atmosphere, which is described – assuming the Sun has an infinitesimal size – by the solar irradiance I_{irr} multiplied by the Dirac delta-function (see, e.g. Thomas and Stamnes 1999). The integral $\int\alpha(s)ds$ along the photon path defines the *optical depth* $\tau(s)$. Integration is performed along the direct solar beam from the surface to the top-of-atmosphere. This equation describes the attenuation of the direct solar or lunar light travelling through the atmosphere. Thus the direct component (7.5) is a good approximation for the radiative transfer when for example directly viewing the Sun or the Moon. Therefore it can be used to describe SCIAMACHY measurements in occultation geometry.

If the diffuse radiation is non-negligible – as for SCIAMACHY nadir and limb measurements – it has to be considered in addition to the direct one. The corresponding radiative transfer Eq. (7.1) for the diffuse component using the scattering source function (7.2) can be written then as follows (details see Liou 2002):

$$\frac{dI_{dif}}{ds} + \alpha\left(I_{dif} - \frac{\omega}{4\pi}\int p(\gamma)\, I_{dif}\, d\Omega\right) = \frac{\omega}{4\pi} p(\gamma_0)\, I_{irr}\,\exp\left(-\int \alpha(s)\, ds\right) \tag{7.6}$$

where γ_o denotes the scattering angle between the direct solar beam and the direction of observation and I_{irr} is the extraterrestrial solar flux, often also termed 'irradiance'. Various standard methods to solve (7.6) are presented in Lenoble (1985) or Liou (2002).

Light is scattered by atoms and molecules as well as by various types of aerosols and clouds. Molecular scattering cross sections are characterised by the Rayleigh law (λ^{-4}), whereas aerosol scattering typically shows a much less pronounced dependence on wavelength with about λ^{-1} (Mie scattering). For cloud-free scenes, molecular scattering usually dominates in the UV spectral range, while aerosols become more important in the VIS and NIR range (see also Fig. 7.3). The molecular scattering consists of two parts: the elastic Rayleigh component which accounts for about 96% of scattering events and the 4% inelastic rotational Raman component, which is responsible for the *Ring* effect, the 'filling in' of solar Fraunhofer lines in the registered earthshine spectra (Grainger and Ring 1962). Further details on atmospheric radiative transfer can be found in, e.g. Liou (2002).

As an example, Fig. 7.2 shows the solar irradiance spectrum and the backscattered radiance at the top-of-atmosphere. When SCIAMACHY nadir radiances at high solar zenith angles SZA $> 75°$ are simulated, the sphericity of the atmosphere must be taken into account when calculating the attenuation of the direct solar light. For simulating SCIAMACHY limb radiances both the direct and the diffuse radiance have to be treated in a spherical atmosphere, including refraction. The numerical solution of the radiative transfer equation is then accomplished, e.g. by an iterative approach (Rozanov et al. 2001). Depending on the scientific application, several radiative transfer models are used in the SCIAMACHY data analysis. These are GOMETRAN/SCIATRAN (Rozanov et al. 1998, 2005a), LIDORT (Linearized Discrete Ordinate Radiative Transfer, Spurr et al. 2001), DAK (Doubling-Adding KNMI, Stammes 2001), SCIARAYS (Kaiser and Burrows 2003), VECTOR (McLinden et al. 2002), and McArtim (Monte Carlo Atmospheric Radiative Transfer Inversion Model, Deutschmann et al. 2010). These radiative transfer models are not only able to simulate the radiance measured by SCIAMACHY, but also to deliver additional parameters to quantify and characterise the geophysical parameters of interest. One of these parameters is the *weighting function*. This function is the derivative of the modelled

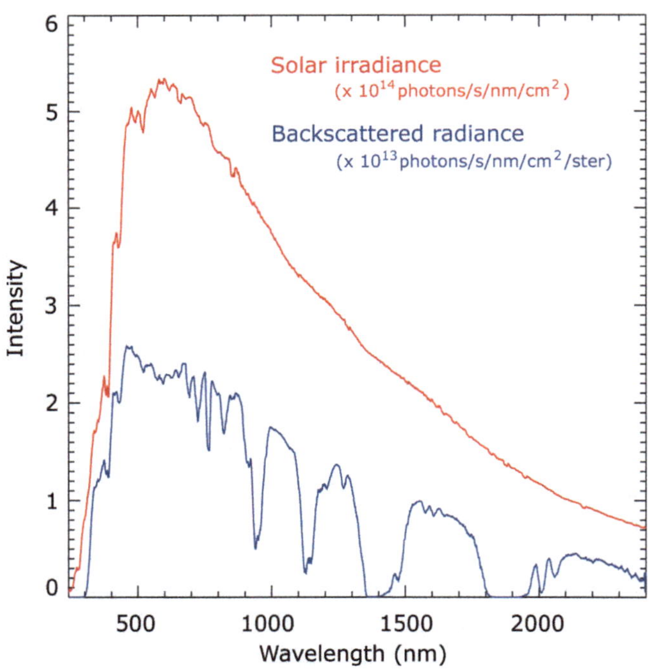

Fig. 7.2 The solar irradiance spectrum (*red*) and Earth radiance spectrum (*blue*) with a shape modified by absorption of trace gases and scattering in the atmosphere (Courtesy: IUP-IFE, University of Bremen).

radiance with respect to a selected parameter, describing how the radiance changes when this parameter is modified. Another important quantity to be delivered by radiative transfer models is the *Air Mass Factor* (AMF, see section "Nadir Trace Gas Retrieval Schemes") which under optically thin conditions (optical depth $\ll 1$) provides a measure for the average photon path length in the atmosphere.

The quality of the radiative transfer simulation results is often assessed in application-specific radiative transfer model inter-comparison exercises. This was done for example for limb radiative transfer applications by Loughman et al. (2004), for the AMF by Wagner et al. (2007a) and for polarised radiative transfer calculations by Kokhanovsky et al. (2010).

7.2 Nadir Trace Gas Retrieval Schemes

DOAS Retrieval

Many molecules of atmospheric relevance have structured absorption spectra in the UV-VIS and NIR spectral range (Fig. 7.3). These can be used to determine the total atmospheric abundance of the species from remote sensing measurements of scattered sunlight using *inverse methods* (see also section "Inversion Theory"), which relate the measured atmospheric spectra to the constituent properties in the atmosphere. One powerful and intuitive approach is the Differential Optical Absorption Spectroscopy (DOAS), originally developed for ground-based measurements using artificial light sources or scattered sunlight (Noxon 1975; Platt et al. 1979; Solomon et al. 1987; Platt and Stutz 2008) and successfully adapted to nadir – and partly limb – measurements from GOME, SCIAMACHY, OMI and GOME-2 (Richter and Wagner 2010). Two main ideas form the basis of the DOAS approach (Fig. 7.4):

- The isolation of high frequency structures of molecular absorbers from broadband scattering features (Rayleigh, Mie) by a high pass filter.
- The separation of spectroscopic retrievals and radiative transfer calculations using the Air Mass Factor (AMF) concept.

Given a measured spectrum $I_{abs}(\lambda)$ containing the absorption of the trace gas of interest, a reference spectrum without the absorption of the trace gas of interest $I_{ref}(\lambda)$ and the absorption cross sections $\sigma_i(\lambda)$ of all relevant species, the optical depth can be written according to Lambert–Beer's law (7.5). Under the assumption of an optically thin atmosphere in the spectral region of interest, this optical depth is approximated as described in (7.7), where it is also assumed that the pressure and temperature dependence of the absorption cross sections can be neglected:

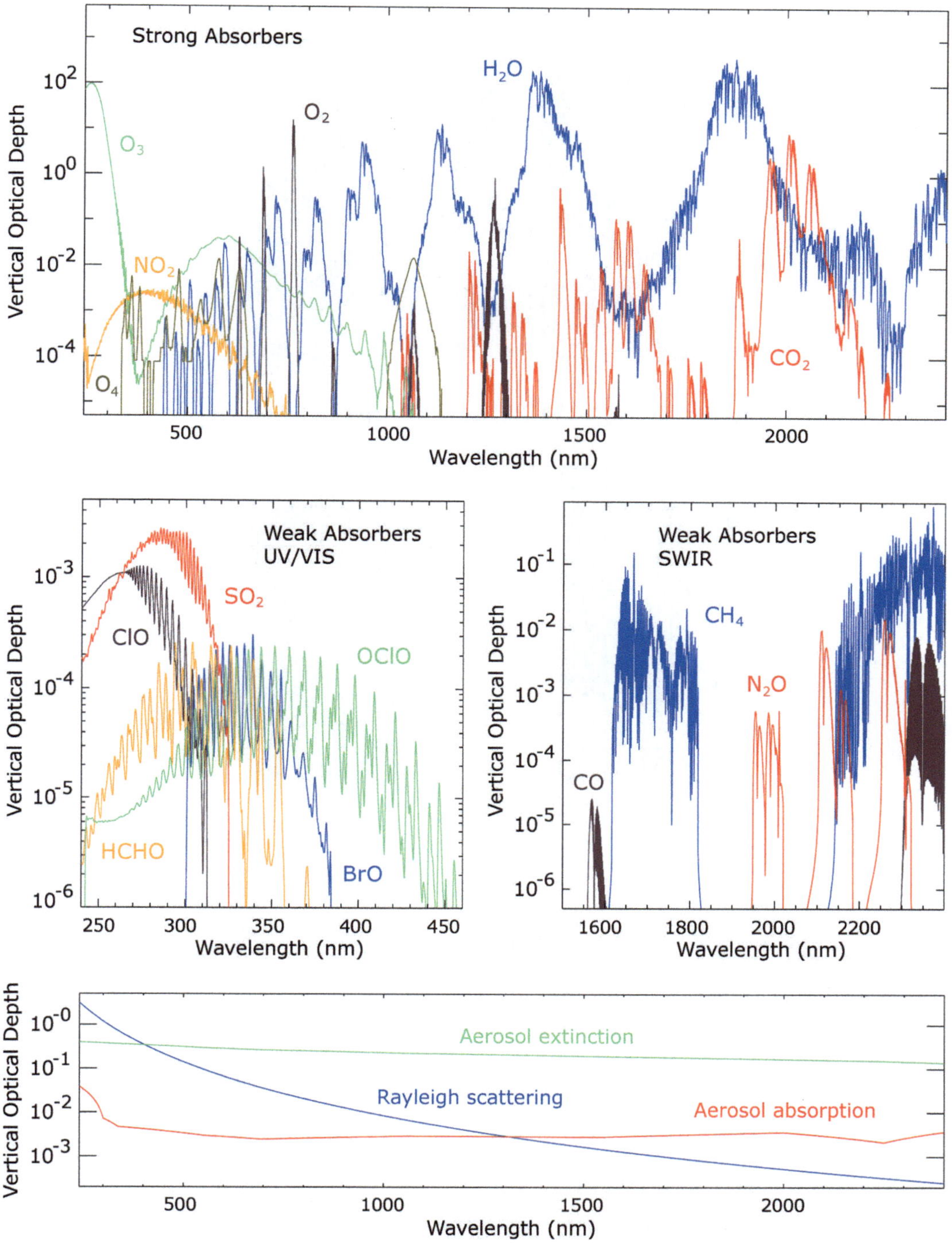

Fig. 7.3 Simulated vertical optical depth of the targeted constituents for 55°N around 10 a.m. The strong absorbers are plotted in the upper part and the relevant weak absorbers in the middle part. In the lower part the vertical optical depth due to Rayleigh scattering, aerosol extinction and absorption is given. Note the large dynamic range of the differential absorption structures used for retrieval of the constituents (Courtesy: IUP-IFE, University of Bremen).

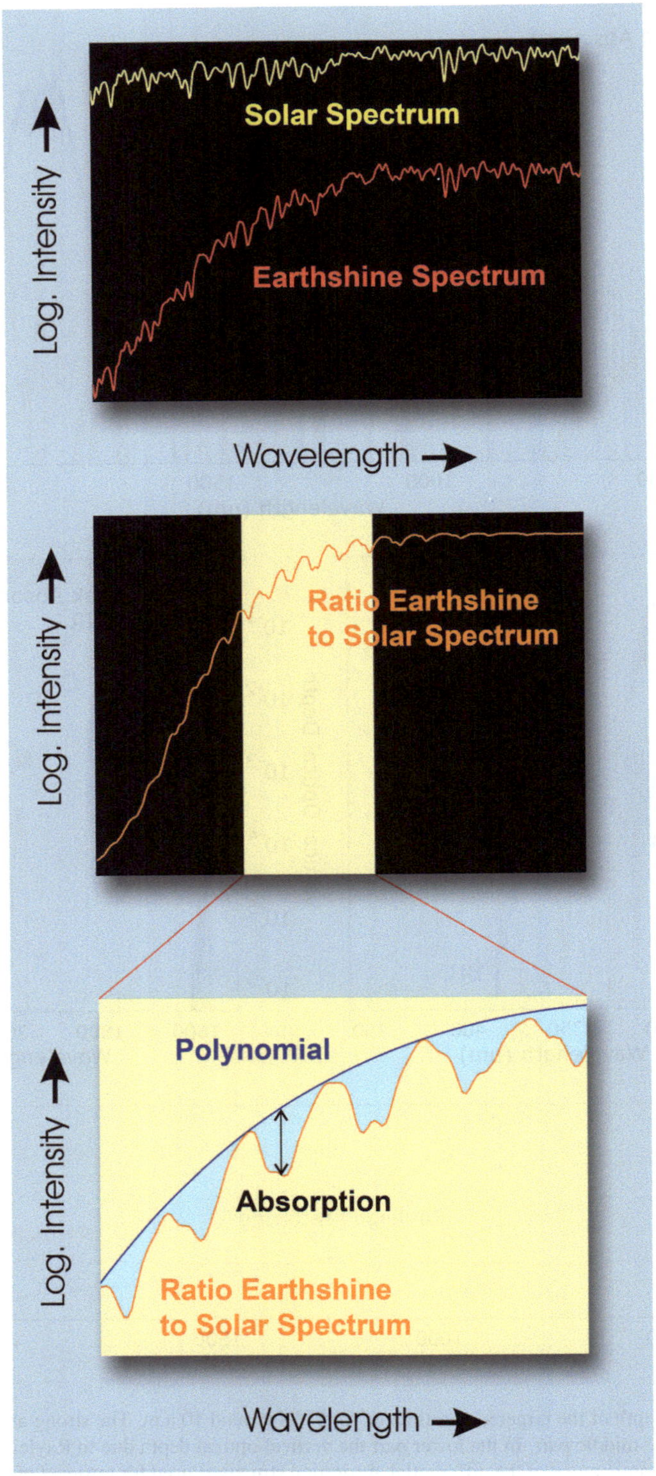

Fig. 7.4 The main steps of the DOAS retrieval. For further details see the text (Courtesy: IUP-IFE, University of Bremen).

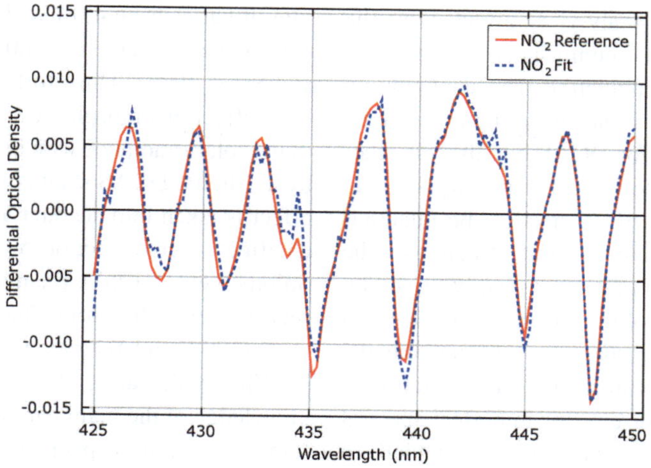

Fig. 7.5 Typical SCIAMACHY NO_2 fit results from a measurement over a polluted area in China on 15 January 2006. The red line is the scaled NO_2 laboratory cross section, the dashed blue line the result of the fit after subtraction of all contributions with the exception of NO_2 (Courtesy: IUP-IFE, University of Bremen).

$$\tau_s(\lambda) = -\ln\left(\frac{I_{abs}(\lambda)}{I_{ref}(\lambda)}\right) \approx \sum_i \sigma_i(\lambda) SCD_i(\lambda) - \sum_j c_j \lambda^j \qquad (7.7)$$

For $I_{abs}(\lambda)$ the nadir or limb radiance $I(\lambda)$ measured by SCIAMACHY is used. For $I_{ref}(\lambda)$ an extra-terrestrial solar spectrum determined from SCIAMACHY's solar measurements is applied. Alternatively, nadir or limb spectra from regions where the absorber of interest shows negligible absorption are used. While molecular absorption cross sections $\sigma_i(\lambda)$ are usually highly structured, absorption by particles, scattering by molecules and particles as well as reflection at the surface show broadband dependences. These broadband modulations are approximated in the DOAS approach by a common polynomial of low order with polynomial coefficients c_j. The slant column density SCD_i is the molecular number density n_i integrated along the light path s:

$$SCD_i(\lambda) = \int n_i(s) ds(\lambda) \qquad (7.8)$$

The spectral fit window is selected in such a way that the trace gas of interest shows strong and characteristic absorption features. The SCD_i is then obtained using a linear least squares fit by minimising the following quadratic form, where the wavelength dependence of the SCD is omitted to simplify the expression to

$$\left\| \tau_s(\lambda) - \sum_i \sigma_i(\lambda) SCD_i + \sum_j c_j \lambda^j \right\|^2 \qquad (7.9)$$

Fit parameters are the slant column densities SCD_i and the polynomial coefficients c_j. Figure 7.5 depicts a typical fit for NO_2. Once the SCD_i have been determined from the measured spectra, the conversion to the desired vertical column densities VCD_i is the last step. This requires for each trace gas a division by the Air Mass Factor (AMF_i) according to

$$VCD_i = \int n_i(z) dz = SCD_i / AMF_i \qquad (7.10)$$

with $n_i(z)$ being the vertical number density profile of the i-th absorber. The AMF describes the sensitivity of the measurement related to the specific properties of the atmospheric light path. Besides the solar elevation, the AMF depends on the measurement geometry and on the vertical profiles of absorbing and scattering constituents in the atmosphere as well as on the surface spectral reflectance. The AMF is calculated with a radiative transfer model and is essentially the ratio of the logarithms of atmospheric radiances simulated including all trace gases and excluding the trace gas of interest (Sarkissian et al. 1995; Burrows et al. 1999; Rozanov and Rozanov 2010). RTM calculations require trace gas profiles as input which are taken from climatologies or chemical transport models. The effect of partial cloudiness of the scene can be accounted for in the AMF calculation (Burrows et al. 1999). Radiative transfer models are also used to extend the AMF concept to a so-called 'box-AMF' concept, allowing for a height-dependent assessment of the sensitivity of DOAS measurements (Wagner et al. 2007a). The box-AMF characterises the ratio of the partial slant column density (SCD_{box}) to the partial vertical column density (VCD_{box}) of an atmospheric layer. It has to be noted, that for optically thin absorbers (optical depth $\ll 1$), the box-AMFs are proportional to the weighting functions.

The quality of the DOAS analysis is affected by the accuracies achieved in the slant column spectral fitting procedure as well as in the AMF calculation. The slant column spectral fitting is, especially for weak absorbing trace gases, a challenge in itself - not only for DOAS but also for other inverse methods described below – as very tiny spectral structures (amplitude of a few 10^{-4}) need to be quantitatively separated from interfering spectral structures. The performance of the slant column fitting is mainly driven by the signal-to-noise ratio (SNR) of the measurement, spectral calibration and knowledge of the instrument slit function, the separation of interfering spectral structures, the separation of the Ring effect and how one corrects for imperfect spectral sampling (Chance 1998). This imperfect sampling, which is a common feature in atmospheric UV-VIS spectra taken by current spaceborne spectrometers, may causes significant systematic spectral structures in the fitting process. For example, undersampled spectra cannot be re-sampled in wavelength without introducing spurious features due to spectral aliasing. Such structures introduced by aliasing can be largely corrected in the case where atmospheric absorptions are optically thin by calculating an undersampling correction using an independent, high resolution, solar reference spectrum (see also Slijkhuis et al. 1999; Chance et al. 2005).

Rayleigh scattering is partially inelastic as manifested via the Ring effect. The inelastic component is Raman scattering by air molecules, predominantly rotational Raman scattering. It varies from 4.0% to 3.4% of the total scattering over the wavelength region of 280–1000 nm (Bates 1984; Chance and Spurr 1997). Thus, to first order, the correction for the Ring effect in atmospheric spectra is simply the convolution of the solar irradiance with the spectrum of Raman scattering by air. In the DOAS slant column fitting procedure the Ring effect is taken into account as an additional 'absorber' with an effective cross section $\sigma_{Ring}(\lambda)$ as originally proposed by Solomon et al. (1987). Examples how to determine this cross section can be found in Chance and Spurr (1997), Vountas et al. (1998) and Wagner et al. (2009).

When calculating the AMF, accuracy depends on the knowledge of the main parameters determining the radiative transfer in the atmosphere, i.e. clouds, aerosols, surface spectral reflectance, surface elevation, height dependence of absorbers and scatterers. Obviously clouds drastically modify the light path through the atmosphere and need to be properly taken into account when calculating the AMF (see section "Cloud and Aerosol Retrieval"). The most important cloud parameters are the cloud fraction and the cloud top altitude, as fractional cloudiness blocks the light path to atmospheric layers below the cloud. The optical thickness of clouds also plays a role.

Using the DOAS algorithm, atmospheric columns of a number of species can be determined, including O_3, NO_2, SO_2, HCHO, CHOCHO, BrO, IO and OClO. For details see Table 7.1 and references therein.

Table 7.1 Atmospheric geophysical parameters and retrieval algorithms – nadir trace gases in the UV-VIS-NIR

Parameter	Spectral window (nm)	Occurrence	Quantity (column)	Retrieval algorithm	Algorithm reference
SO_2	315–327	Troposphere	Total	DOAS	Richter et al. 2006
					Lee et al. 2009
				Operational	Lichtenberg et al. 2010
O_3	325–335	Troposphere, stratosphere	Total	Operational	Lichtenberg et al. 2010
				SDOAS	Lerot et al. 2009
				WFM-DOAS	Weber et al. 2005
				TOSOMI	Eskes et al. 2005
BrO	335–347(51)	Troposphere, stratosphere	Total, tropospheric, stratospheric	DOAS	Afe et al. 2004
	336–351			Operational	Lichtenberg et al. 2010
HCHO	335(28)–347	Troposphere	Total	DOAS	Wittrock et al. 2006
					De Smedt et al. 2008
OClO	363.5–391	Stratosphere	Total	DOAS	Kühl et al. 2006
	365–389	Stratosphere	Total	DOAS	Oetjen et al. 2009
				Operational	Lichtenberg et al. 2010
IO	426–440	Troposphere	Total	DOAS	Saiz-Lopez et al. 2007
	415–430				Schönhardt et al. 2008
NO_2	425–450	Troposphere, stratosphere	Total, tropospheric, stratospheric	Operational	Lichtenberg et al. 2010
				DOAS	Richter et al. 2005
CHOCHO	435–457	Troposphere	Total	DOAS	Wittrock et al. 2006
					Vrekoussis et al. 2009
H_2O	688–700	Troposphere	Total	AMC-DOAS	Noël et al. 2004
				Operational	Lichtenberg et al. 2010

Modified DOAS Methods

One limitation of the classical DOAS technique is the assumption that the atmosphere is optically thin in the wavelength region of interest. In addition, 'line-absorbers' such as H_2O, O_2, CO, CO_2 and CH_4 usually cannot be retrieved precisely by standard DOAS algorithms because their strong absorption also depends on pressure and temperature and their spectra are often not fully spectrally resolved by SCIAMACHY. To overcome these drawbacks, several DOAS-type techniques were developed to account for such effects and to permit successful retrievals of the trace gas species (for a theoretical overview see Rozanov and Rozanov 2010). They are for example the WFM-DOAS (Weighting Function Modified DOAS, Buchwitz et al. 2000, 2004, 2005), AMC-DOAS (Air Mass Corrected DOAS, Noël et al. 2004), TOSOMI (Total Ozone retrieval scheme for SCIAMACHY based on the OMI DOAS algorithm, Eskes et al. 2005), SDOAS (Lerot et al. 2009), IMAP-DOAS (Iterative maximum a posteriori DOAS, Frankenberg et al. 2005a, b), IMLM (Iterative Maximum Likelihood Method DOAS, Gloudemans et al. 2008), FSI-WF-DOAS (Full Spectral Initiation Weighting Function Modified DOAS, Barkley et al. 2006) and BIRRA (Better InfraRed Retrieval Algorithm, Schreier et al. 2009). Figure 7.6 illustrates an example of one day of ozone columns derived from nadir observations with one of these schemes, the TOSOMI algorithm. Deriving optically thicker absorbers or line-absorbers based on such DOAS-type methods was successfully demonstrated for O_3, H_2O, O_2, CO, CO_2 and CH_4 (Richter and Wagner 2010).

Tables 7.1 and 7.2 summarise the DOAS-type retrieval algorithms as applied to SCIAMACHY data including references.

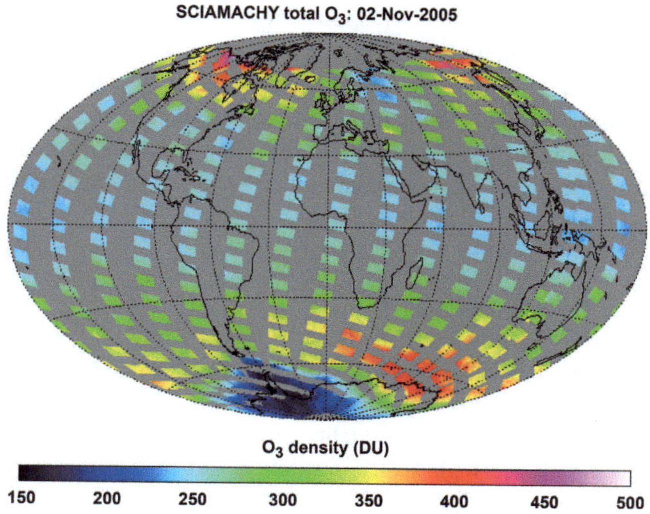

Fig. 7.6 One day of total ozone densities obtained with the TOSOMI algorithm (Courtesy: KNMI/ESA).

7.3 Cloud and Aerosol Retrieval

The primary scientific objective of SCIAMACHY is the measurement of atmospheric trace gases in both the troposphere and stratosphere. However, clouds residing in the troposphere interfere with the retrievals from SCIAMACHY measurements mainly by shielding, atmospheric path enhancement, and albedo effects. Similarly, tropospheric and/or stratospheric aerosols impact the trace gas retrieval, as they alter the light path. Therefore there is a clear need for cloud and aerosol information. In addition, cloud information derived from SCIAMACHY provides relevant data for climate research.

Table 7.2 Atmospheric geophysical parameters and retrieval algorithms – nadir trace gases SWIR

Parameter	Spectral window (nm)	Occurrence	Quantity (column)	Retrieval algorithm	Algorithm reference
CH_4	1627(30)–1671	Troposphere	Total	WFM-DOAS	Schneising et al. 2009
				IMAP	Frankenberg et al. 2005a, 2008
CO_2	1558(63)–1594(85)	Troposphere	Total	WFM-DOAS	Schneising et al. 2008
				IMLM	Houweling et al. 2005
				IMAP	Frankenberg et al. 2005a
				FSI	Barkley et al. 2006
CO	2321(24)–2335	Troposphere	Total	WFM-DOAS	Buchwitz et al. 2007
				IMLM	Gloudemans et al. 2008, 2009
				IMAP	Frankenberg et al. 2005b
				BIRRA/Operational	Schreier et al. 2009
H_2O	2353–2368	Troposphere	Total	IMLM	Schrijver et al. 2009
HDO/H_2O ratio	2354.2–2374.2	Troposphere	Total	IMAP	Frankenberg et al. 2009

In the UV-VIS and NIR spectral regions the solar radiation is strongly scattered by clouds and aerosols, thereby modifying the Earth reflectance spectrum. The presence of clouds and aerosols and their properties can therefore be determined by analysing the top-of-atmosphere reflection function R, often also termed 'top-of-atmosphere reflectance'. It is defined as

$$R = \pi \frac{I_{dif}}{\mu_0 \, I_{irr}} \tag{7.11}$$

where μ_0 is the cosine of the solar zenith angle, I_{irr} is the solar irradiance and I_{dif} is the scattered and reflected radiance in the direction towards the satellite sensor.

Prerequisites for high quality information about aerosols and clouds are high spatial resolution and high calibration accuracy of the reflectance measurements. Typical reflectance spectra for various cloud and surface conditions are shown in Fig. 7.7. Appropriate algorithms are required and available to determine cloud and aerosol properties from SCIAMACHY data. For a summary the reader is referred to Table 7.3.

Cloud Parameters

Information on cloud parameters are of twofold importance in the SCIAMACHY context. Firstly, cloud parameters such as cloud fraction, cloud top pressure (or height) and cloud optical thickness are required to correct the cloud impact on the retrieved trace gas concentrations. Secondly, clouds play

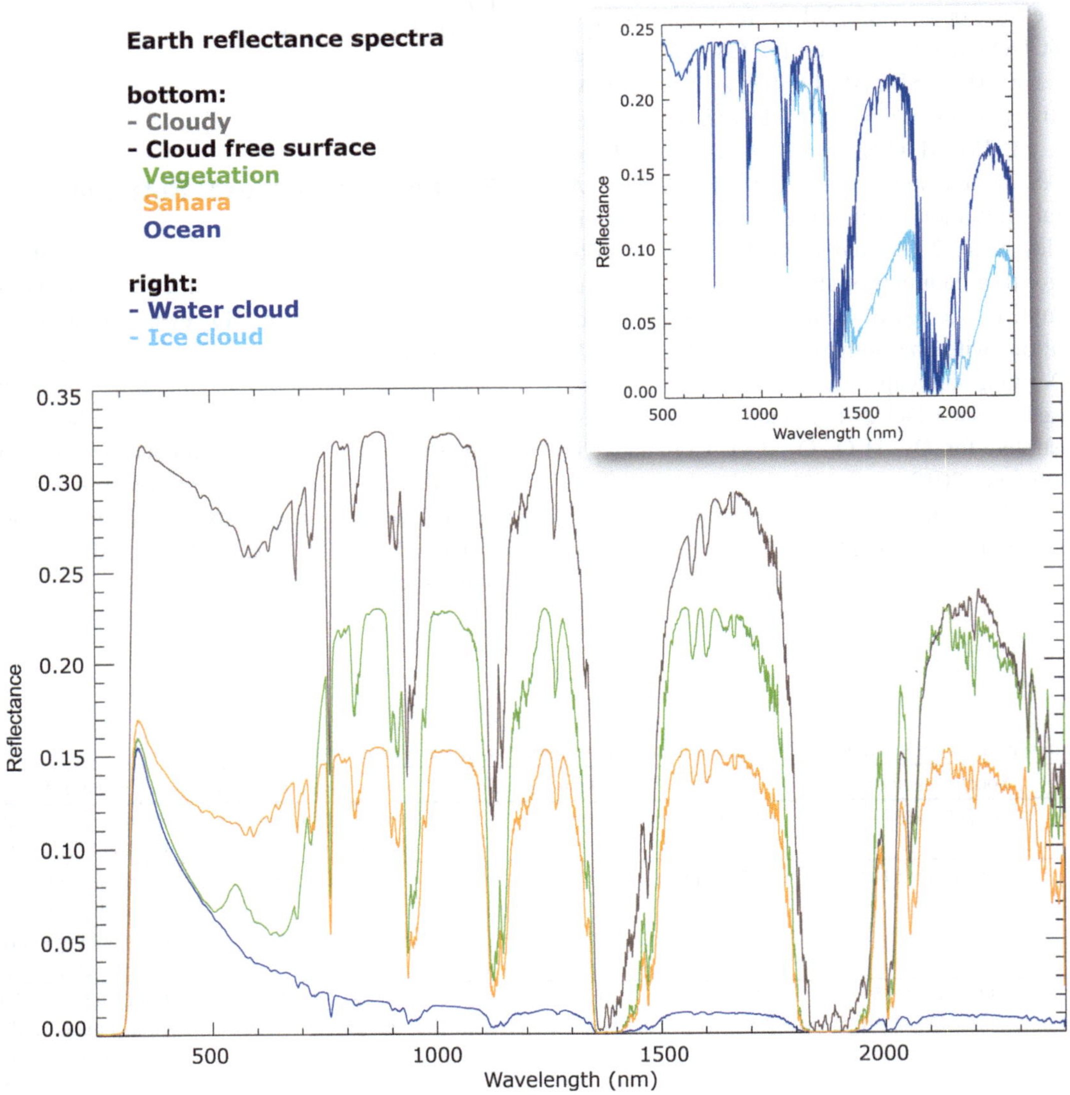

Fig. 7.7 Earth reflectance spectra (Sun normalised intensity) for various cloud and surface conditions. The inset shows the variation in the reflectance spectrum due to changes in the thermodynamic state of water in the cloud from liquid water to ice. The large difference in the reflectance spectrum around 1600 nm for ice and water clouds is used to derive information on the thermodynamical state of water in clouds (Courtesy: IUP-IFE, University of Bremen).

Table 7.3 Atmospheric geophysical parameters and retrieval algorithms – nadir cloud, aerosol and surface parameters. Unnamed algorithms are marked by '*'

Parameter	Spectral window (nm)	Retrieval algorithm	Algorithm reference
CF	PMD (RGB)	OCRA	Loyola 1998
		Operational	Lichtenberg et al. 2010
		HICRU	Grzegorski et al. 2004
		SPCA	Yan 2005
		SPICI	Krijger et al. 2005
	Reflectance near O_2 A-band	FRESCO	Fournier et al. 2006
		SACURA	Kokhanovsky et al. 2006a
	MERIS data	MCFA	Kokhanovsky et al. 2009b
CTH	O_2 A-band	FRESCO	Fournier et al. 2006 Wang et al. 2008
		SACURA	Kokhanovsky et al. 2006 a/b
		Operational	Lichtenberg et al. 2010
	O_2 obsorption at ~630	HICRU	Grzegorski 2009
	~477	DOAS	Acarreta et al. 2004a
COT	Reflectance VIS-NIR	SACURA	Kokhanovsky et al. 2006 a/b
CGT	O_2 A-band	SACURA	Rozanov and Kokhanovsky 2004
$R_{eff,cld}$	Reflectance VIS-NIR	SACURA	Kokhanovsky et al. 2006 a/b
Thermodynamic phase	Reflectance NIR-SWIR	*	Acarreta et al. 2004b
		SACURA	Kokhanovsky et al. 2006b, 2007b
Cloud type	PMDs	SPICS	Lotz et al. 2009
Cloud – surface discrimination		SPICS	Lotz et al. 2009
AAI	340 and 380	*	de Graaf and Stammes 2005
		Operational	Lichtenberg et al. 2010
UVAI	335.5 and 376.5	*	Penning de Vries et al. 2009
AOT	Several wavelengths in the VIS	SCIA-BAER	von Hoyningen-Huene et al. 2003, 2007
	AATSR	SYNAER	Holzer-Popp et al. 2008
Aerosol type	Several wavelengths, combination with AATSR data	SYNAER	Holzer-Popp et al. 2008
Dust altitude	O_2 A-band	*	Kokhanovsky et al. 2010
Phytoplankton groups	340–390, 425–500	PhytoDOAS	Vountas et al. 2007
			Bracher et al. 2009
Vegetation	605–685	DOAS	Wagner et al. 2007b
Surface reflectance	640	*	Grzegorski 2009

an important role in the Earth's climate system. The amount of radiation reflected by the Earth-atmosphere system into outer space depends not only on the cloud cover and the total amount of condensed water in the Earth's atmosphere but also on the size of droplets, the thermodynamic state of water in clouds, and vertical distributions of cloud parameters. The information about microphysical properties and spatial distributions of terrestrial clouds on a global scale can be obtained only with satellite remote sensing systems. SCIAMACHY data, due to its wide spectral range and despite its relatively low spatial resolution, enables the determination of important parameters for climate research like thermodynamic phase or geometrical thickness of clouds.

Cloud Fraction (CF)

In order to correct for the effect of clouds, a fast and reliable cloud fraction algorithm is required for SCIAMACHY. Nearly all cloud fraction algorithms for SCIAMACHY use the PMD measurements, as their higher temporal readout frequency translates into a higher spatial resolution as compared to data from channels 1–8. The basic principle of the algorithms is that the cloud albedo is much higher (except for ice and snow) than the Earth's surface albedo (see Fig. 7.8). Therefore, a pixel which is contaminated by clouds will generally have a higher detector signal than one that is cloud-free. Cloud fractions can therefore be determined

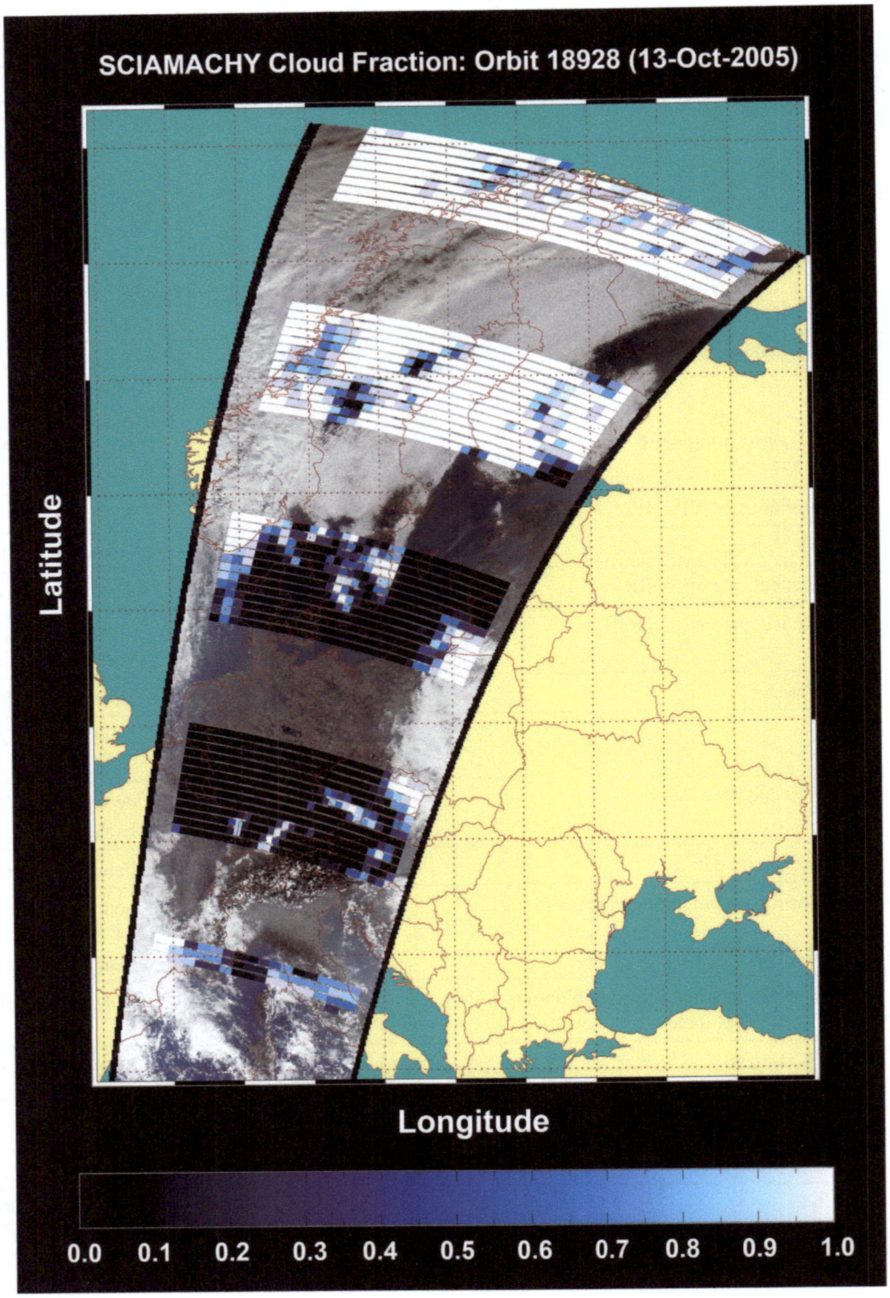

Fig. 7.8 Clouds over Europe on 13 October 2005. Cloud cover as seen in a RGB composite from MERIS with an overlay of the cloud fraction from SCIAMACHY (Courtesy: C. Schlundt, IUP-IFE).

through comparison of PMD intensities. Several derivatives of cloud fraction algorithms using PMD data are currently applied to SCIAMACHY data, such as OCRA (Optical Cloud Recognition Algorithm, Loyola 1998) used in the operational processing, SPCA (SCIAMACHY PMD Cloud Algorithm, Yan 2005) and HICRU (Heidelberg Iterative Cloud Retrieval Utilities, Grzegorski et al. 2004; Grzegorski 2009). In addition, algorithms analysing PMD data were developed and applied to discriminate clouds from ice/snow surface such as, e.g. SPICI (SCIAMACHY PMD Identification of Clouds and Ice/snow, Krijger et al. 2005). Similarly, PMD data can also be used for retrieving further surface classification information (water, sun glint, desert, and vegetation) as demonstrated by SPICS (SCIAMACHY PMD based Identification and Classification of Clouds and Surfaces algorithm, Lotz et al. 2009). The estimation of the cloud fraction and even the distribution of clouds within a SCIAMACHY scene can also be derived from instruments that have a much higher spatial resolution such as MERIS and AATSR on board ENVISAT (Kokhanovsky et al. 2009b; Schlundt et al. 2010).

Cloud Top Height (CTH)

CTH can be estimated from the TOA reflection function by using the changes in the penetration depth of solar photons due to strong changes in the absorption of trace gases with known vertical distributions such as O_2 (see Fig. 7.9). The idea for the CTH retrieval from O_2 A-band absorption was originally proposed by Yamomoto and Wark (1961). For SCIAMACHY two algorithms are currently implemented to derive cloud top height from O_2 A-band measurements. These are FRESCO (Fast Retrieval Scheme for Clouds from the Oxygen A-band, Koelemeijer et al. 2001; Fournier et al. 2006) and SACURA (Semi-Analytical Cloud Retrieval Algorithm, Rozanov and Kokhanovsky 2004). FRESCO was recently improved to FRESCO+ by taking into account single Rayleigh scattering (Wang et al. 2008). SACURA delivers not only CTH but also other cloud parameters like cloud phase index, cloud optical thickness and cloud liquid water path for clouds with an optical thickness larger than 5. A corresponding yearly average cloud top height map derived with SACURA for 2004 is shown in Fig. 7.10. It is worth mentioning that in addition to the O_2 A-band, there is also the option to derive CTH information from the O_2–O_2 absorption (Acarreta et al. 2004a), the Ring effect (Joiner et al. 1995; de Beek et al. 2001) or CH_4 absorption (Gloudemans et al. 2009).

Cloud Geometrical Thickness (CGT)

O_2 A-band absorption can also be used to obtain an estimate of the CGT (Asano et al. 1995). The CGT values represent an estimate of the light absorption inside a cloud and are therefore well suited for reducing uncertainties in the cloud top altitude measurements. This method was further developed by Kokhanovsky and Rozanov (2005).

Cloud Optical Thickness (COT) and Effective Radius ($R_{eff,cld}$)

Measurements of the Earth reflectance spectrum in the VIS or NIR range outside strong gaseous absorption bands permit derivation of the COT and Effective Radius (Nakajima

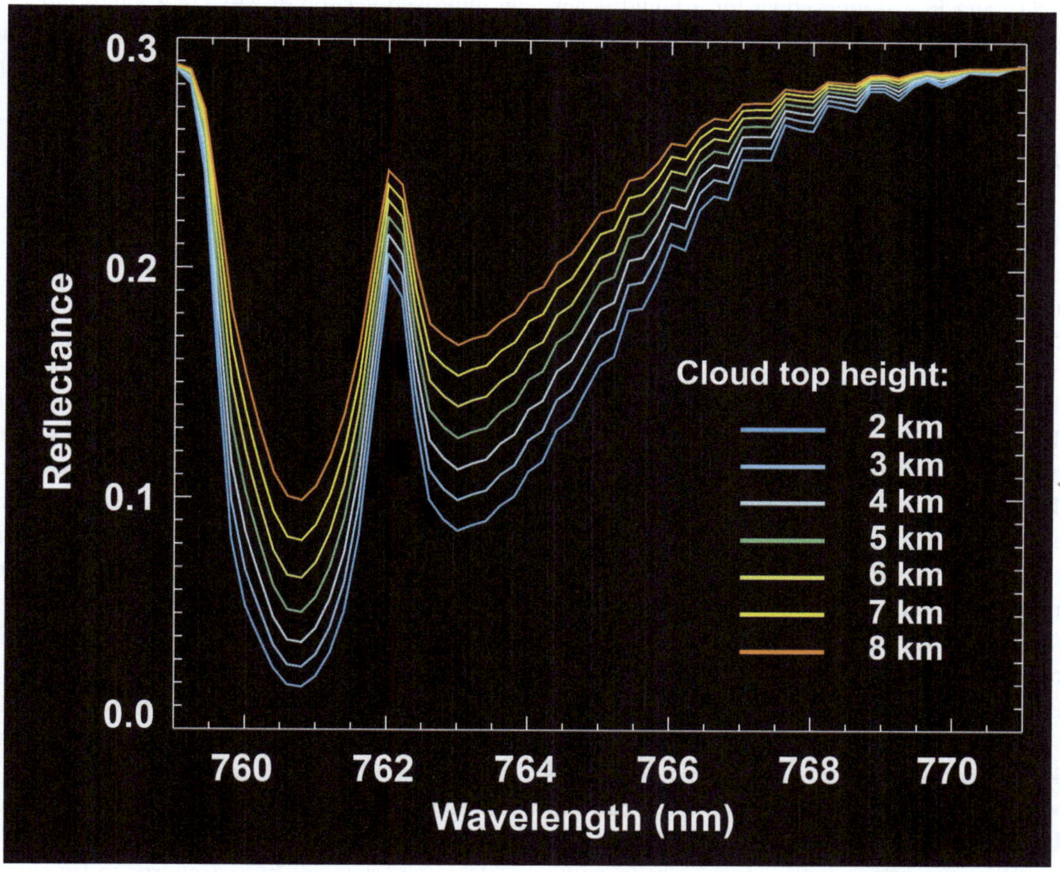

Fig. 7.9 The TOA reflectance in the O_2 A-band as a function of cloud top height (Courtesy: IUP-IFE, University of Bremen).

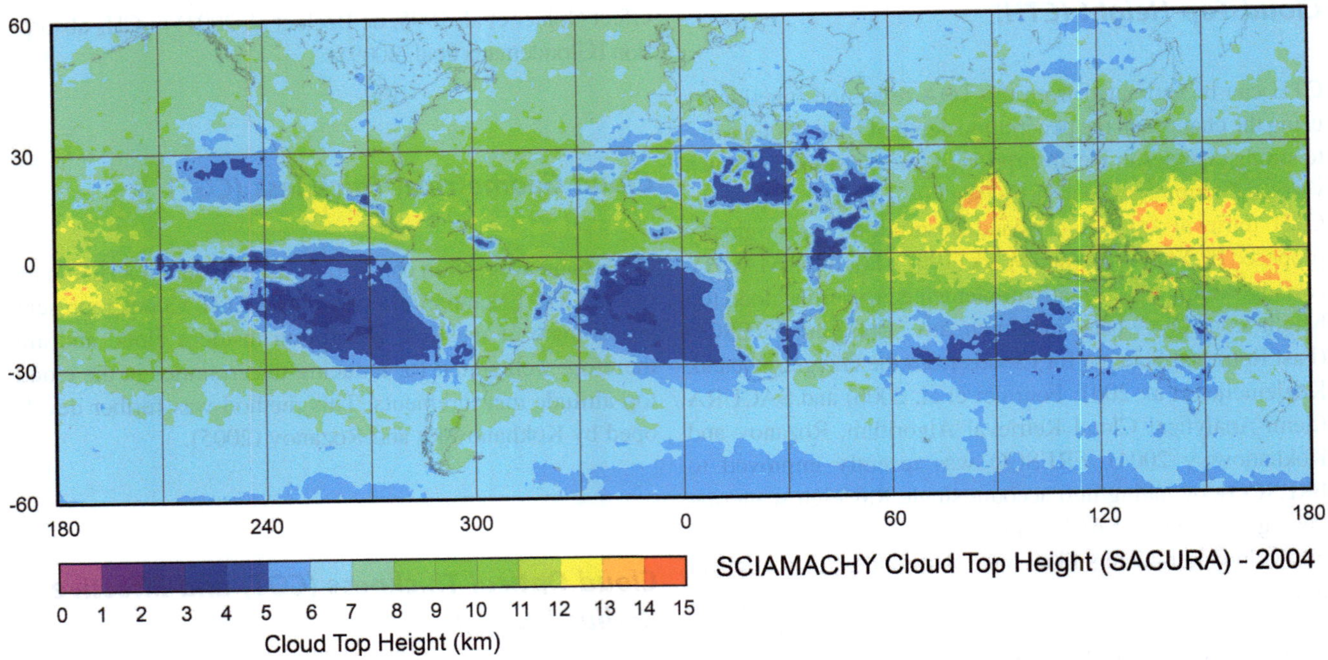

Fig. 7.10 The global distribution of cloud top height as derived with SACURA for 2004 (Courtesy: IUP-IFE, University of Bremen).

and King 1990, Platnick et al. 2003). Kokhanovsky et al. (2006a, b) applied the method to SCIAMACHY data. In cases where COT and $R_{eff,cld}$ are known for a liquid water cloud, the liquid water path can be estimated.

Cloud Phase Index (CPI)

SWIR reflectance measurements are suitable for obtaining the cloud phase index CPI (Knap et al. 2002; Acarreta et al. 2004b; Kokhanovsky et al. 2007a), as the difference in liquid water and ice absorption around 1.6 μm results in detectable differences in the TOA reflectance (see Fig. 7.7). A discrimination between both is feasible when the reflectance ratio at $\lambda = 1550$ nm and 1670 nm is analysed. This can be further extended by simultaneously using thermal IR data from AATSR and SCIAMACHY measurements to detect regions with mixed clouds and clouds containing supercooled water (Kokhanovsky et al. 2007b).

Of the various cloud retrieval algorithms described above the GOME/SCIAMACHY data processing environment mainly exploits two. One is FRESCO, which uses measurements inside and outside the O_2 A-band (758–778 nm). FRESCO is developed for cloud correction of trace gas retrievals, like O_3 and NO_2. It simultaneously retrieves an effective cloud fraction (CF_{eff}) and an effective cloud top height (CTH_{eff}) assuming that the cloud can be represented as a bright Lambertian surface with a fixed albedo value of 0.8. The second is the combination of OCRA with SACURA. OCRA delivers cloud fraction as input for SACURA to determine CTH and COT. In addition to that, an improved version of SACURA yields cloud geometrical thickness, effective radius, liquid water path, and the cloud thermodynamic state in the case of optically thick clouds (Kokhanovsky et al. 2006a, b).

Tropospheric Aerosol Parameters

Similar to clouds, aerosols play two different roles from SCIAMACHY's perspective. They are both substantial error sources in the retrieval of tropospheric trace gases and provide important geophysical information: aerosol particles affect the radiative budget of the atmosphere directly by absorbing and scattering radiation, and indirectly by influencing cloud properties such as droplet size and hence cloud brightness. Aerosols are characterised by high spatial and temporal variability and can consist of mixtures of chemically and physically different particles, depending on their origin (Pöschl 2005). They are typically roughly divided into five categories: (1) sea spray: large particles, no absorption of UV-VIS radiation, (2) mineral dust: large particles, UV-absorbing, (3) biomass burning smoke: small to large particles, weak to strong absorption, (4) volcanic ash: small to large particles, weak to strong absorption, and (5) sulphate droplets: small particles, weak to no absorption. Detection of aerosols by spaceborne instruments makes use of the effect of aerosols on the reflected solar radiation observed at the top of the atmosphere. Contributions from Earth's surface reflections (Fig. 7.7) and atmospheric gases (Fig. 7.3) need to be separated using available information on the surface properties and the effects of gases in combination with radiative transfer simulations. Thus the signatures of aerosols can be derived and can be used to retrieve aerosol properties.

Most of the currently existing aerosol retrieval algorithms are aimed at determining the Aerosol Optical Thickness (AOT), i.e. columnar extinction by aerosol particles and its spectral behaviour (Kokhanovsky and de Leeuw 2009). AOT is extracted by fitting modelled reflectance spectra to the measured reflectance spectra. This approach needs careful constraints based on the knowledge of molecular scattering, absorption and the surface reflectance (von Hoyningen-Huene et al. 2003). Some microphysical parameters, such as aerosol number density and effective particle radius, can be inferred from the magnitude and spectral dependence of the AOT. Other aerosol properties, of which aerosol absorption and layer altitude are of the most interest, need to be derived separately. Although aerosol measurements are not the primary focus of SCIAMACHY, its large wavelength range can be exploited for the study of spectrally dependent aerosol properties in the UV-VIS range and of surface or cloud properties in the near-IR range (von Hoyningen-Huene et al. 2003, 2007). A more standardised and less complicated way of obtaining information on aerosol properties is offered by application of the UV Aerosol Index (UVAI) algorithm (Herman et al. 1997; Torres et al. 1998; de Graaf and Stammes 2005). The UVAI takes advantage of the strong spectral dependence of molecular Rayleigh scattering in the UV range to obtain semi-quantitative information on aerosol optical thickness, layer altitude, and absorption. The UVAI is most sensitive to elevated layers of aerosols that absorb UV radiation, thus causing a positive UVAI, also widely known as Absorbing Aerosol Index (AAI). The AAI mainly indicates the presence of mineral dust or biomass burning smoke (Fig. 7.11). Initially developed as an error indicator for ozone retrieved from TOMS data (Herman et al. 1997), the AAI is an aerosol quantity with a long data record. The AAI is derived as the positive residual between the measured reflectance from an atmosphere enriched with aerosols and the simulated reflectance of an atmosphere with only Rayleigh scattering, absorption by molecules, plus surface reflection and absorption. Such an algorithm using SCIAMACHY data at 340 nm and 380 nm delivers meaningful AAI values even in the presence of clouds and over bright surfaces (de Graaf and Stammes 2005; de Graaf et al. 2007, as shown in Fig. 7.11). In addition, the concept of the UVAI can also be used for the detection of weakly absorbing ('scattering') aerosols, such as sulphate and secondary organic aerosols, which cause a negative UVAI or Scattering Index (SCI – Penning de Vries et al. 2009). Figure 7.12 shows seasonal averages of the UV Aerosol Index data for July–September 2005, after application of a cloud filter. The colour scale is chosen so that strongly absorbing aerosols, i.e. mineral dust and mainly biomass burning smoke, show up in blue tones, whereas scattering aerosols, such as sulphate and secondary organic aerosols, but also clouds, show up in yellow and red tones. The high positive UVAI values over the global dust belt, defined by the world's largest deserts at around 25°N, indicate large amounts of elevated mineral dust which are transported westward across the Atlantic ocean. High, positive values near the Angolan coast and over the Amazon rainforest are caused by massive biomass burning in the local dry season. Areas influenced by scattering aerosols are not as easily identified. However, when pixels with cloud fractions exceeding 5% were removed from the average, high amounts of scattering aerosols are clearly seen to occur in the South-Eastern US, Eastern China, and in the northern part of South America.

Further aerosol related retrievals concern the aerosol layer height and the aerosol extinction vertical profile. Both methods use the O_2 A-band (Corradini and Cervino 2006; Kokhanovsky and Rozanov 2010). Finally even deriving aerosol types is possible by combining SCIAMACHY spectral information with high resolution images from AATSR on ENVISAT. Holzer-Popp et al. (2008) followed this approach by applying the synergistic method SYNAER. A summary of aerosol property retrieving algorithms is given in Table 7.3.

7.4 Surface Parameter Retrieval

The measured earthshine spectra are also affected by surface reflection and absorption, i.e. the broadband ground albedo, and narrowband spectral structures from the surface. Dedicated algorithms were developed to extract surface information from SCIAMACHY data, since the unique spectral coverage in combination with the high spectral resolution offers the possibility to determine new parameters from space. An important prerequisite for the successful application of such algorithms is a good cloud screening (see section "Cloud and Aerosol Retrieval") to avoid artefacts introduced by clouds.

Surface Albedo

The spectral surface albedo is an important parameter, which determines the sensitivity of satellite observations towards the lower troposphere. It can be derived from TOA reflectance under cloud-free conditions, using a radiative transfer model, as, e.g. pursued by Grzegorski (2009).

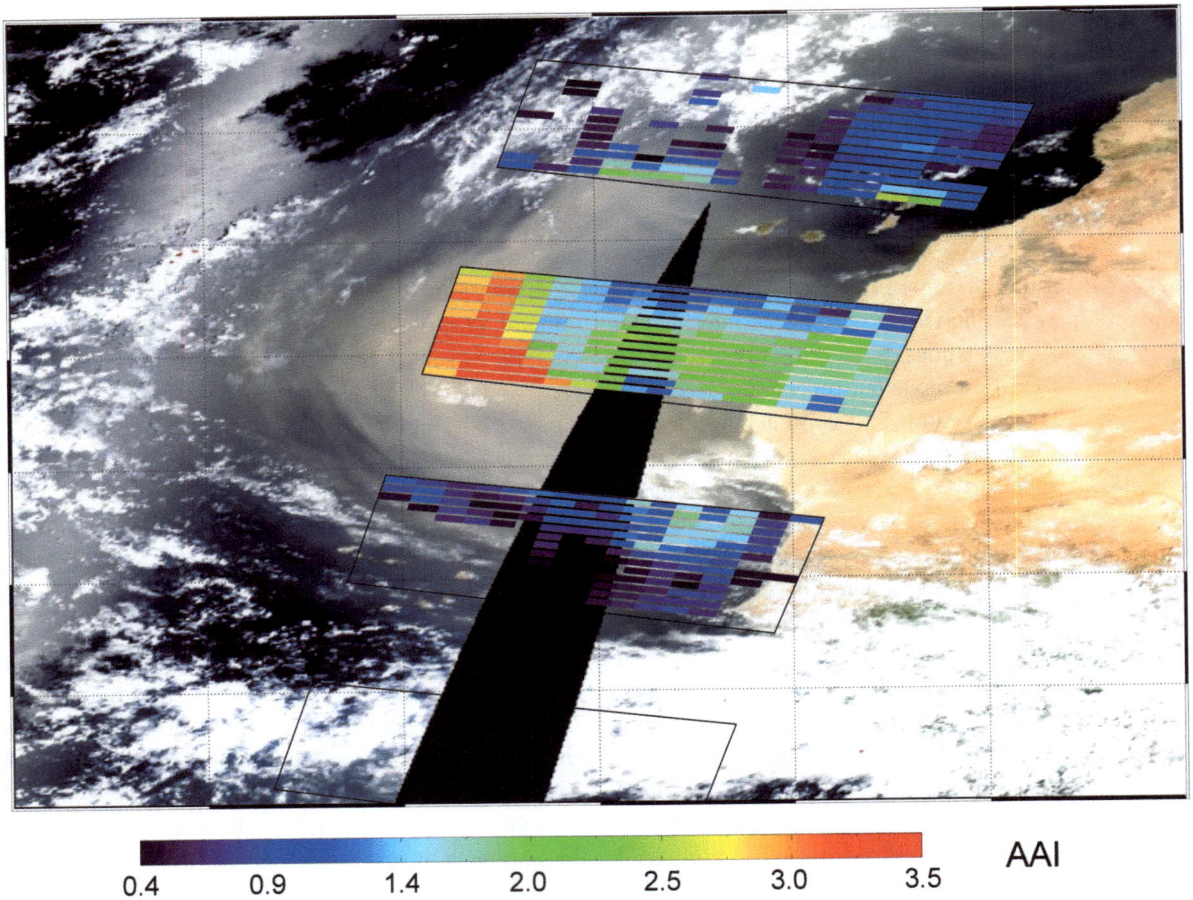

Fig. 7.11 Saharan desert dust outbreak to the Atlantic on 25 July 2004. Shown is the SCIAMACHY AAI at 9:15 UTC of that day overlaid on a MODIS RGB picture, acquired around 11:10 UTC (*right side of the plot*) and 12:50 UTC (*left side of the plot*). High SCIAMACHY AAI values coincide with the dust plume, visible as a yellow haze on the MODIS image (Courtesy: de Graaf et al. 2007).

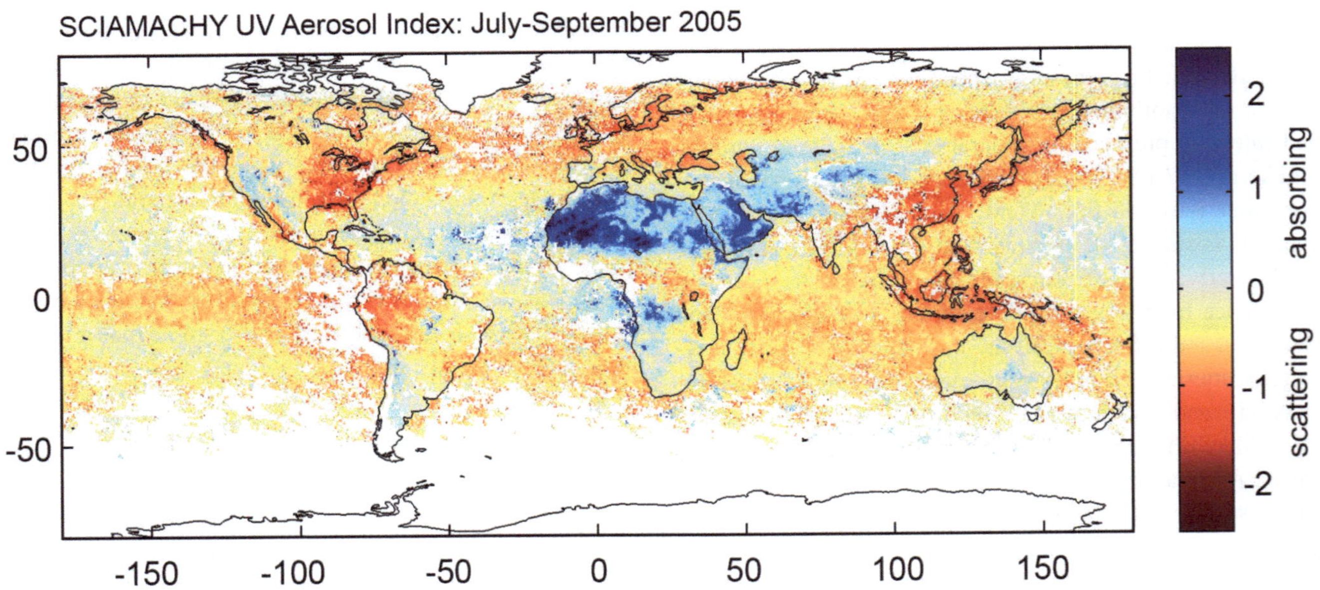

Fig. 7.12 Seasonal average of SCIAMACHY UV Aerosol Index for July–September, 2005. Blue indicates absorbing aerosols (mineral dust, biomass burning smoke); red/orange indicates weakly absorbing but scattering aerosols (sulphate and secondary organic aerosols) and remaining clouds. Only pixels with cloud fraction smaller than 5% are included in the average (Courtesy: M. Penning de Vries; MPI for Chemistry, Mainz).

Land Vegetation

Light reflected from the Earth's surface towards SCIAMACHY contains the spectral signature of the surface constituents. In case of land, vegetation has a clearly identifiable spectral signature at about 600–680 nm, which can be used to derive vegetation information from measured reflectance spectra (Wagner et al. 2007b). In contrast to existing algorithms, which rely on the strong changes of the reflectivity in the red and near infrared spectral region, the new method analyses weak narrowband – only a few nanometers wide – absorption and reflectance structures of vegetation in the abovementioned red spectral range. This method is based on the DOAS approach, which is usually applied for the analysis of atmospheric trace gas concentrations. Since the spectra of atmospheric absorption as well as vegetation absorption and reflectance are simultaneously included in the analysis, the effects of atmospheric absorptions are automatically corrected, in contrast to other algorithms using hyperspectral data. Another effect of including vegetation spectra is the significant improvement of the results of the trace gas retrievals. Improved sets of vegetation spectra might lead to more accurate and more specific identification of vegetation types in future.

Oceanic Biological Activity

The DOAS methodology can also be used to derive information about biological activity, i.e. chlorophyll in the ocean. Fitting a modelled Vibrational Raman Scattering (VRS) spectrum of liquid water (Vassilkov et al. 2002) and in situ measured phytoplankton absorption reference spectra to optical depths measured by SCIAMACHY permits identifying marine areas with enhanced chlorophyll (Vountas et al. 2007). Phytoplankton absorption reference spectra were obtained during two cruises of the research vessel *Polarstern* (see section "Correlative Measurements" of Chapter 9). They have been acquired in regions where different species – cyanobacteria and diatoms – dominate the phytoplankton composition. It was demonstrated that the distinct absorption characteristics of the two phytoplankton groups enable their unambiguous identification in SCIAMACHY spectra such that their global surface distribution can be retrieved (Bracher et al. 2009). Combining the phytoplankton absorption with light penetration depth estimated from the inelastic scattering in the liquid water, globally distributed pigment concentrations for both phytoplankton groups can be determined. As in the case of vegetation, accounting for particular spectral signatures from an ocean surface component leads to improved atmospheric trace gas DOAS retrievals over sea.

7.5 Inversion Theory

The forward modelling described in section 7.1 "Radiative Transfer in the Earth's Atmosphere" is commonly employed to simulate a measured quantity, e.g. intensity of the radiation, for a predefined state of the atmosphere. Contrary to this, the objective of inversion problems is to retrieve certain characteristics of the atmospheric state – for example trace gas concentration profiles – based on the measured quantities. These can be the solar radiance transmitted through the Earth's atmosphere as observed in the occultation geometry or the radiance scattered by the Earth's atmosphere and/or reflected by the surface as observed in the limb or nadir geometry. In this sense the DOAS method as described above is a special case to solve the inverse problem. Here we will summarise a more general approach to describe the inverse problem and possible solutions. The parameters to be retrieved from the measurements are represented by a *model state vector* x. For example, for trace gas vertical profile retrieval, the model state vector contains the number densities of atmospheric constituents defined at discrete altitude levels. Each state vector can be mapped to the measurement space by means of the forward model operator F to obtain the corresponding measurement vector y, i.e., for each atmospheric state described by vector x an appropriate measured quantity y can be simulated using a radiative transfer model (Fig. 7.13).

In the case of SCIAMACHY occultation or limb measurements, the measured quantities are represented by a set of intensities measured at different tangent heights in selected spectral windows, in case of the nadir geometry the measured quantity is represented by the intensity measured in selected spectral windows. Taking into account that measurements are made to a finite accuracy, a measurement error ε has to be considered which is commonly assumed to be normally distributed with mean zero and known error covariance matrix S_y. Thus, the relationship between the model state vector and the measurement vector can be written as

$$y = F(x) + \varepsilon \qquad (7.12)$$

It is interesting to note that the DOAS approach can be derived from (7.12) and is thus only a special case among the methods addressing the inverse problem (see, e.g. Rozanov and Rozanov 2010). In order to solve the inverse problem, the generally non-linear relationship in (7.12) can be linearised expanding the forward model operator, F, as a Taylor series about an initial guess state x_0 (also referred as *a priori* state) which is the best beforehand estimation of the solution. Ignoring the higher-order terms one obtains

$$F(x) \approx F(x_0) + \left.\frac{\partial F}{\partial x}\right|_{x_0} (x - x_0) = y_0 + K_0(x - x_0) \qquad (7.13)$$

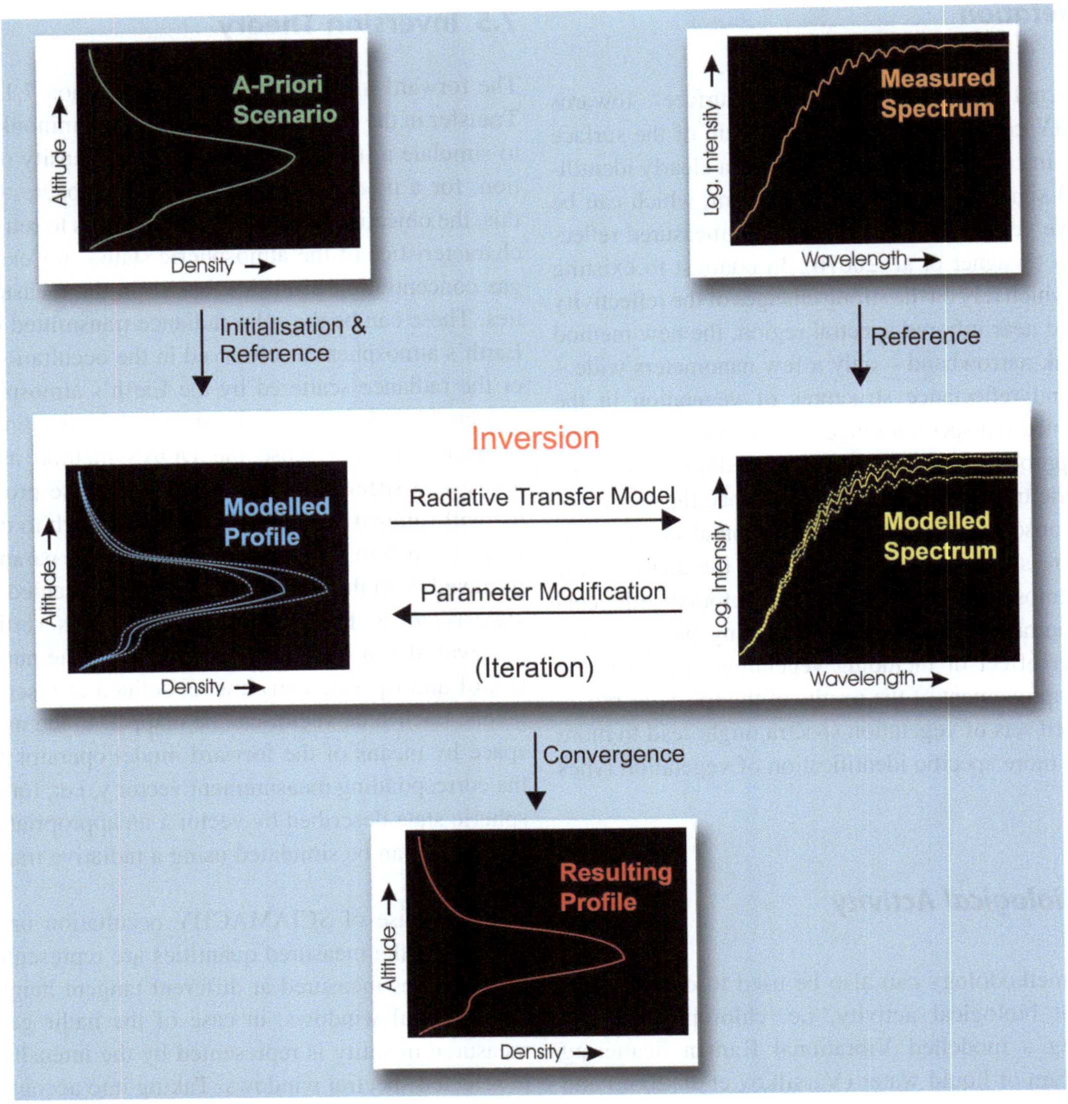

Fig. 7.13 The principle of inversion for the retrieval of geophysical parameters. For further details see the text (Courtesy: IUP-IFE, University of Bremen).

Here, K_0 is the linearised forward model operator. In the discrete representation the linearised forward model operator is given by the weighting function matrix describing the sensitivity of the measured quantities to variations of the atmospheric parameters at different altitude levels. This weighting function matrix is calculated with the radiative transfer model.

Atmospheric inversion problems are usually 'ill-posed', i.e., they do not have a unique solution. Thus, additional constraints need to be introduced to determine a geophysical solution from the set of mathematically allowed solutions and an iterative method has to be used. Most commonly, methods of statistical regularisation, e.g. the *maximum a posteriori* estimate as described by Rodgers (2000), are employed to solve the inversion problem. Using the maximum *a posteriori* estimate the solution is found by minimising the following quadratic form

$$\| (y - y_0) - K_0(x - x_0) \|^2_{S_y^{-1}} + \|x - x_0\|^2_{S_a^{-1}} \rightarrow \min \quad (7.14)$$

where S_a is the a priori covariance matrix of the solution and S_y is the measurement covariance matrix. The latter is usually assumed to be diagonal, i.e. no correlation between measurement errors at different wavelengths or different tangent heights is considered. The linear solution results in

$$x = x_0 + \left(S_a^{-1} + K_0^T S_y^{-1} K_0\right)^{-1} K_0^T S_y^{-1} (y - y_0) \quad (7.15)$$

The non-linearity of the inverse problem can be accounted for employing the iterative approach as follows:

$$x_{n+1} = x_0 + \left(S_a^{-1} + K_n^T S_y^{-1} K_n\right)^{-1} K_n^T S_y^{-1} \times [(y - y_n) - K_n(x_0 - x_n)] \tag{7.16}$$

where subscripts n and $n + 1$ denote the number of the iteration.

In the retrieval of the vertical distributions of the atmospheric species the *a priori* covariance matrix S_a is commonly chosen as a block diagonal matrix, i.e. vertical distributions of different atmospheric trace gases are assumed to be uncorrelated. The diagonal elements of S_a represent the variances of atmospheric trace gases σ which can be derived from, e.g. a climatology.

The quality of the obtained solution is characterised by the *a posteriori* covariance matrix

$$S = \left(K^T S_y^{-1} K + S_a^{-1}\right)^{-1} \tag{7.17}$$

and by the averaging kernels

$$A = \frac{\partial x}{\partial x_{true}} = \left(K^T S_y^{-1} K + S_a^{-1}\right)^{-1} K^T S_y^{-1} K \tag{7.18}$$

characterising the response of the retrieved solution to the variation of the true atmospheric state. The square roots of the diagonal elements of the *a posteriori* covariance matrix are referred to as *theoretical precisions*.

Employing the averaging kernels, the retrieved solution x can be related to the true solution x_{true} as

$$x = x_0 + A(x_{true} - x_0) \tag{7.19}$$

If the model state vector represents a vertical profile of an atmospheric trace gas, the retrieved values at each altitude are expressed as the sum of the a priori value at this altitude and of the deviation of the true profile from the a priori profile smoothed with the associated row of the averaging kernel matrix. For an ideal observing system, A is a unit matrix. In reality, the rows of the averaging kernel matrix are peaked with a finite width, which can be regarded as a measure of the vertical resolution of the retrieved profile.

7.6 Application of Inversion Theory to Limb Retrieval

For the retrieval of the vertical distributions of atmospheric species from the measurements the so called 'Global Fit' technique is an effective method for inversion. The measurement vector contains the logarithms of the radiances at all selected spectral points and at all line-of-sight angles, both for limb and occultation geometry, referenced to an appropriate extraterrestrial spectrum. In the limb viewing geometry, the extraterrestrial spectrum can be replaced by a limb measurement at an upper tangent height with negligible absorption. Due to this normalisation the retrieval is relatively robust with respect to the radiometric calibration. In addition, the normalisation significantly reduces the sensitivity of the retrievals with respect to ground albedo and cloud cover. Commonly, before the main inversion step, a preprocessing is performed intended to correct for possible misalignment in the wavelength calibration and to account for known atmospheric corrections such as the Ring effect. If required a polynomial can be subtracted from all relevant spectra that account for both missing or inappropriate instrument calibration and unknown scattering properties of the atmosphere.

To illustrate the limb inversion in practice, Fig. 7.14 shows an example of averaging kernels, weighting functions at 338.6 nm, and theoretical precision typical for BrO vertical profile retrievals from SCIAMACHY limb measurements. These results were obtained using the limb measurement at a tangent height of 38.5 km as a reference spectrum. As can be seen, the peak values of the averaging kernels are close to 1.0 only at tangent heights above 15 km. The peak value of about 0.55 at 12 km altitude indicates an increased dependence of the retrieved BrO concentration at this altitude on the BrO concentration at neighbouring altitude levels and the *a priori* information. Looking at the width of the averaging kernels, the height resolution of the measurements can be estimated to be about 3 km, close to the geometrical resolution of the instrument. The weighting functions in the middle panel exhibit relatively sharp peaks near the tangent height down to 18 km tangent height, whereas at all lower tangent heights the weighting functions peak at about 18 km altitude. Nevertheless, BrO amounts down to 12 km can be retrieved due to the different shapes of the corresponding weighting functions. In accordance with the averaging kernels, the theoretical precision of the BrO vertical profile retrieval shown in the right panel has reasonable values of 10–40% only above 15 km altitude and rapidly decreases below indicating the low information content in the measurements below 15 km.

To date, such types of retrieval algorithms and associated derivates have been used to obtain stratospheric profiles of O_3, NO_2 (Bracher et al. 2005; Rozanov et al. 2005b), BrO (Rozanov et al. 2005b) and H_2O (Rozanov et al. 2010) from SCIAMACHY limb scattering profiles. They were also applied to derive trace gas concentrations from lunar (Amekudzi et al. 2005a, b) and solar (Meyer et al. 2005) occultation measurements. It is worth noting that a similar approach as described above can also be applied to retrieve trace gas information from nadir measurements, as for

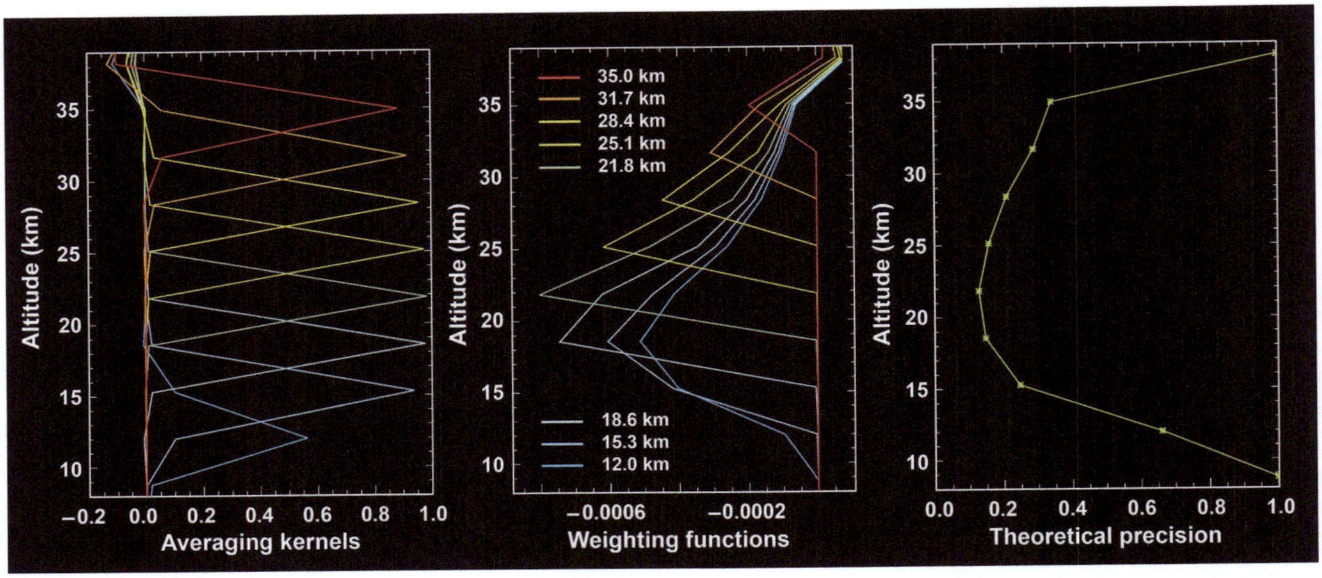

Fig. 7.14 Averaging kernels (*left*), weighting functions at 338.6 nm (*middle*), and theoretical precision (*right*) for BrO vertical profile retrievals from SCIAMACHY limb measurements (Courtesy: IUP-IFE, University of Bremen).

example demonstrated for the ozone profile retrieval from GOME nadir measurements (Munro et al. 1998; Hoogen et al. 1999). For a complete list see Table 7.4.

All the applications listed above use a continuous spectral range to derive the trace gas information. Another group of limb inversion algorithms employs discrete spectral points in and outside strong trace gas absorption bands for the retrieval of vertical profiles. Retrieval algorithms utilising the difference in absorption between the centre and wings of the ozone Chappuis and Huggins bands were devised by Flittner et al. (2000). In a first step, the limb radiance profiles are normalised with respect to a reference tangent height between 40 and 45 km. For the SCIAMACHY O_3 Chappuis band retrieval (von Savigny et al. 2005a), the normalised limb radiance profiles in the centre of the Chappuis band are divided by the mean of two normalised limb radiance profiles at non-absorbing wavelengths outside the Chappuis band and then analysed in an optimal estimation scheme to retrieve the stratospheric O_3 profiles. Retrievals in the O_3 Chappuis band allow extraction of stratospheric O_3 profiles for altitudes between about 12 and 40 km. The retrieval range is limited at lower altitudes because the atmosphere becomes optically thick with respect to Rayleigh scattering and/or O_3 absorption. Above about 40 km the O_3 absorption signatures become too weak to be observed. However, normalised limb radiance profiles in the O_3 Hartley and Huggins bands can also be used without further wavelength pairing for the derivation of ozone profiles in the upper stratosphere and lower mesosphere (Rohen et al. 2006). This type of O_3 retrieval was recently extended to cover the lower stratosphere up to the mesosphere (at least 65 km) by combining the absorption in the Hartley and Huggins bands with those in the Chappuis band (Sonkaew et al. 2009). In addition, limb retrieval algorithms can derive information about Noctilucent Clouds (NLC), Polar Stratospheric Clouds (PSC) and the temperature at the mesopause ($T_{mesopause}$). References to the algorithms are given in Table 7.4.

A third group of limb inversion algorithms is based on a combination of DOAS and vertical inversion and performed in two steps (Sioris et al. 2004; Sioris et al. 2006; Kühl et al. 2008). In the first step slant column densities, i.e. the integrated concentration of the absorbers along the light path, are derived from the SCIAMACHY limb spectra by DOAS. As a reference spectrum, a measurement at a tangent height is used, where the absorption of the considered trace gas is small. Remaining small abundances of the considered absorbers, which appear at the tangent height of the reference spectrum, are estimated by a latitude-dependent a-priori value and are added to the retrieved slant column densities (SCD). In the second step, the trace gas SCD are converted into vertical concentration profiles applying a radiative transfer model and an inversion scheme, for example the optimal estimation method (Rodgers 2000) or Chahine-type inversion (Chahine 1970). Like for nadir DOAS, the limb DOAS retrieval is valid only for weak absorptions and was successfully applied to NO_2, BrO and OClO retrieval (Kühl et al. 2008). Recently Puķīte et al. (2009) extended the limb DOAS concept to absorbers in an optically thicker atmosphere improving the retrieval of BrO.

The above mentioned algorithms typically assume that the atmosphere is horizontally homogeneous, which is not fulfilled under all conditions. For example, substantial gradients exist across the Arctic polar vortex boundary. Puķīte et al. (2010) investigated the effect of horizontally inhomogeneous distributions of trace gases along the flight/viewing

Table 7.4 Atmospheric geophysical parameters and retrieval algorithms – limb and occultation

Parameter	Spectral window (nm)	Occurrence	Quantity	Retrieval algorithm reference
O_3	240–310 (selected wavelengths)	Mesosphere	Profile limb	Rohen et al. 2008
	240–675	Stratosphere, mesosphere	Profile limb	Sonkaew et al. 2009
	525, 600, 675	Stratosphere	Profile limb	von Savigny et al. 2005a
	525–590	Stratosphere	Profile limb	Doicu et al. 2002
				Rozanov et al. 2007
				Lichtenberg et al. 2010
	520–595	Stratosphere	Profile sun occultation	Meyer et al. 2005
	510–560	Stratosphere	Profile moon occultation	Amekudzi et al. 2005a
BrO	338–356	Stratosphere	Profile limb	Rozanov et al. 2005b
				Lichtenberg et al. 2010
	338–357			Kühl et al. 2008
	344.1–360			Sioris et al. 2006
OClO	363.5–391	Stratosphere	Profile limb	Kühl et al. 2008
NO_2	425–450(70)	Stratosphere	Profile limb	Rozanov et al. 2005b
				Sioris et al. 2004
				Doicu et al. 2002
				Lichtenberg et al. 2010
	420–450	Stratosphere	Profile limb	Kühl et al. 2008
	420–460	Stratosphere	Profile sun occultation	Meyer et al. 2005
	430–460	Stratosphere	Profile moon occultation	Amekudzi et al. 2005a
NO_3	610–680	Stratosphere	Profile moon occultation	Amekudzi et al. 2005b
H_2O	928–968	Stratosphere	Profile sun occultation	Noël et al. 2010
	1375–1390	Lower stratosphere	Profile limb	Rozanov et al. 2010
NLC	265–300	Mesosphere	Indicator, particle radius	von Savigny et al. 2004a
PSC	750, 1090	Stratosphere	Indicator	von Savigny et al. 2005b
$T_{mesopause}$	1515–1550	Mesosphere	Nighttime temperature at mesopause	von Savigny et al. 2004b
Mg, Mg+	275–290	Mesosphere		Scharringhausen et al. 2008

direction on the retrieval of profiles. They introduced a method called 'limb 2D tomography' to correct for this effect by combining consecutive limb scanning sequences and using the overlap in their measurement sensitivity regions. The retrieval was applied to synthetic as well as to real SCIAMACHY data (Fig. 7.15). It was found that if horizontal inhomogeneity is not properly accounted for, typical errors of 20% for NO_2 and up to 50% for OClO around the altitude of the profile peak can arise for measurements close to the Arctic polar vortex boundary in boreal winter.

To derive meteoric metal content at different levels of the atmosphere from limb and nadir measurements, Scharringhausen et al. (2008) developed and applied a limb-nadir 2D tomographic retrieval scheme based on optimal estimation to evaluate SCIAMACHY limb and nadir emissions simultaneously for every SCIAMACHY orbit. The inversion algorithm gives the total content as well as partial columns in the thermosphere and in the mesosphere. This retrieval successfully distinguished the mesospheric from the thermospheric column abundances of MgII and MgI for the first time.

The wide variety of retrieval algorithms currently applied to SCIAMACHY limb and occultation data are summarised in Table 7.4, together with their corresponding references.

7.7 Derivation of Tropospheric Information

Distributions of trace gases in the troposphere are of prime scientific, as well as public interest. Two cases need to be distinguished:

- Constituents with the majority of the atmospheric amount residing in the lower troposphere (e.g. CO, CH_4, CO_2, HCHO, SO_2, H_2O): The total column derived from UV-VIS and SWIR solar backscatter measurements with the techniques described in section "Nadir Trace Gas Retrieval Schemes" directly represents the tropospheric column amount including the boundary layer under cloud-free conditions.

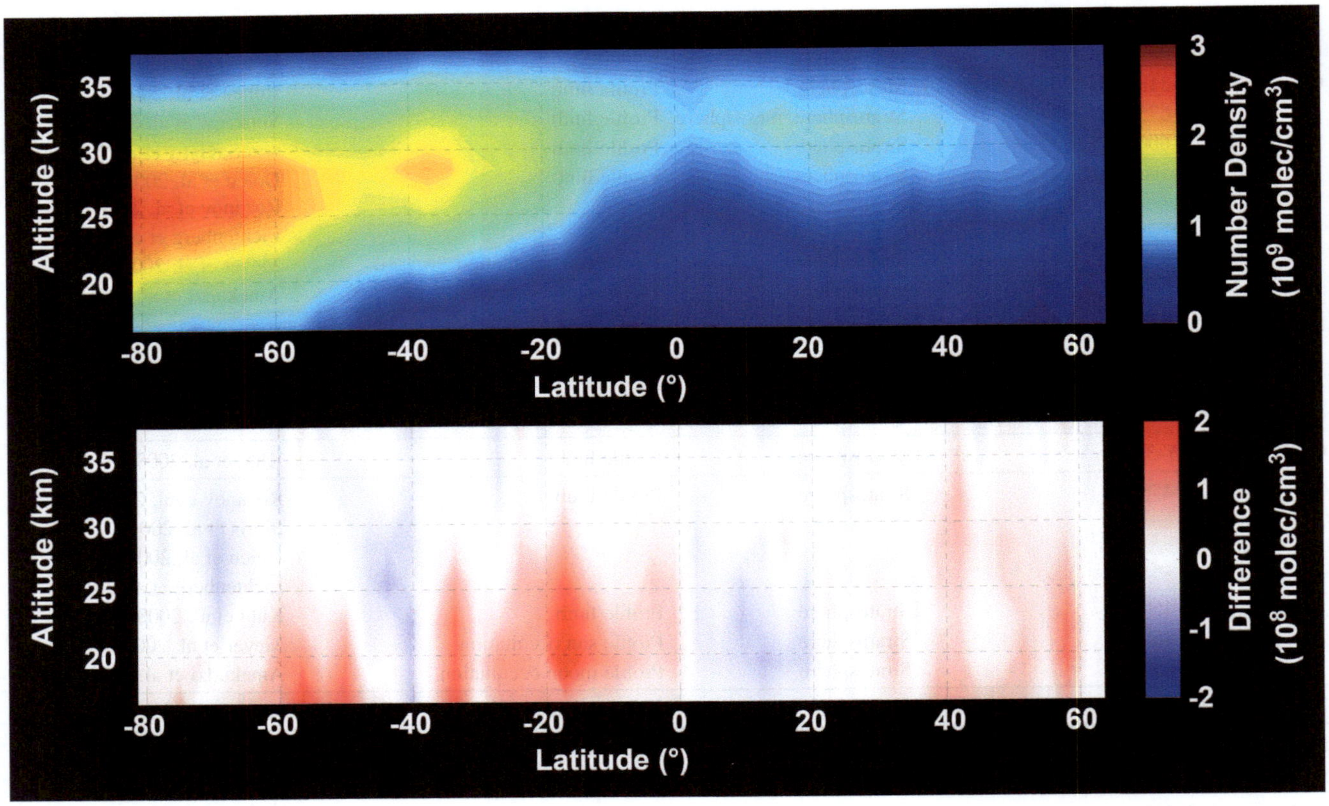

Fig. 7.15 NO_2 densities (*top panel*) as derived from SCIAMACHY limb measurements using the limb 2D tomography method for the orbit 35499 (14 December 2008) consisting of limb states only. In the bottom panel the difference to results from the less accurate 1D approach is illustrated (Courtesy: J. Puķīte; MPI for Chemistry, Mainz).

– Trace gases with comparable column amounts in the troposphere and stratosphere (e.g. BrO, NO_2) or with the stratospheric amount dominating the total column (e.g. O_3).

For the latter case, dedicated techniques are required to separate the tropospheric and the stratospheric contributions. One approach is to use measurements over a clean-air region as a background, the so-called 'Reference Sector Method' (see below). In addition, SCIAMACHY's unique limb/nadir matching capabilities (see Chapter 4) provide a nearly simultaneous stratospheric profile measurement for each nadir measurement. In that respect, SCIAMACHY is clearly superior to GOME on ERS-2, which obtained measurements of the same species in the UV-VIS range but only in nadir geometry. An error analysis for approaches to derive tropospheric NO_2 has been given by Boersma et al. (2004) providing error components and retrieval uncertainties.

Reference Sector Method

The Reference Sector Method (RSM), also referred to as *Tropospheric Residual Method* (Velders et al. 2001; Richter and Burrows 2002; Martin et al. 2002) allows the separation of tropospheric and stratospheric contributions to the total NO_2 content under the assumption of negligible tropospheric NO_2 concentrations over the clean, free Pacific Ocean. It is assumed that the stratospheric NO_2 distribution is homogeneous with longitude. Then the tropospheric NO_2 is primarily the difference between the total column measured over a polluted area and the total column measured over the clean Pacific Ocean. This technique needs no stratospheric profile information and can therefore be applied to generate a consistent GOME – SCIAMACHY tropospheric NO_2 data set (see Chapter 10). A similar idea is also pursued in the retrieval of tropospheric HCHO and SO_2 concentrations, to correct for artificial biases.

Limb/Nadir Matching

Especially for regions with low or moderate pollution levels, the errors in the retrieved tropospheric columns introduced by the simple Reference Sector Method can become important. To reduce the error, it is required to use the measured stratospheric column above the ground scene of interest. SCIAMACHY with its limb/nadir matching measurement

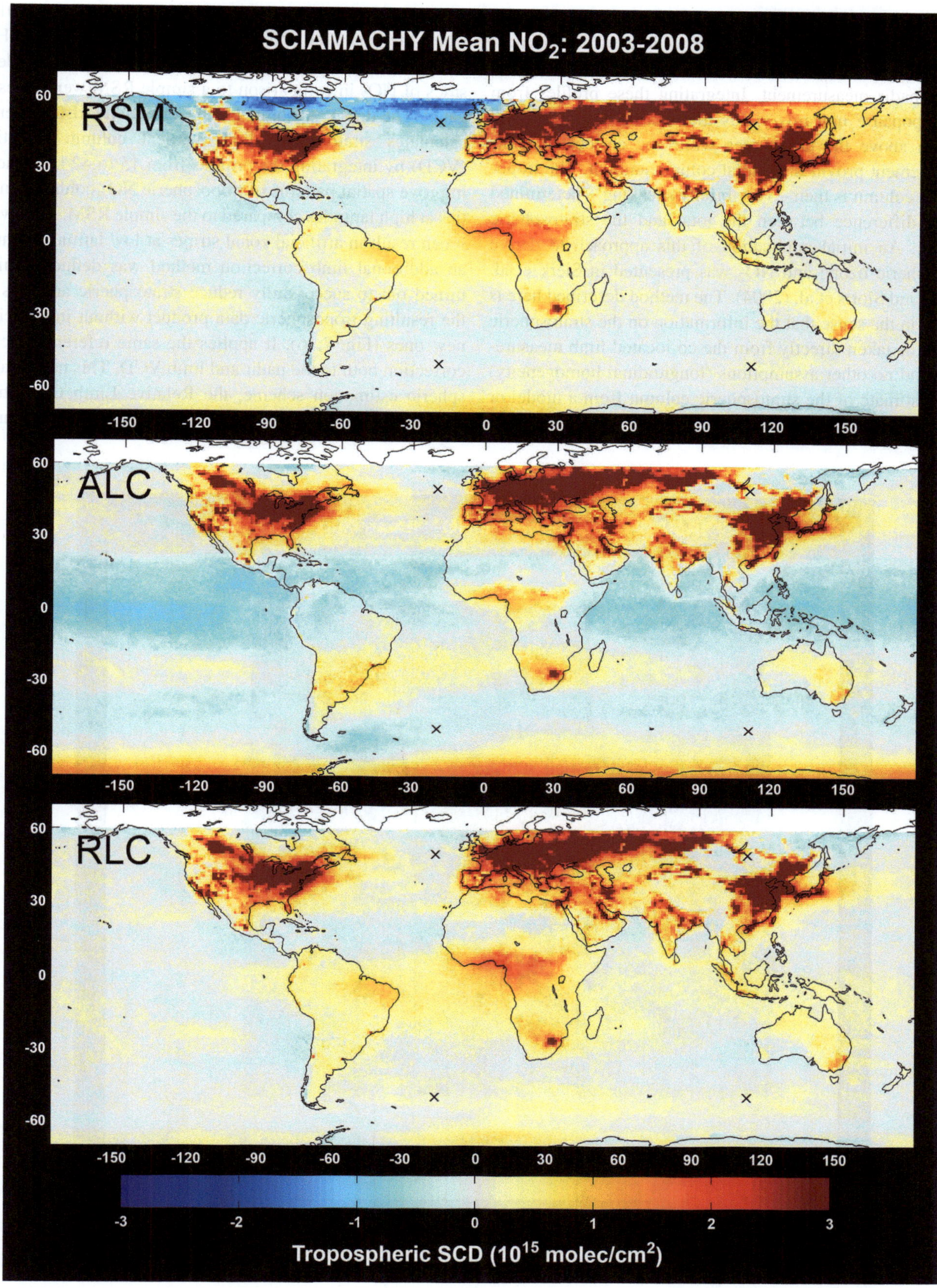

Fig. 7.16 Mean tropospheric NO_2 slant column densities as derived for the period 2003–2008 using the Reference Sector Method (*RSM – top*), Absolute Limb Correction (*ALC – middle*) and Relative Limb Correction (*RLC – bottom*) (Courtesy: S. Beirle, MPI for Chemistry, Mainz).

mode provides radiances from the same volume of air in limb and nadir geometry. This allows inference of vertical stratospheric concentration profiles directly over the region of the nadir measurement. Integrating these profiles from the tropopause upwards yields the measured stratospheric column above the target area while the co-located nadir measurement provides the total column amount. The tropospheric column is then – very briefly speaking – determined as the difference between the total and the stratospheric column. An initial application of this approach to derive tropospheric ozone and NO_2 was presented in Sierk et al. (2006) and Sioris et al. (2004). The method described here is unique in the sense that the information on the stratospheric content is taken directly from the co-located limb measurement and no other assumptions (longitudinal homogeneity) or an estimate of the stratospheric column from a model or from data assimilation are necessary.

Further progress was recently made by Beirle et al. (2010) when the potential and the limitations of SCIAMACHY limb measurements for estimating stratospheric column densities of NO_2 in comparison to a simple RSM were investigated. The direct, Absolute Limb Correction (ALC) scheme calculates stratospheric limb vertical column densities (VCD) by integrating the profiles from 15 to 42 km. These improve spatial patterns of tropospheric NO_2 column densities at high latitudes compared to the simple RSM. However, it can result in artificial zonal stripes at low latitudes. Thus, an additional limb correction method was defined, which turned out to successfully reduce stratospheric artefacts in the resulting tropospheric data product without introducing new ones (Fig. 7.16). It applies the same reference sector correction both to the nadir and limb VCD. This new stratospheric estimation scheme, the Relative Limb Correction (RLC), improves monthly mean tropospheric slant column

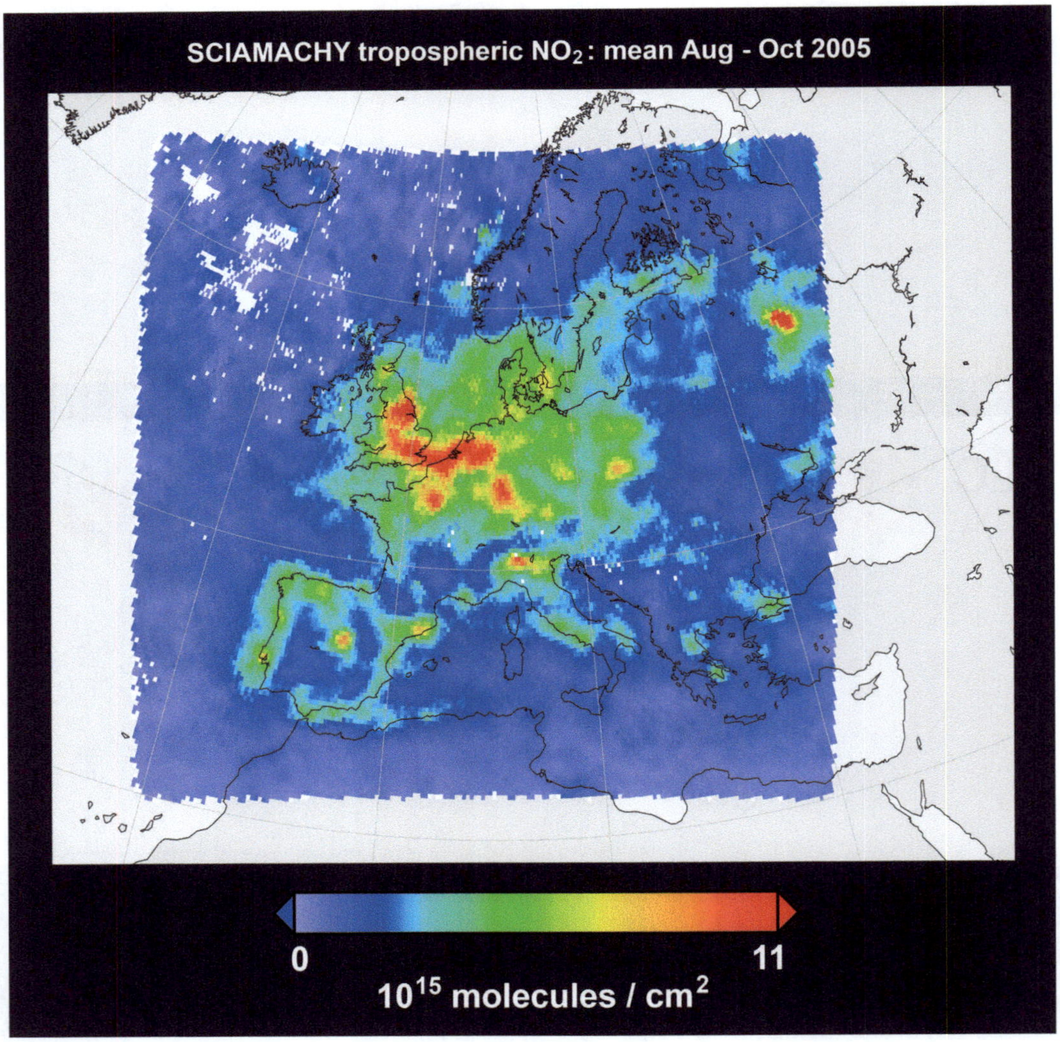

Fig. 7.17 Mean tropospheric NO_2 vertical column densities over Europe as derived from SCIAMACHY for August to October 2005 with the DLR-DFD assimilation approach (Courtesy: T. Erbertseder, DLR-DFD).

densities significantly, e.g. from a bias of -1×10^{15} molecules/cm^2 (using a simple RSM) to nearly zero in the Atlantic ocean, and from $+1 \times 10^{15}$ molecules/cm^2 to nearly zero over Siberia, at 50°N in January.

Data Assimilation – The NO_2 Example

Data assimilation (see section "Scientific and Value-Added Products" of Chapter 8) provides another possibility for the separation of tropospheric and stratospheric contributions. NO_2 permanently resides in the stratosphere and shows significant amounts in the troposphere near source areas. Firstly, the stratospheric and tropospheric parts of the column need to be separated and subsequently, a tropospheric airmass factor has to be applied to the tropospheric slant column. At KNMI, in collaboration with BIRA-IASB, a data assimilation system was applied to NO_2 to derive the stratospheric part of the slant column by assimilation of observed slant columns in a chemistry-transport model (Eskes et al. 2003). This resulted in a stratospheric analysis consistent with the observations and with the variations observed in the stratosphere that are due to atmospheric dynamics and chemical reactions. The tropospheric NO_2 slant column was then extracted by subtracting the assimilated stratospheric slant column from the retrieved total slant column.

In a similar way the stratospheric NO_2 slant column density, exactly for the SCIAMACHY overpass time, was derived from the stratospheric chemistry transport model ROSE at DLR-DFD. To avoid a bias, the modelled stratospheric slant columns were scaled to 'clean conditions' over the Pacific Ocean. The tropospheric NO_2 slant column was then again extracted by subtracting the modelled stratospheric slant column from the retrieved total slant column. Figure 7.17 shows the resulting tropospheric NO_2 distribution over Europe. The tropospheric NO_2 distributions could be further improved by using NO_2 profile shapes estimated with air quality models like EURAD. In this case also properties of clouds, aerosols and the surface need to be taken into account.

References

Acarreta JR, de Haan JF, Stammes P (2004a) Cloud pressure retrieval using O2–O2 absorption band at 477 nm. J Geophys Res 109: D05204. doi:10.1029/2003JD003915

Acarreta JR, Stammes P, Knap WH (2004b) First retrieval of cloud phase from SCIAMACHY spectra around 1.6 micron. Atmos Res 72:89–105

Afe OT, Richter A, Sierk B, Wittrock F, Burrows JP (2004) BrO emission from volcanoes: A survey using GOME and SCIAMACHY measurements. Geophys Res Lett 31:L24113. doi:10.1029/2004GL020994

Amekudzi LK, Bracher A, Meyer J, Rozanov A, Bovensmann H, Burrows JP (2005a) Lunar occultation with SCIAMACHY: first retrieval results. Adv Space Res 36:906–914

Amekudzi LK, Sinnhuber B-M, Sheode NV, Meyer J, Rozanov A, Lamsal L, Bovensmann H, Burrows JP (2005b) Retrieval of stratospheric NO_3 vertical profiles from SCIAMACHY lunar occultation measurement over the Antarctic. J Geophys Res 110:D20304. doi:10.1029/2004JD005748

Asano S, Shiobara M, Uchiyama A (1995) Estimation of cloud physical parameters from airborne solar spectral reflectance measurements for stratocumulus clouds. J Atmos Sci 52:3556–3576

Barkley MP, Frieß U, Monks PS (2006) Measuring atmospheric CO_2 from space using Full Spectral Initiation (FSI) WFM-DOAS. Atmos Chem Phys 6:3517–3534

Bates DR (1984) Rayleigh scattering by air. Planet Space Sci 32:785–790

Beirle S, Kühl S, Pukite J, Wagner T (2010) Retrieval of tropospheric column densities of NO_2 from combined SCIAMACHY nadir/limb measurements. Atmos Meas Tech 3:283–299

Boersma KF, Eskes HJ, Brinksma EJ (2004) Error analysis for tropospheric NO_2 retrieval from space. J Geophys Res 109:D04311. doi:10.1029/2003JD003962

Bracher A, Sinnhuber M, Rozanov A, Burrows JP (2005) NO_2 modelling used for the comparison of NO_2 Satellite Measurements at different Solar Zenigh Angles. Atmos Chem Phys 4:5515–5548

Bracher A, Vountas M, Dinter T, Burrows JP, Röttgers R, Peeken I (2009) Cyanobacteria and diatoms from space using PhytoDOAS on SCIAMACHY data. Biogeosciences 6:751–764

Buchwitz M, Rozanov VV, Burrows JP (2000) A near-infrared optimized DOAS method for the fast global retrieval of atmospheric CH_4, CO, CO_2, H_2O, and N_2O total column amounts from SCIAMACHY Envisat-1 nadir radiances. J Geophys Res 105:15231–15245

Buchwitz M, de Beek R, Bramstedt K, Noël S, Bovensmann H, Burrows JP (2004) Global carbon monoxide as retrieved from SCIAMACHY by WFM-DOAS. Atmos Chem Phys 4:1954–1960

Buchwitz M, de Beek R, Burrows JP, Bovensmann H, Warneke T, Notholt J, Meirink JF, Goede APH, Bergamaschi P, Körner S, Heimann M, Schulz A (2005) Atmospheric methane and carbon dioxide from SCIAMACHY satellite data: Initial comparison with chemistry and transport models. Atmos Chem Phys 5:941–962

Buchwitz M, Khlystova I, Bovensmann H, Burrows JP (2007) Three years of global carbon monoxide from SCIAMACHY: comparison with MOPITT and first results related to the detection of enhanced CO over cities. Atmos Chem Phys 7:2399–2411

Burrows JP, Weber M, Buchwitz M, Rozanov V, Ladstätter-Weißenmayer A, Richter A, DeBeek R, Hoogen R, Bramstedt K, Eichmann K-U, Eisinger M, Perner D (1999) The Global Ozone Monitoring Experiment (GOME): mission concept and first scientific results. J Atmos Sci 56:151–175

Chahine MT (1970) Inverse problems in radiative transfer: determination of atmospheric parameters. J Atmos Sci 27:960967

Chance K (1998) Analysis of BrO measurements from the Global Ozone Monitoring Experiment. Geophys Res Lett 25:3335–8

Chance K, Spurr RJD (1997) Ring effect studies: Rayleigh scattering, including molecular parameters for rotational Raman scattering, and the Fraunhofer spectrum. Appl Opt 36:5224–5230

Chance K, Kurosu TP, Sioris CE (2005) Undersampling correction for array-detector based satellite spectrometers. Appl Opt 44:1296–1304

Corradini S, Cervino M (2006) Aerosol extinction coefficient profile retrieval in the oxygen A-band considering multiple scattering atmosphere. Test case: SCIAMACHY nadir simulated measurements. J Quant Spectrosc Radiat Transfer 97:354–380

Deutschmann T, Beirle S, Frieß U, Grzegorski M, Kern C, Kritten L, Platt U, Pukite J, Wagner T, Werner B, Pfeilsticker K (2010) The Monte Carlo Atmospheric Radiative Transfer Model McArtim: Introduction and Validation of Jacobians and 3D Features, accpeted for publication in J Quant Spectr Rad Transf

de Beek R, Vountas M, Rozanov VV, Richter A, Burrows JP (2001) The Ring effect in the cloudy atmosphere. Geophys Res Lett 28:721–724

de Graaf M, Stammes P (2005) SCIAMACHY Absorbing Aerosol Index – calibration issues and global results from 2002-2004. Atmos Chem Phys 5:2385–2394

de Graaf M, Stammes P, Aben EAA (2007) Analysis of reflectance spectra of UV-absorbing aerosol scenes measured by SCIAMACHY. J Geophys Res 112:D02206. doi:10.1029/2006JD007249

De Smedt I, Müller J-F, Stavrakou T, van der A R, Eskes H, Roozendael M (2008) Twelve years of global observations of formaldehyde in the troposphere using GOME and SCIAMACHY sensors. Atmos Chem Phys 8:4947–4963

Doicu A, Schreier F, Hess M (2002) Iteratively regularized Gauss–Newton method for atmospheric remote sensing. Comp Phys Commun 148:214–226

Dorf M, Bösch H, Butz A, Camy-Peyret C, Chipperfield MP, Engel A, Goutail F, Grunow K, Hendrick F, Hrechanyy S, Naujokat B, Pommereau J-P, Van Roozendael M, Sioris C, Stroh F, Weidner F, Pfeilsticker K (2006) Balloon-borne stratospheric BrO measurements: comparison with Envisat/SCIAMACHY BrO limb profiles. Atmos Chem Phys 6:2483–2501

Eskes HJ, van Velthoven PFJ, Valks PJM, Kelder HM (2003) Assimilation of GOME total ozone satellite observations in a three-dimensional tracer transport model. Quart J R Meteorol Soc 129:1663–1681

Eskes HJ, van der A RJ, Brinksma EJ, Veefkind JP, de Haan JF, Valks PJM (2005) Retrieval and validation of ozone columns derived from measurements of SCIAMACHY on Envisat. Atmos Chem Phys Discuss 5:4429–4475

Flittner DE, Bhartia PK, Herman BM (2000) O_3 profiles retrieved from limb scatter measurements: theory. Geophys Res Lett 27:2061–2064

Fournier N, Stammes P, de Graaf M, van der A RJ, Piters A, Grzegorski M, Kokhanovsky A (2006) Improving cloud information over deserts from SCIAMACHY oxygen A-band measurements. Atmos Chem Phys 6:163–172

Frankenberg C, Platt U, Wagner T (2005a) Iterative maximum a posteriori (IMAP)-DOAS for retrieval of strongly absorbing trace gases: Model studies for CH_4 and CO_2 retrieval from near infrared spectra of SCIAMACHY onboard ENVISAT. Atmos Chem Phys 5:9–22

Frankenberg C, Platt U, Wagner T (2005b) Retrieval of CO from SCIAMACHY onboard ENVISAT: detection of strongly polluted areas and seasonal patterns in global CO abundances. Atmos Chem Phys 5:1639–1644

Frankenberg C, Bergamaschi P, Butz A, Houweling S, Meirink JF, Notholt J, Petersen AK, Schrijver H, Warneke T, Aben I (2008) Tropical methane emissions: a revised view from SCIAMACHY onboard ENVISAT. Geophys Res Lett 35:L15811. doi:doi:10.1029/2008GL034300

Frankenberg C, Yoshimura K, Warneke T, Aben I, Butz A, Deutscher N, Griffith D, Hase F, Notholt J, Schneider M, Schrijver H, Röckmann T (2009) Dynamic processes governing lower-tropospheric HDO/H2O ratios as observed from space and ground. Science 325:1374

Gloudemans AMS, Schrijver H, Hasekamp OP, Aben I (2008) Error analysis for CO and CH_4 total columns retrieved from SCIAMACHY 2.3 μm spectra. Atmos Chem Phys 8:3999–4017

Gloudemans AMS, de Laat ATJ, Schrijver H, Aben I, Meirink JF, van der Werf GR (2009) SCIAMACHY CO over land and oceans: 2003-2007 interannual variability. Atmos Chem Phys 9:3799–3813

Grainger JF, Ring J (1962) Anomalous Fraunhofer line profiles. Nature 193:762–762

Grzegorski M (2009) Cloud retrieval from UV/VIS satellite instruments (SCIAMACHY and GOME). Ph.D. thesis, University of Heidelberg

Grzegorski M, Frankenberg C, Platt U, Wenig M, Fournier N, Stammes P, Wagner T (2004) Determination of cloud parameters from SCIAMACHY data for the correction of tropospheric trace gases. Proceedings of the ENVISAT & ERS Symposium, Salzburg, Austria, ESA SP-572

Herman JR, Bhartia PK, Torres O, Hsu C, Seftor C, Celarier EA (1997) Global distributions of UV-absorbing aerosols from NIMBUS 7/TOMS data. J Geophys Res 102:16911–16922

Holzer-Popp T, Schroedter-Homscheidt M, Breitkreuz H, Martynenko D, Klüser L (2008) Improvements of synergetic aerosol retrieval for ENVISAT. Atmos Chem Phys 8:7651–7672

Hoogen R, Rozanov VV, Burrows JP (1999) Ozone profiles from GOME satellite data: algorithm description and first validation. J Geophys Res 104:8263–8280

Houweling S, Hartmann W, Aben I, Schrijver H, Skidmore J, Roelofs G-J, Breon F-M (2005) Evidence of systematic errors in SCIAMACHY-observed CO_2 due to aerosols. Atmos Chem Phys 5:3003–3013

Joiner J, Bhartia PK, Cebula RP, Hilsenrath E, McPeters RD, Park H (1995) Rotational Raman scattering (Ring effect) in satellite backscatter ultraviolet measurements. Appl Opt 34:4513–4525

Kaiser JW, Burrows JP (2003) Fast weighting functions for retrievals from limb scattering measurements. J Quant Spectrosc Radiat Transfer 77:273–283

Knap WH, Stammes P, Koelemeijer RBA (2002) Cloud thermodynamic phase determination from near-infrared spectra of reflected sunlight. J Atmos Sci 59:83–96

Koelemeijer RBA, Stammes P, Hovenier JW, de Haan JF (2001) A fast method for retrieval of cloud parameters using oxygen A band measurements from the Global Ozone Monitoring Experiment. J Geophys Res 106:3475–3490

Kokhanovsky AA, G. de Leeuw (2009) Satellite Aerosol Remote Sensing Over Land, Berlin: Springer-Praxis.

Kokhanovsky AA, Rozanov VV (2005) Cloud bottom altitude determination from a satellite. IEEE Trans Geosci Remote Sens Lett 2:280–283

Kokhanovsky AA, Rozanov VV (2010) The determination of the dust cloud altitude from a satellite using hyperspectral measurements in the gaseous absorption band. Int. J. Remote Sens. 31:2729–2744

Kokhanovsky AA, Rozanov VV, Nauss T, Reudenbach C, Daniel JS, Miller HL, Burrows JP (2006a) The semianalytical cloud retrieval algorithm for SCIAMACHY: I. Validation. Atmos Chem Phys 6:1905–1911

Kokhanovsky AA, von Hoyningen-Huene W, Rozanov VV, Noël S, Gerilowski K, Bovensmann H, Bramstedt K, Buchwitz M, Burrows JP (2006b) The semianalytical cloud retrieval algorithm for SCIAMACHY II. The application to MERIS and SCIAMACHY data. Atmos Chem Phys 6:4129–4136

Kokhanovsky AA, Vountas M, Rozanov VV, Lotz W, Bovensmann H, Burrows JP (2007a) Global cloud top height and thermodynamic phase distribution as obtained by SCIAMACHY on ENVISAT. Int J Remote Sens 28:4499–4507

Kokhanovsky AA, Jourdan O, Burrows JP (2007b) The cloud phase discrimination from a satellite. IEEE Trans Geosci Remote Sens Lett 3:103–106

Kokhanovsky AA, Deuzé JL, Diner DJ, Dubovik O, Ducos F, Emde C, Garay MJ, Grainger RG, Heckel A, Herman M, Katsev IL, Keller J, Levy R, North PRJ, Prikhach AS, Rozanov VV, Sayer AM, Ota Y, Tanré D, Thomas GE, Zege EP (2010) The inter-comparison of major satellite aerosol retrieval algorithms using simulated intensity and polarization characteristics of reflected light. Atmos Meas Tech 3:909–932

Kokhanovsky AA, von Hoyningen-Huene W, Burrows JP (2009b) The determination of the cloud fraction in the SCIAMACHY ground scene using spectral MERIS measurements. Int J Remote Sens 30:6151–6167

Krijger JM, Aben I, Schrijver H (2005) Distinction between clouds and ice/snow covered surfaces in the identification of cloud-free observations using SCIAMACHY PMDs. Atmos Chem Phys 5:2729–2738

Kühl S, Wilms-Grabe W, Frankenberg C, Grzegorski M, Platt U, Wagner T (2006) Comparison of OClO nadir measurements from SCIAMACHY and GOME. Adv Space Res 37:2247–2253

Kühl S, Pukite J, Deutschmann T, Platt U, Wagner T (2008) SCIAMACHY limb measurements of NO_2, BrO and OClO. Retrieval of vertical profiles: Algorithm, first results, sensitivity and comparison studies. Adv Space Res 42:1747–1764

Lee C, Martin RV, van Donkelaar A, O'Byrne G, Krotkov N, Richter A, Huey G, Holloways JS (2009) Retrieval of vertical columns of sulfur dioxide from SCIAMACHY and OMI: air mass factor algorithm development and validation. J Geophys Res 114:D22303. doi:10.1029/2009JD012123

Lenoble J (ed) (1985) Radiative transfer in scattering and absorbing atmospheres. Standard computational procedures. A. Deepak Publishing, Hampton, VA

Lerot C, Van Roozendael M, van Geffen J, van Gent J, Fayt C, Spurr R, Lichtenberg G, von Bargen A (2009) Six years of total ozone column measurements from SCIAMACHY nadir observations. Atmos Meas Tech 2:87–98

Lichtenberg G, Bovensmann H, Van Roozendael M, Doicu A, Eichmann K-U, Hess M, Hrechanyy S, Kokhanovsky A, Lerot C, Noël S, Richter A, Rozanov A, Schreier F, Tilstra LG (2010) SCIAMACHY Offline level 1b-2 processor algorithm theoretical baseline document (ENV-ATB-QWG-SCIA-0085). Technical Document, SQWG/DLR-IMF. Available at http://atmos.caf.dlr.de/sciamachy/documentation.html

Liou KN (2002) An introduction to atmospheric radiation, vol 84, International geophysical series. Academic, Burlington, MA

Lotz WA, Vountas M, Dinter T, Burrows JP (2009) Cloud and surface classification using SCIAMACHY polarization measurement devices. Atmos Chem Phys 9:1279–1288

Loughman RP, Griffioen E, Oikarinen L, Postylyakov OV, Rozanov A, Flittner DE, Rault DF (2004) Comparison of radiative transfer models for limb-viewing scattered sunlight measurements. J Geophys Res 109(D6):D06303. doi:10.1029/2003JD003854

Loyola D (1998) A new cloud recognition algorithm for optical sensors. IEEE International Geoscience and Remote Sensing Symposium IGARSS'98 Digest, II:572–574

Martin RV, Chance K, Jacob DJ, Kurosu TP, Spurr RJD, Bucsela E, Gleason JF, Palmer PI, Bey I, Fiore AM, Li Q, Yantosca RM, Koelemeijer RBA (2002) An improved retrieval of tropospheric nitrogen dioxide from GOME. J Geophys Res 107:4437–4456. doi:10.1029/2001JD001027

McLinden CA, McConnell JC, Griffioen E, and McElroy CT 2002 A vector radiative transfer model for the Odin/OSIRIS project, Can. J. Phys., 80:375–393,

Meyer J, Bracher A, Rozanov A, Schlesier AC, Bovensmann H, Burrows JP (2005) Solar occultation with SCIAMACHY: algorithm description and first validation. Atmos Chem Phys 5:1589–1604

Montoux N, Hauchecorne A, Pommereau J-P, Lefèvre F, Durry G, Jones RL, Rozanov A, Dhomse S, Burrows JP, Morel B, Bencherif H (2009) Evaluation of balloon and satellite water vapour measurements in the Southern tropical and subtropical UTLS during the HIBISCUS campaign. Atmos Chem Phys 9:5299–5319

Munro R, Siddans R, Reburn WJ, Kerridge B (1998) Direct measurement of tropospheric ozone from space. Nature 392:168–171

Nakajima T, King MD (1990) Determination of the optical thickness and effective particle radius of clouds from reflected solar radiation measurements. Part I: Theory. J Atmos Sci 47:1878–1893

Noël S, Buchwitz M, Bovensmann H, Burrows JP (2004) First retrieval of global water vapour column amounts from SCIAMACHY measurements. Atmos Chem Phys 4:111–125

Noël S, Bramstedt K, Rozanov A, Bovensmann H, Burrows JP (2010) Water vapour profiles from SCIAMACHY solar occultation measurements derived with an onion peeling approach. Atmos Meas Tech 3:523–535

Noxon JF (1975) Nitrogen dioxide in the stratosphere and troposphere measured by ground-based absorption spectroscopy. Science 189:547

Oetjen H, Wittrock F, Richter A, Chipperfield MP, Medeke T, Sheode N, Sinnhuber B-M, Sinnhuber M, Burrows JP (2009) Evaluation of stratospheric chlorine chemistry for the Arctic spring 2005 using modelled and measured OClO column densities. Atmos Chem Phys Discuss 9:26539–26575

Penning de Vries M, Beirle S, Wagner T (2009) UV aerosol indices from SCIAMACHY: introducing the SCattering Index (SCI). Atmos Chem Phys 9:9555–9567

Platnick S, King MD, Ackerman SA, Menzel WP, Baum BA, Rièdi JC, Frey RA (2003) The MODIS cloud products: algorithms and examples from Terra. IEEE Trans Geosci Remote Sens 41:459–473

Platt U, Stutz J (2008) Differential optical absorption spectroscopy: principles and applications. Springer, Physics of earth and space environments. ISBN 978-3-540-21193-8

Platt U, Perner D, Pätz HW (1979) Simultaneous measurements of atmospheric CH_2, O_3 and NO_2 by differential optical absorption. J Geophys Res 84:6329–6335

Pöschl U (2005) Atmospheric aerosols: composition, transformation, climate and health effects. Angew Chem Int Ed 44:7520–7540

Puķīte J, Kühl S, Deutschmann T, Platt U, Wagner T (2009) Extending differential optical absorption spectroscopy for limb measurements in the UV. Atmos Meas Tech 3:631–653

Puķīte J, Kühl S, Deutschmann T, Dörner S, Jöckel P, Platt U, Wagner T (2010) The effect of horizontal gradients and spatial measurement resolution on the retrieval of global vertical NO_2 distributions from SCIAMACHY measurements in limb only mode. Atmos Meas Tech 3:1155–1174, doi:10.5194/amt-3-1155-2010

Richter A, Burrows JP (2002) Retrieval of tropospheric NO_2 from GOME measurements. Adv Space Res 29:1673–1683

Richter A, Wagner T (2011) Solar backscattered radiation: UV, visible and near IR – trace gases. In: Burrows JP, Platt U, Borrell P (eds.) (2011) The remote sensing of Tropospheric Composition from Space. Springer Heidelberg Dordrecht London New York, ISBN 978-3-642-14790-6, DOI: 10.1007/978-3-642-14791-3

Richter A, Burrows JP, Nüß H, Granier C, Niemeier U (2005) Increase in nitrogen dioxide over China observed from space. Nature 437:129–132

Richter A, Wittrock F, Burrows JP (2006) SO_2 measurements with SCIAMACHY. Proceedings of the ESA atmospheric science conference, Frascati, Italy, ESA SP-628

Rodgers CD (2000) Inverse methods for atmospheric sounding: theory and practice. Series on atmospheric, oceanic and planetary physics, vol 2. World Scientific

Rohen GJ, von Savigny C, Eichmann K-U, Llewellyn EJ, Bracher A, Burrows JP (2006) Retrieval of mesospheric ozone profiles from SCIAMACHY limb scattering observations: theory, first validation results and ozone depletion during the Oct./Nov. 2003 solar proton event. Adv Space Res 37:2263–2268. doi:10.1016/j.asr.200503.160

Rohen GJ, von Savigny C, Kaiser JW, Llewellyn EJ, Froidevaux L, Lopez-Puertas M, Steck T, Palm M, Winkler H, Sinnhuber M, Bovensmann H, Burrows JP (2008) Ozone profile retrieval from limb scatter measurements in the HARTLEY bands: further retrieval details and profile comparisons. Atmos Chem Phys 8:2509–2517

Rozanov A, Rozanov V, Burrows JP (2001) A numerical radiative transfer model for a spherical planetary atmosphere: Combined differential-integral approach involving the Picard iterative approximation. J Quant Spectrosc Radiat Transfer 69:491–512

Rozanov A, Rozanov V, Buchwitz M, Kokhanovsky A, Burrows JP (2005a) SCIATRAN 2.0 – A new radiative transfer model for geophysical applications in the 175–2400 nm spectral region. Adv Space Res 36:1015–1019

Rozanov A, Bovensmann H, Bracher A, Hrechany S, Rozanov V, Sinnhuber M, Stroh F, Burrows JP (2005b) NO_2 and BrO vertical profile retrieval from SCIAMACHY limb measurements: sensitivity studies. Adv Space Res 36:846–854

Rozanov A, Eichmann K-U, von Savigny C, Bovensmann H, Burrows JP, von Bargen A, Doicu A, Hilgers S, Godin-Beekmann S, Leblanc T, McDermid IS (2007) Comparison of the inversion algorithms applied to the ozone vertical profile retrieval from SCIAMACHY limb measurements, Atmos. Chem. Phys 7:4763–4779

Rozanov A, Weigel K, Bovensmann H, Dhomse S, Eichmann K-U, Kivi R, Rozanov V, Vömel H, Weber M, Burrows JP (2010) Retrieval of water vapor vertical distributions in the upper troposphere and the lower stratosphere from SCIAMACHY limb measurements, Atmos Meas Tech Discuss 3:4009–4057, doi:10.5194/amtd-3-4009-2010

Rozanov VV, Kokhanovsky AA (2004) Semi-analytical cloud retrieval algorithm as applied to the cloud top altitude and the cloud geometrical thickness determination from top-of-atmosphere reflectance measurements in the oxygen A-band. J Geophys Res 109:D05202. doi:10.1029/2003JD004104

Rozanov VV, Rozanov AV (2010) Differential optical absorption spectroscopy (DOAS) and air mass factor concept for a multiply scattering vertically inhomogeneous medium: theoretical consideration. Atmos Meas Tech 3:751–780

Rozanov VV, Kurosu T, Burrows JP (1998) Retrieval of atmospheric constituents in the UV-visible: a new quasi-analytical approach for the calculation of weighting functions. J Quant Spectrosc Radiat Transfer 60:277–299

Saiz-Lopez A, Chance K, Liu X, Kurosu TP, Sander SP (2007) First observations of iodine oxide from space. Geophys Res Lett 34: L12812. doi:10.1029/2007GL030111

Sarkissian A, Roscoe HK, Fish DJ (1995) Ozone measurements by zenith-sky spectrometers: an evaluation of errors in Air Mass Factors calculated by radiative transfer models. J Quant Spectrosc Radiat Transfer 54:471–480

Scharringhausen M, Aikin AC, Burrows JP, Sinnhuber M (2008) Global column density retrievals of mesospheric and thermospheric Mg I and Mg II from SCIAMACHY limb and nadir radiance data. J Geophys Res 113:D13303. doi:10.1029/2007JD009043

Schlundt C, Kokhanovsky AA, von Hoyningen-Huene W, Dinter T, Istomina L, Burrows JP (2010) Synergetic cloud fraction determination from SCIAMACHY using MERIS. Atmospheric Measurements Techniques Discussions 3:3601–3642

Schneising O, Buchwitz M, Burrows JP, Bovensmann H, Reuter M, Notholt J, Macatangay R, Warneke T (2008) Three years of greenhouse gas column-averaged dry air mole fractions retrieved from satellite – Part 1: Carbon dioxide. Atmos Chem Phys 8:3827–3853

Schneising O, Buchwitz M, Burrows JP, Bovensmann H, Bergamaschi P, Peters W (2009) Three years of greenhouse gas column-averaged dry air mole fractions retrieved from satellite – Part 2: Methane. Atmos Chem Phys 9:443–465

Schönhardt A, Richter A, Wittrock F, Kirk H, Oetjen H, Roscoe HK, Burrows JP (2008) Observations of iodine monoxide (IO) columns from satellite. Atmos Chem Phys 8:637–653

Schreier F, Gimeno-Garcia S, Hess M, Doicu A, Lichtenberg G (2009) Carbon monoxide vertical column density retrieval from SCIAMACHY infrared nadir observations. In: Nakajiama T, Yamasoe MA (eds) IRS 2008: current problems in atmospheric radiation. American Institute of Physics. doi:10.1063/1.3116983

Schrijver H, Gloudemans AMS, Frankenberg C, Aben I (2009) Water vapour total columns from SCIAMACHY spectra in the 2.36 μm window. Atmos Meas Tech 2:561–571

Sierk B, Richter A, Rozanov A, von Savigny C, Schmoltner AM, Buchwitz M, Bovensmann H, Burrows JP (2006) Retrieval and Monitoring of atmospheric trace gas concentrations in nadir and limb geometry using the space-borne SCIAMACHY instrument. Environ Mon Assess. doi:10.1007/s10661-005-9049-9

Sioris CE, Kurosu TP, Martin RV, Chance K (2004) Stratospheric and tropospheric NO_2 observed by SCIAMACHY: first results. Adv Space Res 34:780–785

Sioris CE, Kovalenko LJ, McLinden CA, Salawitch RJ, Van Roozendael M, Goutail F, Dorf M, Pfeilsticker K, Chance K, von Savigny C, Liu X, Kurosu TP, Pommereau J-P, Bösch H, Frerick J (2006) Latitudinal and vertical distribution of bromine monoxide in the lower stratosphere from Scanning Imaging Absorption Spectrometer for Atmospheric Chartography limb scattering measurements. J Geophys Res 111:D14301. doi:10.1029/2005JD006479

Slijkhuis S, von Bargen A, Thomas W, Chance K (1999) Calculation of undersampling correction spectra for DOAS spectral fitting. In: Proceedings of the atmospheric measurements from space – ESAMS '99, Noordwijk, The Netherlands, WPP-161 563-569

Solomon S, Schmeltekopf AL, Sanders RW (1987) On the interpretation of zenith sky absorption measurements. J Geophys Res 92:8311–8319

Sonkaew T, Rozanov VV, von Savigny C, Rozanov A, Bovensmann H, Burrows JP (2009) Cloud sensitivity studies for stratospheric and lower mesospheric ozone profile retrievals from measurements of limb-scattered solar radiation. Atmos Meas Tech 2:653–678

Spurr RJD, Kurosu TP, Chance KV (2001) A linearized discrete ordinate radiative transfer model for atmospheric remote-sensing retrieval. J Quant Spectrosc Radiat Transfer 68:689–735

Stammes P (2001) Spectral radiance modelling in the UV-Visible range. In: Smith WL, Timofeyev YM (eds) IRS 2000: current problems in atmospheric radiation. A. Deepak Publication, Hampton, VA, pp 385–388

Thomas GE, Stamnes K (1999) Radiative transfer in the atmosphere and ocean. Cambridge: Cambridge University Press

Torres O, Bhartia PK, Herman JR, Ahmad Z, Gleason J (1998) Derivation of aerosol properties from satellite measurements of backscattered UV radiation: theoretical basis. J Geophys Res 103(D14):17099–17110

Vassilkov AP, Joiner J, Gleason J, Bhartia P (2002) Ocean Raman scattering in satellite backscatter UV measurements. Geophys Res Lett 10:1837–1840

Velders GJM, Granier C, Portmann RW, Pfeilsticker K, Wenig M, Wagner T, Platt U, Richter A, Burrows JP (2001) Global tropospheric NO_2 column distributions: comparing 3-D model calculations with GOME measurements. J Geophys Res 106:12643–12660

von Hoyningen-Huene W, Freitag M, Burrows JP (2003) Retrieval of aerosol optical thickness over land surfaces from top-of-atmosphere radiance. J Geophys Res 108. doi:10.1029/2001JD002018

von Hoyningen-Huene W, Kokhanovsky AA, Wuttke M, Buchwitz M, Noël S, Gerilowski K, Burrows JP, Latter B, Siddans R, Kerridge BJ (2007) Validation of SCIAMACHY top-of-atmosphere reflectance for aerosol remote sensing using MERIS L1 data. Atmos Chem Phys 7:97–106

von Savigny C, Kokhanovsky A, Bovensmann H, Eichmann K-U, Kaiser JW, Noël S, Rozanov AV, Skupin J, Burrows JP (2004a)

NLC detection and particle size determination: first results from SCIAMACHY on ENVISAT. Adv Space Res 34:851–856

von Savigny C, Eichmann K-U, Llewellyn EJ, Bovensmann H, Burrows JP, Bittner M, Höppner K, Offermann D, Steinbrecht W, Winkler P, Taylor MJ, Cheng Y (2004b) First near-global retrieval of OH rotational temperatures from satellite-based Meinel band emission measurements. Geophys Res Lett 31:L15111. doi:10.1029/2004GL

von Savigny C, Rozanov A, Bovensmann H, Eichmann K-U, Noël S, Rozanov VV, Sinnhuber B-M, Weber M, Burrows JP (2005a) The ozone hole break-up in September 2002 as seen by SCIAMACHY on ENVISAT. J Atmos Sci 62:721–734

von Savigny C, Ulasi EP, Eichmann K-U, Bovensmann H, Burrows JP (2005b) Detection and mapping of polar stratospheric clouds using limb scattering observations. Atmos Chem Phys 5:3071–3079

Vountas M, Rozanov V, Burrows J (1998) Ring effect: impact of rotational Raman scattering on radiative transfer in earth's atmosphere. J Quant Spectrosc Radiat Transfer 60:943–961

Vountas M, Dinter T, Bracher A, Burrows JP, Sierk B (2007) Spectral studies of ocean water with space-borne sensor SCIAMACHY using Differential Optical Absorption Spectroscopy (DOAS). Ocean Sci 3:429–440

Vrekoussis M, Wittrock F, Richter A, Burrows JP (2009) Temporal and spatial variability of glyoxal as observed from space. Atmos Chem Phys 9:4485–4504

Wagner T, Burrows JP, Deutschmann T, Dix B, von Friedeburg C, Frieß U, Hendrick F, Heue K-P, Irie H, Iwabuchi H, Kanaya Y, Keller J, McLinden CA, Oetjen H, Palazzi E, Petritoli A, Platt U, Postylyakov O, Pukite J, Richter A, van Roozendael M, Rozanov A, Rozanov V, Sinreich R, Sanghavi S, Wittrock F (2007a) Comparison of box-air-mass-factors and radiances for Multiple-Axis Differential Optical Absorption Spectroscopy (MAX-DOAS) geometries calculated from different UV/visible radiative transfer models. Atmos Chem Phys 7:1809–1833

Wagner T, Beirle S, Deutschmann T, Grzegorski M, Platt U (2007b) Satellite monitoring of different vegetation types by differential optical absorption spectroscopy (DOAS) in the red spectral Range. Atmos Chem Phys 7:69–79

Wagner T, Beirle S, Deutschmann T (2009) Three-dimensional simulation of the Ring effect in observations of scattered sun light using Monte Carlo radiative transfer models, Atmos. Meas. Tech 2:113–124

Wang P, Stammes P, van der A R, Pinardi G, Van Roozendael M (2008) FRESCO+: an improved O_2 A-band cloud retrieval algorithm for tropospheric trace gas retrievals. Atmos Chem Phys 8:6565–6576

Weber M, Lamsal LN, Coldewey-Egbers M, Bramstedt K, Burrows JP (2005) Pole-to-pole validation of GOME WFM-DOAS total ozone with groundbased data. Atmos Chem Phys 5:1341–1355

Wittrock F, Richter A, Oetjen H, Burrows JP, Kanakidou M, Myriokefalitakis S, Volkamer R, Beirle S, Platt U, Wagner T (2006) Simultaneous global observations of glyoxal and formaldehyde from space. Geophys Res Lett 33:L16804. doi:10.1029/2006GL026310

Yamamoto GA, Wark DQ (1961) Discussion of the letter by R.A. Hanel: determination of cloud altitude from satellite. J Geophys Res 66:3596

Yan X (2005) A fast SCIAMACHY PMD Cloud Algorithm (SPCA). Master Thesis, University of Bremen. Available at http://www.iup.uni-bremen.de/doas/paper/yang_thesis_spca_0509.pdf

Chapter 8
Data Processing and Products

G. Lichtenberg, K-U. Eichmann, C. Lerot, R. Snel, S. Slijkhuis, S. Noël, R. van Hees, B. Aberle, K. Kretschel, M. Meringer, D. Scherbakov, H. Weber, and A. von Bargen

Abstract SCIAMACHY data processing generates products on various levels as required by the user community. These products are either produced in the ENVISAT Payload Data Segment as *operational* products or by science institutes as *scientific* products or *value-added* products. Operational processing occurs in two steps – level 0–1b and level 1b–2. In both steps different timescales may apply – near-realtime, fast delivery or offline. Level 0–1b processing generates geolocated and calibrated radiances from the raw atmospheric measurements, as well as from measurements for calibration and instrument monitoring. The algorithms convert measured signals into calibrated radiances. Therefore a sequence of calibration steps has to be applied starting with correcting the memory effect and non-linearity and ending with applying the radiometric instrument response. Particular attention has to be given to the correction of polarisation and degradation. The goal of level 1b–2 processing is to provide geophysical parameters such as column densities and profiles from trace gas species as well as cloud and aerosol parameters. Nadir measurements permit retrieving total column densities or cloud and aerosol parameters. From limb observations height resolved profiles of atmospheric parameters can be inferred. Together with the scientific and value-added products the SCIAMACHY data can serve a wide range of applications.

Keywords Operational products • Scientific products • Value-added products • Level 0–1b processing • Level 1b–2 processing

The processing of SCIAMACHY measurement data has to serve the requirements of different user communities. Therefore, products, differing in content, are provided at various levels. In addition, product generation may occur in two environments. Whereas the *operational* data processing follows the specific guidelines and rules of the ENVISAT Payload Data Segment (PDS) architecture and organisation, *scientific* product generation is based on science group algorithms which have not been developed within the framework of the PDS. This twofold approach, when fully implemented, has the advantage that standard products with well-defined requirements can be generated in a configuration controlled but also flexible processor regime. It relieves users from providing significant resources for basic processing steps, e.g. calibrating the measurement data. On the other hand, scientific products can utilise most recent, often advanced algorithms since the stringent requirements of the operational ground segment need not to be fulfilled. Also new geophysical parameters which are not part of the operational chains will first be introduced as scientific products demonstrating the potential of SCIAMACHY's measurement data. Later these algorithms may be implemented in the operational environment, thus providing users with a continuous and up-to-date global dataset. This feedback between scientific and operational products ensures that algorithm improvements are rapidly and continuously fed into the operational system.

As outlined in Chapter 2, the SCIAMACHY products follow the general ENVISAT data product rules and requirements. Thus for each orbit a level 0, level 1b and level 2 file is generated. The SCIAMACHY specific definitions include:

G. Lichtenberg (✉), S. Slijkhuis, B. Aberle, K. Kretschel, M. Meringer, D. Scherbakov, and H. Weber
Remote Sensing Technology Institute, German Aerospace Center (DLR-IMF), Oberpfaffenhofen, 82234 Wessling, Germany
e-mail: guenter.lichtenberg@dlr.de

K-U. Eichmann and S. Noël
Institute of Environmental Physics/Institute of Remote Sensing (IUP-IFE), University of Bremen, Otto-Hahn-Allee 1, 28359 Bremen, Germany

C. Lerot
Belgian Institute for Space Aeronomy (BIRA-IASB), 3 Avenue Circulaire, 1180 Brussels, Belgium

R. Snel and R. van Hees
SRON, Netherlands Institute for Space Research, Sorbonnelaan 2, 3584 CA Utrecht, The Netherlands

A. von Bargen
German Aerospace Center, Space Agency, Königswinterer Str. 522-524, 53227 Bonn, Germany

M. Gottwald and H. Bovensmann (eds.), *SCIAMACHY - Exploring the Changing Earth's Atmosphere*,
DOI 10.1007/978-90-481-9896-2_8,

- Level 1b: The level 0 content plus all data needed for calibration together with the geolocation of the ground pixel measured
- Level 2: Geophysical quantities such as total trace gas columns, trace gas profiles and cloud and aerosol information

Level 1b and level 2 products are distributed directly to the users. For calibration additional products were defined:

- Calibrated spectra (level 1c): The user has the choice to apply either the entire or only a limited set of calibrations (see below). Applying the full calibration results in Earth radiances as a function of wavelength.
- Auxiliary Data Files (ADF): Data needed for calibration and geolocation are usually not measured as frequently as the Earth radiances. During the level 0–1b processing the appropriate ADF are extracted and archived. Combining appropriate ADF and Earth measurement data yields a level 1b product.

Since the level 1b data contain all information required for calibration, including the relevant ADF, level 1c products can be generated with the specific software tool *SciaL1c*, which is available to the users. It permits filtering level 1b data and processing these to level 1c, i.e. calibrated radiances.

The operational processing is done in two steps starting with level 0 to level 1b and ending with level 1b to level 2. Each day of SCIAMACHY operations generates 14 or 15 level 1b and level 2 products, one for each orbit. ESA's responsibility for data generation, dissemination and archiving ends with level 2. This does not exclude experienced science users from processing their own scientific products of the same levels. It is up to the individual user how to specify and process products higher than level 2, known as *value-added* (VA) products. In the VA environment, geophysical parameters are often gridded on a global scale, i.e. these products are well suited to provide the interested public with SCIAMACHY's view of the Earth's atmosphere.

8.1 Operational Processing Overview

Operational processing of SCIAMACHY measurement data occurs in the PDS as part of the ENVISAT mission. It covers:

- Facilities to serve the need of generating, archiving, and disseminating data products including the interface to the users, e.g. Processing and Archiving Centres (PAC)
- Software to generate the data products, e.g. data processors and calibration facilities which run within the processing centres
- Data products on level 0, level 1b, and level 2
- Tools for monitoring and controlling the quality of the operational production chain

Basic processing related information is provided via the file names of the products. A filename includes the sequence

- Product ID: 'SCI_NL__0P', 'SCI_NL__1P', 'SCI_NL__2P' and 'SCI_OL__2P' for level 0, 1b or 2 data
- Processing status flag: 'N' for NRT products, letters between 'N' and 'V' for consolidated products
- Originator ID: 'PDK', 'PDE', 'LRA', 'DPA' for PDHS-K, PDHS-E, LRAC or D-PAC
- Start date: year, month and day of measurement start
- Start time: hours, minutes and seconds of measurement start in UTC
- Duration of product coverage in seconds
- ENVISAT mission phase
- ENVISAT cycle number
- Relative orbit number within cycle
- Absolute orbit number
- File counter
- File extension: 'N1' corresponds to ENVISAT

This permits an unambiguous identification of files for a given orbit. A filename example is provided in Fig. 8.1. All ENVISAT files have two headers – the Main Product Header (MPH) that contains information common to all instruments of the platform such as orbit data and the Specific Product Header (SPH) which holds instrument specific information.

Level 1b Product

Additionally to the MPH and SPH the level 1 file contains the Measurement Data Sets (MDS) and the Annotated Data Sets (ADS) with supplemental information needed for, e.g. calibration. Each type of measurement has its own data structure, giving four different MDS:

- Nadir MDS
- Limb MDS
- Occultation MDS
- Monitoring MDS

The atmospheric measurements for nadir, limb and occultation can be found in the corresponding MDS in raw format, i.e. uncorrected binary units. The MDS also contains calibration parameters such as, e.g. size of memory effect, amount of stray light, polarisation parameters, for each measurement. Calibration measurements using the Sun, the Moon or the calibration lamps are stored in the monitoring MDS. Individual dark measurements are not stored in the product thus avoiding too large file sizes. However, for each

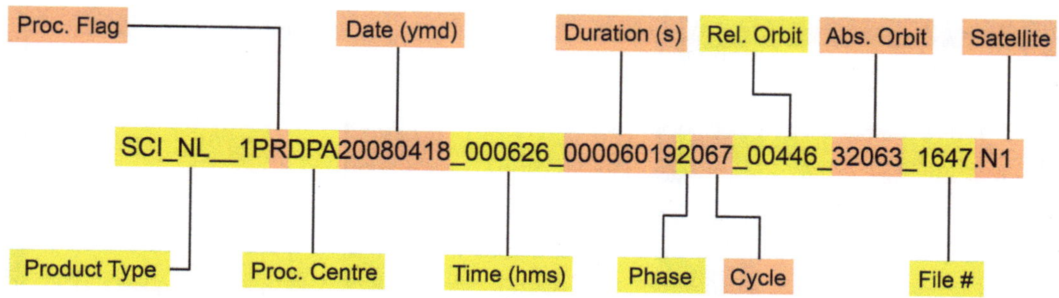

Fig. 8.1 An example for a SCIAMACHY level 1b product filename illustrating the individual filename fields (Courtesy: DLR-IMF).

dark state the average values and standard deviations are available.

The ADS contain the calibration data, quality indicators, geolocation and instrument configuration parameters from the level 0 files. Two types of ADS exist, *global* ADS with data from the auxiliary data files (ADF, see above) and the *newly calculated* ADS: if calibration data are measured in a given orbit, these are stored in the newly calculated ADS. The global ADS are used in the standard calibration of the spectra with the exception of the dark correction in the short-wave infrared (SWIR) channels, for which the dark measurements performed during the same orbit are always used (see also Fig. 8.3). The experienced user also has the possibility to apply the newly calculated ADS, but only if the corresponding measurements were performed in the same orbit.

Level 1c Product

The level 1c file has the same structure as the level 1b file. It contains the data which were filtered and calibrated with the SciaL1c tool. Additionally the applied calibration options are provided in an ADS for the sake of traceability.

Level 2 Product

In the level 2 product four basic types of MDS are available:

- Cloud and aerosol MDS
- Nadir MDS
- Limb MDS
- Limb cloud MDS

The cloud and aerosol MDS combines cloud and aerosol information from nadir measurements. It is followed by multiple MDS corresponding to the different trace gas products retrieved from nadir and limb geometries, one for each trace gas. The limb cloud MDS contains height resolved information about clouds in the limb view. Additionally, the level 2 product contains several ADS with auxiliary information, e.g. data quality and geolocation.

Detailed information about SCIAMACHY's operational data products is provided via ESA's websites for the ENVISAT missions. The product descriptions can be found under http://envisat.esa.int/instruments/sciamachy/data-app/ while product status, including disclaimers, are given via http://envisat.esa.int/dataproducts/availability/. Finally, a listing of the product performance on a daily and bi-monthly basis is provided by http://earth.esa.int/pcs/envisat/sciamachy/reports/.

Processing Chains

Remote sensing data are used for different applications with different requirements. While forecasting requires data availability shortly after sensing, climate research and other longer-term scientific applications require the most accurate data calibration and retrieval. In order to satisfy both requirements, two separate processing chains were set up: the near-realtime (NRT) level 0–1b processing together with the fast delivery (FD) of level 2 products and the offline (OL) level 0–2 processing chain (see Fig. 8.2). While the processors used for both chains are identical, the input data are different. The tight temporal requirements for the NRT chain do not allow to process and inventory all the calibration data, including calculation of attitude corrections. However experience shows that the content of NRT and OL products differs only by small amounts. Main differences between the processing chains comprise:

- Delivery time: NRT level 1b products are delivered a few hours after sensing, FD products within 24 h after sensing. Offline data are delivered typically 1 week after sensing.
- Start/Stop times: OL products contain one orbit always starting at the ascending node crossing (ANX). NRT/FD data can contain data from different parts of the orbit.

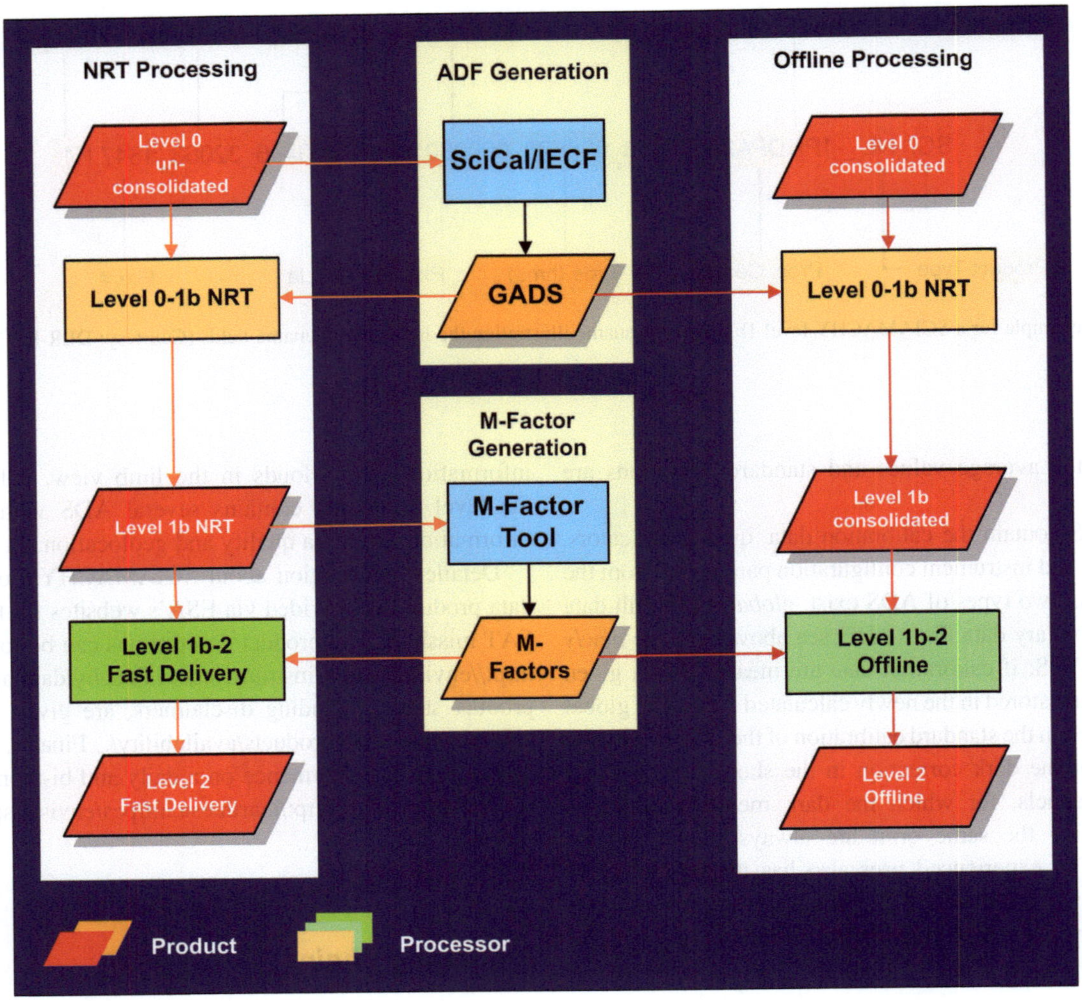

Fig. 8.2 General processing flow in NRT/FD (*left*) and OL (*right*). Both chains receive the calibration information from the same auxiliary flow (Courtesy: DLR-IMF).

- Geolocation: The geolocation differs slightly since the restituted attitude information for the ENVISAT platform is only available a few days after sensing.
- Calibration Data: In the NRT/FD chain the latest available in-flight calibration data are used while in the OL chain the most recent calibration data made before the measurement are applied.

All operational data products are subject to quality monitoring under the responsibility of the Payload Data Control Centre (PDCC). Its goal is to screen all products at the time of generation in order to identify anomalies or deviations from expected results. Quality monitoring includes content and consistency checks, e.g. formal correctness of the product or parameter limits. In the case of detected anomalies, data shall be flagged to initiate further actions. The PDCC executes quality monitoring activities on various timescales ranging from daily to multi-monthly. Level 0 products receive particular attention since they are the basis for all further products. The national agencies providing SCIAMACHY receive copies of the complete level 0 dataset for their own purposes, including instrument monitoring activities. In support of such functions, DLR as one of the instrument providers has generated a master data archive from the consolidated level 0 product stream. This repository, maintained throughout the mission, is free of occasional consolidation flaws; it keeps for each orbit a single consolidated level 0 product which reflects the instrument measurement sequences as best as possible. Whenever the PDS schedules reprocessing campaigns – after major algorithm and processor updates have become operational – the content of the master archive forms the applicable level 0 product basis.

8.2 Operational Level 0–1b Processing

SCIAMACHY level 1b data products comprise geolocated and calibrated radiances of the scientific measurements, as well as measurements for calibration and instrument monitoring. The algorithms used in operational level 0–1b processing are primarily driven by the scientific need to convert measured signals into calibrated radiances. However constraints, particularly imposed by the operational data processing environment or sometimes by instrument operations, may force use of different strategies as would be employed for a ground-based instrument. The wish to obtain a level 1b data product which is not excessively large and complicated imposes an additional constraint. The principle processing cycle starting with instrument level 0 data and ending with the level 1b product is outlined in Fig. 8.2. A major requirement from the ENVISAT PDS architecture rules is that there may be only one output product per processing chain, in this case the level 1b product. As a consequence, the level 1b product must not only hold processed science data but also calibration measurements and instrument monitoring data, as well as newly calculated calibration parameters (see above).

In the operational processing from level 0–1b, all necessary calibration constants for each science measurement are processed from the input calibration data using both ground-based and in-flight measurements. The level 1b data product contains the raw detector signals of the science measurements plus calibration constants, mainly coded as single byte integers. In addition to measurement specific calibration constants, e.g. stray light or atmospheric polarisation, lookup tables are generated for globally applicable calibration constants, e.g. instrument polarisation sensitivity as a function of scan angle. The Instrument Engineering Calibration Facility (IECF) copies the calibration data from the level 1b product into the calibration database. The calibration data files themselves are generated by the *SciCal* tool using level 0 data as an input. In order to allow users processing level 1b products to fully calibrated data, the calibration application tool *SciaL1c* is provided (see above). For the user's convenience certain calibrations can be optionally omitted at extraction, or a subset of data can be filtered out.

Figure 8.3 provides an overview of the steps for level 1b–1c calibration of scientific measurements (nadir, limb and solar occultation) and the used measurement data and ADS. The individual calibration related steps as listed below reflect the calibration characteristics as described in Chapter 5 and specified in detail by Slijkhuis (2004) in the Algorithm Theoretical Baseline Document (ATBD).

Memory Effect and Non-linearity

All channel detector readouts have to be corrected for the memory effect (in channels 1–5) or non-linearity (channels 6–8). The memory effect correction requires the value of the previous readout. If this is not available, e.g. for co-added measurements, an estimate has to be made. In case of co-adding, the value of each readout before co-adding is estimated on the basis of the PMD intensity which is sampled at a higher rate than the detector. Non-linearity correction is implemented similarly for co-added signals. A larger uncertainty in memory effect correction exists for limb measurements: at each new tangent height a reset readout of the detectors with an integration time of 31.25 ms is performed. This reset readout is discarded on-board, i.e. it is not available in the data. Similarly, the memory effect for the very first readout in a state can only be estimated, because before the first readout the detector picks up an unknown signal as the mirrors are driven to their start position. For both cases an estimation of the memory effect is based on the measured signal and state setup parameters.

Dark Signal

The dark signal can be separated into three components:

- Analogue offset or fixed pattern noise: It is added by the electronics to each pixel to avoid negative or near zero signals. Thus no dependence on the thermal environment or exposure times exists.
- Leakage current: It comprises the time dependent part of the dark signal mainly caused by thermally created electron-hole pairs in the detector. Its orbital phase independent component is derived from the orbital measurements on the dark side of the Earth. Different illumination conditions along the orbit with corresponding small thermal perturbations result in an orbital dependent part which is derived as a function of 12 orbit segments from the monthly dark measurements. This variable leakage current is only relevant for the SWIR channels (Fig. 8.4).
- Stray light: Near sunrise, the limb port of the instrument is directly exposed to sunlight leading to a significant amount of spatial stray light. This stray light is characterised during the monthly dark orbits. It is only relevant for orbit phases around sunrise, i.e. at high latitudes over the northern hemisphere. The stray light information is used for monitoring purposes only. For an actual correction of spatial stray light the spatial coverage of the data is insufficient.

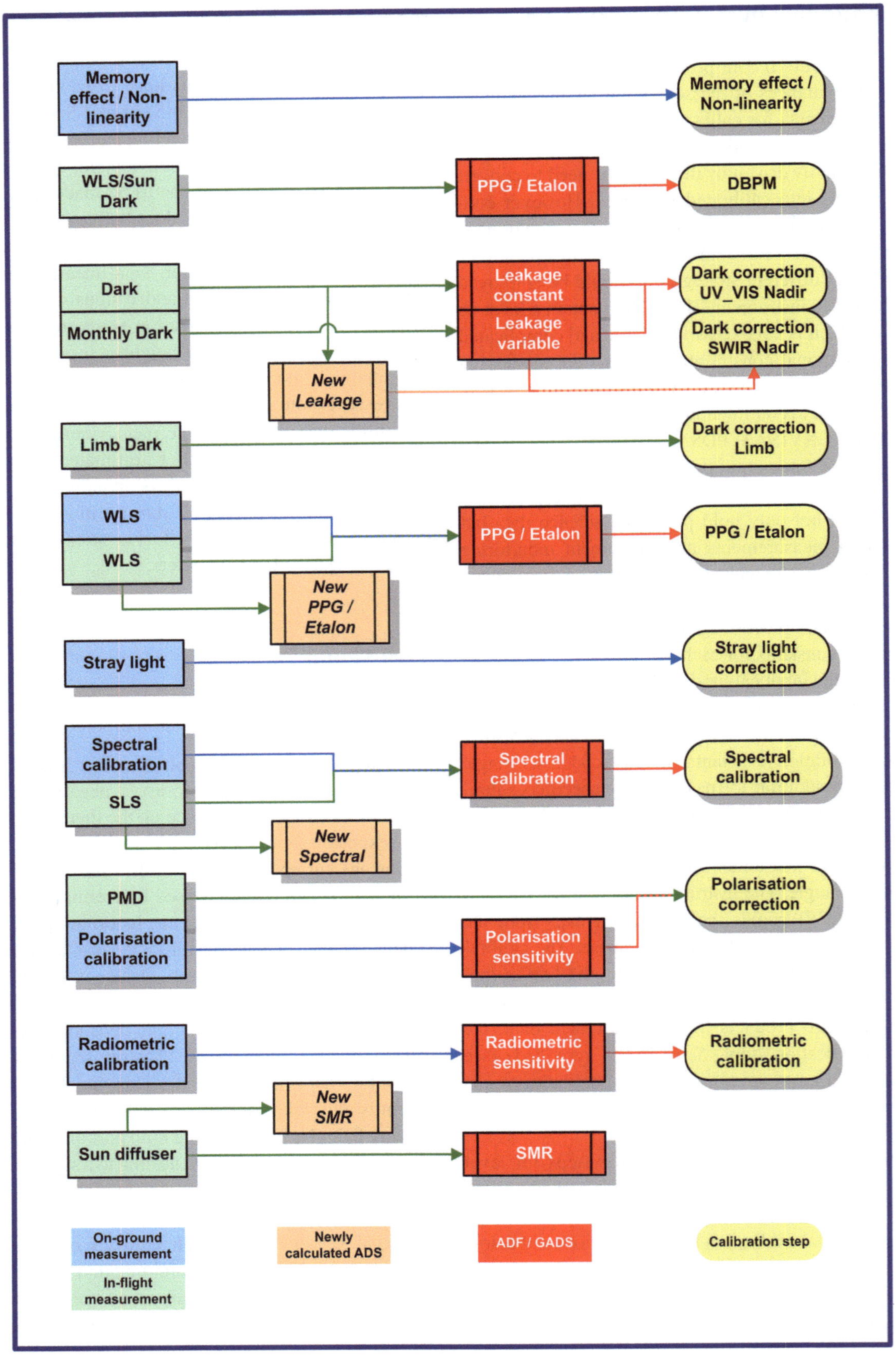

Fig. 8.3 Relations between calibration measurements (*blue: on-ground, green: in-flight*), ADF/GADS (*red*), newly calculated ADS (*orange*) and calibration steps (*yellow*). The calibration information is stored either directly in the newly calculated ADS of the level 1b file or written into an ADF and subsequently copied into the GADS of the level 1b file. Some calibration parameters, e.g. the stray light fractions, are contained in the measurement datasets of the level 1b file. With the exception of the *NEW LEAKAGE* ADS, the newly calculated ADS are not used in the standard calibration (Courtesy: DLR-IMF).

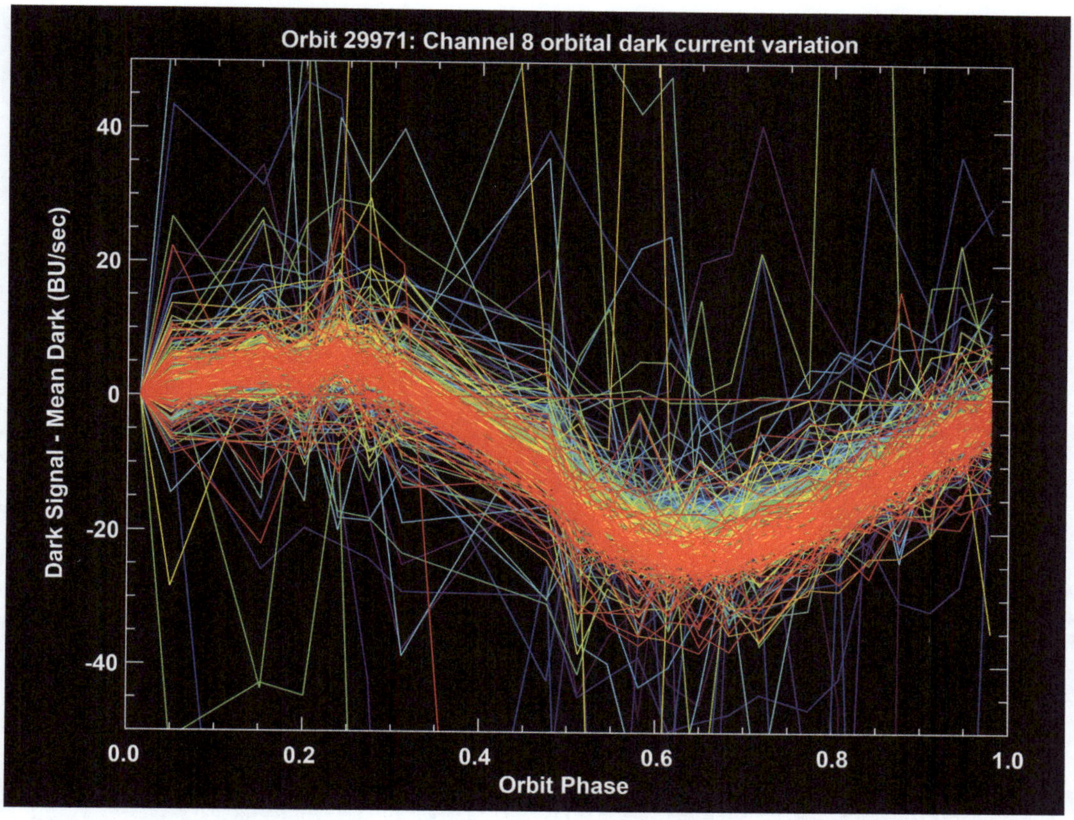

Fig. 8.4 Orbital variation of the thermally induced dark signal for each pixel in channel 8. Orbit phase = 0 corresponds to eclipse entry. The *Mean Dark* is the dark signal per pixel averaged over a complete dark current state close to phase 0. It permits to normalise the orbital variation at phase 0 (Courtesy: DLR-IMF).

All orbital phase independent dark parameters are part of the *LEAKAGE CONSTANT* ADS. The *LEAKAGE VARIABLE* ADS contains the orbital phase dependent components of the dark signal. Both ADS are used in the dark correction for nadir measurements of channels 1–5. After launch it turned out that the original dark signal correction using *LEAKAGE CONSTANT* ADS and *LEAKAGE VARIABLE* ADS had to be adjusted for limb data and the SWIR channels:

- Limb Data: At the end of each limb state the Instantaneous Field of View (IFoV) is set to 250 km and a dedicated dark measurement is executed. This dark measurement is used for the dark correction of limb data.
- SWIR channels 6–8: Since these channels are sensitive to small changes of temperature and the ice layer in channels 7 and 8 leads to changes of the dark signal from one orbit to the next, the dark parameters from the same orbit (from the *LEAKAGE NEW* ADS*)* are used. The orbit phase dependence is taken into account by additionally exploiting data from the *LEAKAGE VARIABLE* ADS.

Generally, no calibration measurements are used which fall into the South Atlantic Anomaly (SAA, Fig. 8.5). This is to reduce contamination of the spectrum from particle radiation which shows as spikes on the detector signal. For orbital dark measurements this rule would imply that for many orbits per day the dark signal is not updated. Therefore, measurements from the SAA are incorporated into the orbital dark calibration for channels 6–8, even though this may imply a small degradation in quality. A 'hot pixel detection algorithm', which analyses a sequence of measurements for each detector pixel to detect and reject spikes in the data, will filter out large deviations. The individual dark measurements are not stored on the level 1b product, but state averaged mean values (after removal of spikes) and their standard deviations are stored in the *DARK AVERAGE* ADS.

Pixel-to-Pixel Gain and Etalon

The Pixel-to-Pixel Gain (PPG) correction aims at correcting relative gain changes with time. It is monitored in-flight using the White Light Source (WLS) measurements. However, observations show that the PPG does not change significantly in-orbit, thus the correction is currently not applied.

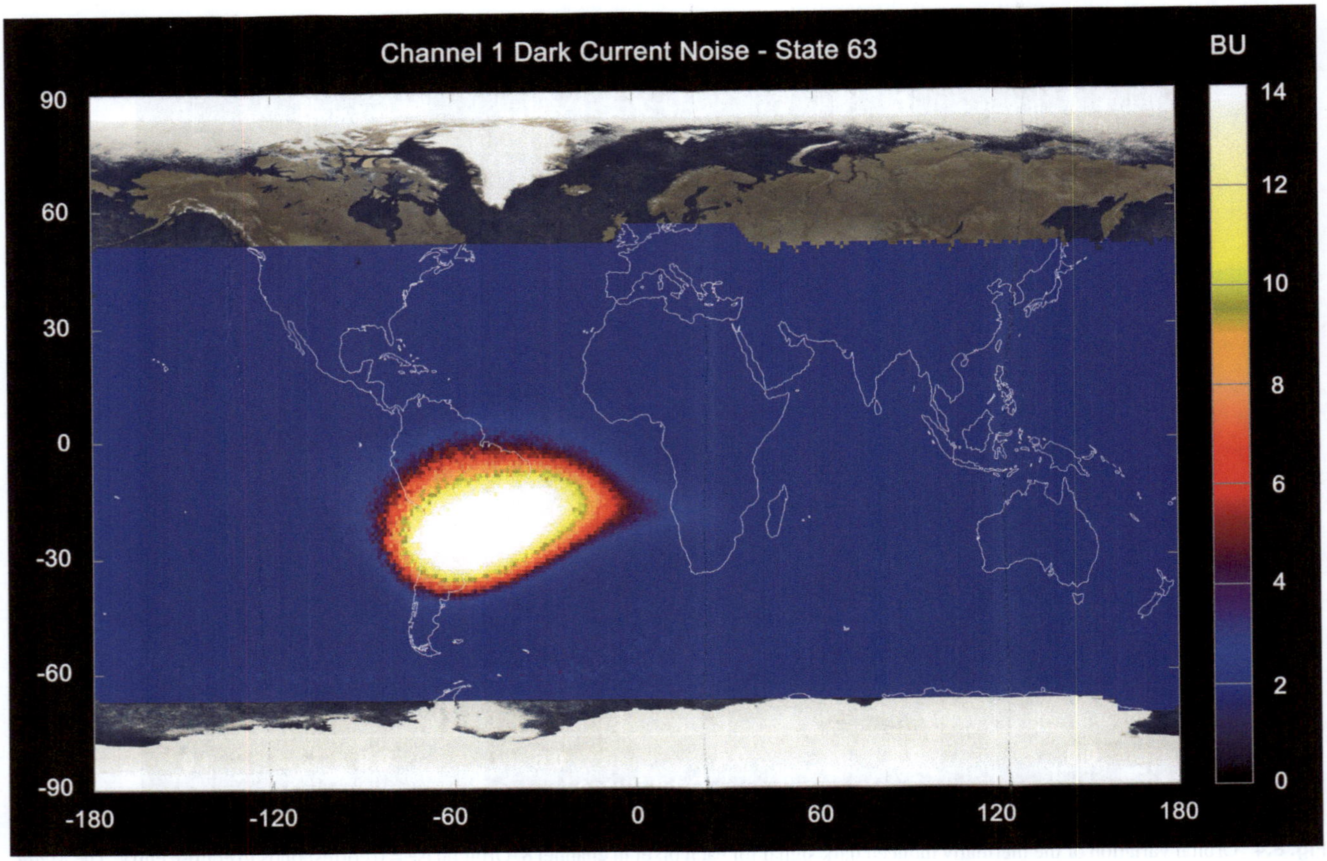

Fig. 8.5 The SAA as derived from SCIAMACHY dark current measurements (Courtesy: SRON).

The detector etalon effect is caused by multiple reflections at the boundaries of the protective silicon oxide layer on top of the light detecting silicon. It leads to interference pattern in the measured spectrum. If material is deposited on top of the detector, the optical path and thus also the interference pattern changes compared to the situation during the on-ground calibration. This change is corrected using data from the weekly WLS measurements and is only relevant for channels 1–5. Both PPG correction and the detector etalon correction are stored in the *PPG ETALON* ADS.

Dead and Bad Pixel Mask

The SWIR channels 6+, 7 and 8 suffer from a rising number of bad or even dead pixels which are not – or only to a small extent – suitable for retrieval. The reason is a lattice constant mismatch between the substrate material and the light detecting material of the detectors (Kleipool et al. 2007). Such dead and bad pixels are detected using dark, WLS and Sun measurements. Several thresholds were determined to separate dead and bad from good pixels. The number of dead and bad pixels is not constant (Fig. 8.6). It increases with in-orbit operations time due to proton impacts. Therefore a dead and bad pixel mask is calculated for each orbit and stored in the *PPG ETALON* ADS.

Spectral Stray Light

The corrections for spectral stray light are derived from each science measurement using calibration constants measured on-ground. Contrary to the situation on-ground, this correction has to cope with the clustered readout of the detectors (see section "Optical Assembly" of Chapter 3) in flight. The stray light at one detector pixel is the sum of stray light originating from wavelengths corresponding to all other detector pixels in a channel (Fig. 8.7), sometimes even from a neighbouring channel. If only part of the channel, i.e. one or more clusters, is read out, the intensity at other wavelengths is not known and must be estimated. For level 0–1b processing, the stray light calculation algorithm is applied after the spectrum covering the complete SCIAMACHY wavelength range has been estimated. Such a spectrum is calculated for each integration time (IT) in the state. After each longest IT, a complete spectrum has been measured.

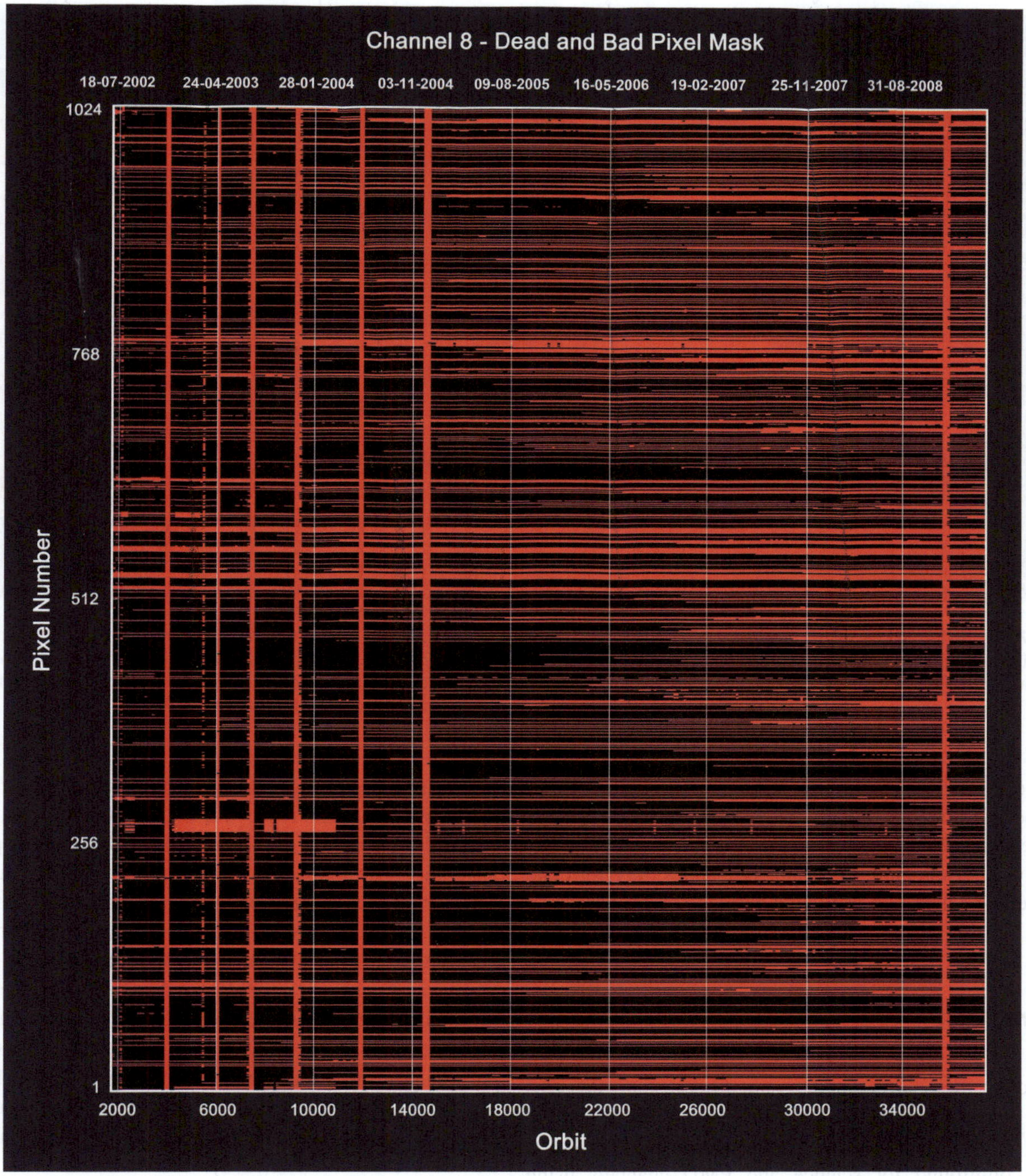

Fig. 8.6 DBPM for channel 8 between August 2002 and March 2009. Each row corresponds to a single pixel. When a pixel turns red it becomes either dead or bad. The vertical stripes correspond to decontamination periods (Courtesy: DLR-IMF).

For shorter IT, intensity from missing clusters is estimated by assuming that the spectrum for shorter IT may only have a different intensity but retains the spectral shape of the longest integration time measurement, i.e. a complete spectrum for each IT is estimated by scaling the complete spectrum at the longest IT to the shorter IT. The stray light correction values are stored directly in each MDS of the level 1b product.

Wavelength Calibration

The wavelength calibration uses on-ground and in-flight data for channels 1–6. From weekly and monthly in-flight measurements of the spectral line source, the difference to the on-ground calibration is determined. The line positions are derived using the Falk algorithm (Falk 1984). Subsequently

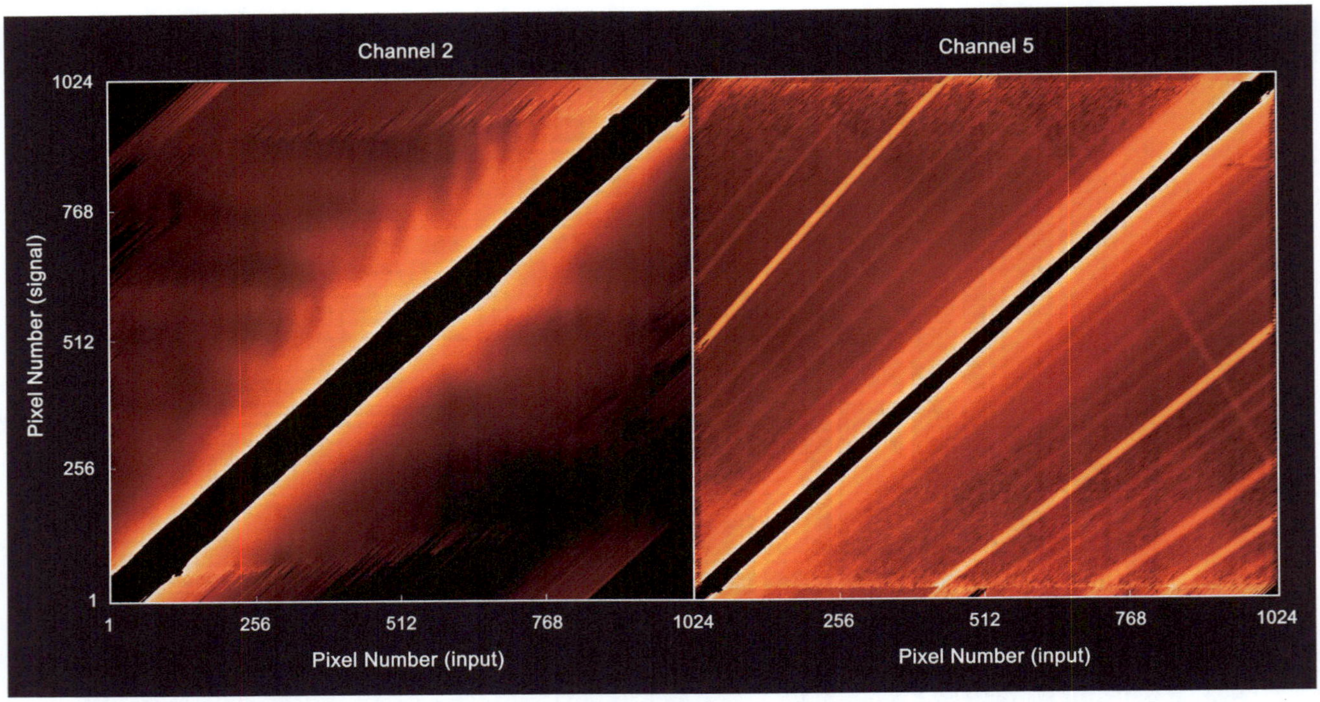

Fig. 8.7 Spectral stray light matrix (1,024 × 1,024) in channel 2 (*left*) and channel 5 (*right*). Each row shows the stray light spectrum (signal) due to unit input in the corresponding pixel (horizontal axis) with bright regions displaying the strongest stray light contribution. The region around the diagonal is defined as being free from stray light. The lines parallel and normal to the diagonal in the plot for channel 5 are reflected and refocused ghosts. These are absent in channel 2, which has a holographic instead of a classically ruled grating (Courtesy: DLR-IMF).

a wavelength for each pixel is stored in the product. Since the spectral line source does not have enough lines in the infrared and dead and bad pixels interfere with the line position determination, for channels 7 and 8 the on-ground wavelength calibration is applied without any changes.

Polarisation

The polarisation correction is required because the detected signal depends on the polarisation state of the incoming light. The instrument's polarisation response was determined on-ground for science and PMD channels using polarised calibration sources and is stored in the calibration Key Data. They contain the ratio of the PMD signal to the science channel signal for several instrument configurations and polarisation directions. As science and PMD channels are performing polarisation sensitive measurements in flight, from the obtained signals the elements Q and U of the Stokes vector are derived. By using the polarisation information from the Key Data a linear polarisation correction factor is calculated for all wavelengths. A complication arises from the fact that the PMD channels are sampled independently and at a different rate from the channel detectors. Using the time stamps of both types of measurements, the PMD channels are resampled on a time grid synchronised with the detector readout, at twice the BCPS frequency with the BCPS corresponding to the maximum detector data packet generation rate of 62.5 ms. For channels 1–5 the time of detector readout depends on the pixel number, as each pixel is read out sequentially. The synchronisation is roughly performed for the part of the detector corresponding to the mean wavelength of each PMD bandpass. All data necessary to calculate the polarisation correction factors are contained in the level 1b product. The polarisation sensitivities from on-ground measurement are stored in the *POLARISATION SENSITIVITY* ADS for nadir, limb and occultation while the polarisation fractions and the integrated PMD values for each observation are directly stored in the MDS. Raw PMD measurements – which are not further used in operational processing – are saved in the *PMD* ADS.

Radiometric Correction

The final calibration step concerns the radiometric calibration, i.e. applying the radiometric instrument response to the measured and corrected signals. Globally valid radiance calibration constants are taken from the calibration Key Data measured on-ground. The radiometric sensitivity for each observation is stored directly in the MDS of the level 1b product.

Sun Mean Reference Spectrum

For the purpose of trace gas retrieval in level 1b–2 processing based on reflectances, the level 0–1b processing generates for each state of solar observations a Sun Mean Reference (SMR) spectrum. Absolute irradiance calibration on-ground was only performed for the ESM diffuser mode. The irradiance calibration of all other solar observation modes is based on a rough estimate which leads to an incorrect absolute irradiance level and may leave residual instrument polarisation features in the spectrum. However, test trace gas retrievals showed good results with radiometrically uncalibrated ASM diffuser spectra. Therefore it was decided to generate two SMR for each diffuser state: one with all calibrations applied and one without radiometric calibration. For monitoring purposes, SMR are also generated for the other solar observations, i.e. occultation and monitoring. All SMR are stored in the *SUN MEAN REFERENCE* ADS.

Degradation

Although correcting the degradation with the m-factors is implemented in the level 0–1b processing, it is not applied operationally at present. However the *SciaL1c* tool provides the capability for applying the radiometric degradation correction and thus producing degradation corrected level 1c data. Because the m-factors impact the science and PMD channels differently the concept for calculating and providing them pursues a two-step approach. In a simple solution only the readouts of the science channels are corrected which comprises only multiplicative calculations. Thus their application can be switched 'on' and 'off' as needed. The current baseline is to apply the simple correction during operational level 1b–2 processing for the products requiring radiometrically calibrated data (such as the Absorbing Aerosol Index). The full solution would include correcting the PMD readouts as well, a processor feature which is presently not activated. Since the calculation of m-factors is decoupled from the operational level 0–1b processing, updates of the external m-factor database are possible without interfering with operational PDS requirements. While all auxiliary data necessary to generate a calibrated level 1c product are stored in the level 1b product itself, the m-factors can be accessed via a separate location at http://www.iup.uni-bremen.de/sciamachy/mfactors/.

Here they are generated and maintained as part of the SCIAMACHY optical performance monitoring.

8.3 Operational Level 1b–2 Data Processing

The goal of level 1b–2 data processing is to provide geophysical parameters such as column densities and profiles from atmospheric constituents as well as cloud and aerosol parameters. These data are given in the MDS of the level 2 product in combination with geolocation and additional auxiliary information (state geolocation, quality flagging, etc.) in the appropriate ADS.

The operational level 1b–2 data processing occurs in OL mode and fast delivery mode. Both chains use the same processor, i.e. SCIAMACHY Ground Processor (SGP) L12 OL V5. Since the level 2 products in the fast delivery chain are based on NRT level 1b data they cannot utilise the most accurate auxiliary and in-flight calibration data. The operational level 1b–2 product suite is comprised of retrievals from the nadir and limb observation geometry. Operational processing of occultation measurements is currently not foreseen, but the existing system can be extended to cope with it at a later stage. Table 8.1 summarises the level 2 product content.

Figure 8.8 (nadir DOAS) and Fig. 8.9 (limb) outline the general processing chain for the level 1b–2 data processors. It is implemented as a multi-processor Linux cluster system. Since scientific progress in retrieval techniques continues, an outline of level 1b–2 algorithms can only provide a snapshot. The reader is referred to the ATBD (Lichtenberg et al. 2010) for up-to-date algorithm information. Particularly experience gained from scientific product generation will be reflected in the operational chain stepwise, i.e. algorithm improvements that are validated will be introduced into the processor. Level 2 products will be generated anew whenever significant algorithm improvements are implemented into the processor. Even beyond the in-orbit mission lifetime, an ambitious atmospheric science project like SCIAMACHY requires continuing retrieval algorithm research and reprocessing.

Column Densities from Nadir Observations

UV-VIS data processing uses the DOAS approach for the retrieval of slant column densities from launch onwards. The DOAS concept, together with the derivation of vertical column densities is described in section "Nadir Trace Gas Retrieval Schemes" of Chapter 7. The operational DOAS algorithm is based on the approach originally implemented for the GOME data processor (GDP, Thomas and Spurr 1999). Since DOAS algorithm development undergoes per-

Table 8.1 Operational SCIAMACHY level 2 product suite

Spectral range	Nadir (FD& OL)		Limb (FD&OL)
UV-VIS-NIR	O_3 (vertical column)[a]	HCHO (vertical column)[b]	O_3 (profile)
	NO_2 (vertical column)[a]	Cloud cover	NO_2 (profile)
	BrO (vertical column)	Cloud optical thickness	BrO (profile)
	SO_2 (vertical column for anthropogenic and volcanic scenario)	Cloud top height	
	OClO (slant column)	AAI	
	H_2O (vertical column)	CHOCHO (vertical column)[b]	
SWIR	CO (vertical column)	CH_4 (vertical column)[b]	Cloud parameters

[a] Constitute METEO products available for the meteorological community.
[b] Planned for the next processor version (SGP V.6).

manent improvements – currently GDP 4.0 is operational (Spurr et al. 2004; Van Roozendael et al. 2006) – also SCIAMACHY's processor SGP L12 follows these changes (Lerot et al. 2009). For O_3 the GDP 4.0 implementation yields vertical columns which are derived in an iterative way by successively updating the air mass factor calculated using the vertical column provided by the previous iteration. It replaces the early standard DOAS approach where a distinction between AMF extraction, slant column density derivation, and the vertical column density calculation was made. Other UV-VIS products of the weak absorbing gases can rely on this standard method while the water vapour retrieval is based instead on a further DOAS variant, the AMC-DOAS approach (Noël et al. 1999; Noël et al. 2004).

In the SWIR range the cross sections of line absorbers strongly depend on pressure, temperature and wavelength. Therefore, retrieval algorithms from the UV-VIS-NIR cannot be applied here. The BIRRA (Beer InfraRed Retrieval Algorithm, Schreier et al. 2009) retrieval scheme, which takes into account the characteristics of SWIR absorption spectroscopy, handles the retrieval of CO. Further trace gases to be determined from the SWIR range in the near future shall rely on BIRRA as well. BIRRA performs a nonlinear least squares fit of molecular vertical column densities together with some auxiliary parameters. When fitting the observed near infrared intensity/radiance spectrum, being essentially described by Beer's law with the Sun as source, the molecular absorption features are modelled by means of line-by-line calculations.

Cloud and Aerosol Parameters in the UV-VIS-NIR for Nadir Observations

The parameter cloud fraction is determined by the threshold algorithm OCRA (Loyola 2000). PMD measurements identified to be free of clouds from the red, green, and blue PMD are stored as RGB composite. The distance between the white point in the RGB reflectance space (= fully clouded) and the actually determined PMD reflectance is taken to determine the cloud fraction. Heuristically derived scaling factors allow the appropriate representation of cloud fraction between zero (cloud-free) and one (fully cloudy). In the initial processor versions cloud top height or cloud top pressure was extracted from the ISCCP (International Satellite Cloud Climatology Project) database but it is replaced from version 3.0 onwards by an algorithm based on evaluation of the O_2 A-band in the visible spectral region (SACURA, Rozanov and Kokhanovsky 2004). This yields improved cloud top height and cloud optical thickness parameters since in the previous implementation only one value for optical thickness served as fixed input.

The Absorbing Aerosol Index (AAI) relies on the ratio of the reflectances between a spectral band which is not covered by spectral absorptions and a spectral band with ozone absorptions. The AAI is calculated from the difference between logarithmic ratios of measured and modelled reflectances. It indicates the presence of absorbing aerosols and depends strongly on an accurate calibration of the reflectance (Tilstra et al. 2007).

Profiles from Limb Observations

Retrieving trace gas profiles from limb measurements is a challenging task. The algorithm currently used for limb retrievals, the Iteratively Regularized Gauss–Newton Method (Doicu et al. 2002; Doicu 2005 – see also Section "Application of Inversion Theory to Limb Retrieval" of Chapter 7), consists of mainly two parts, a forward model calculation followed by the inversion, using a Tikhonov regularisation with a variable regularisation parameter. In the forward model calculation the sequence of steps includes

- Calculation of the geometrical parameters describing the limb scans
- Calculation of the optical properties for each layer

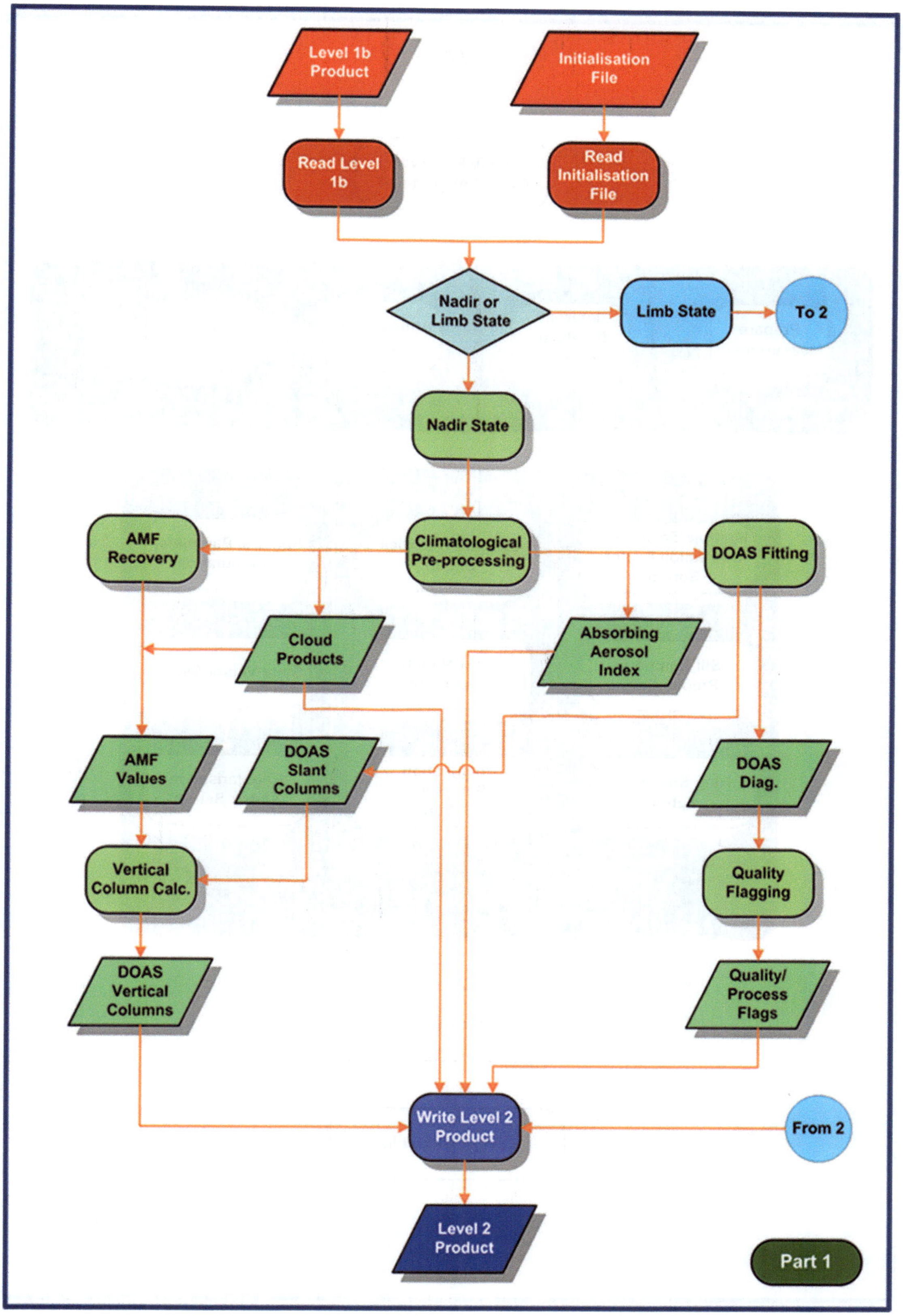

Fig. 8.8 Level 1b–2 data processing flow diagram illustrating the preprocessing and nadir DOAS algorithm. Limb algorithms are continued in Fig. 8.9 (Courtesy: DLR-IMF).

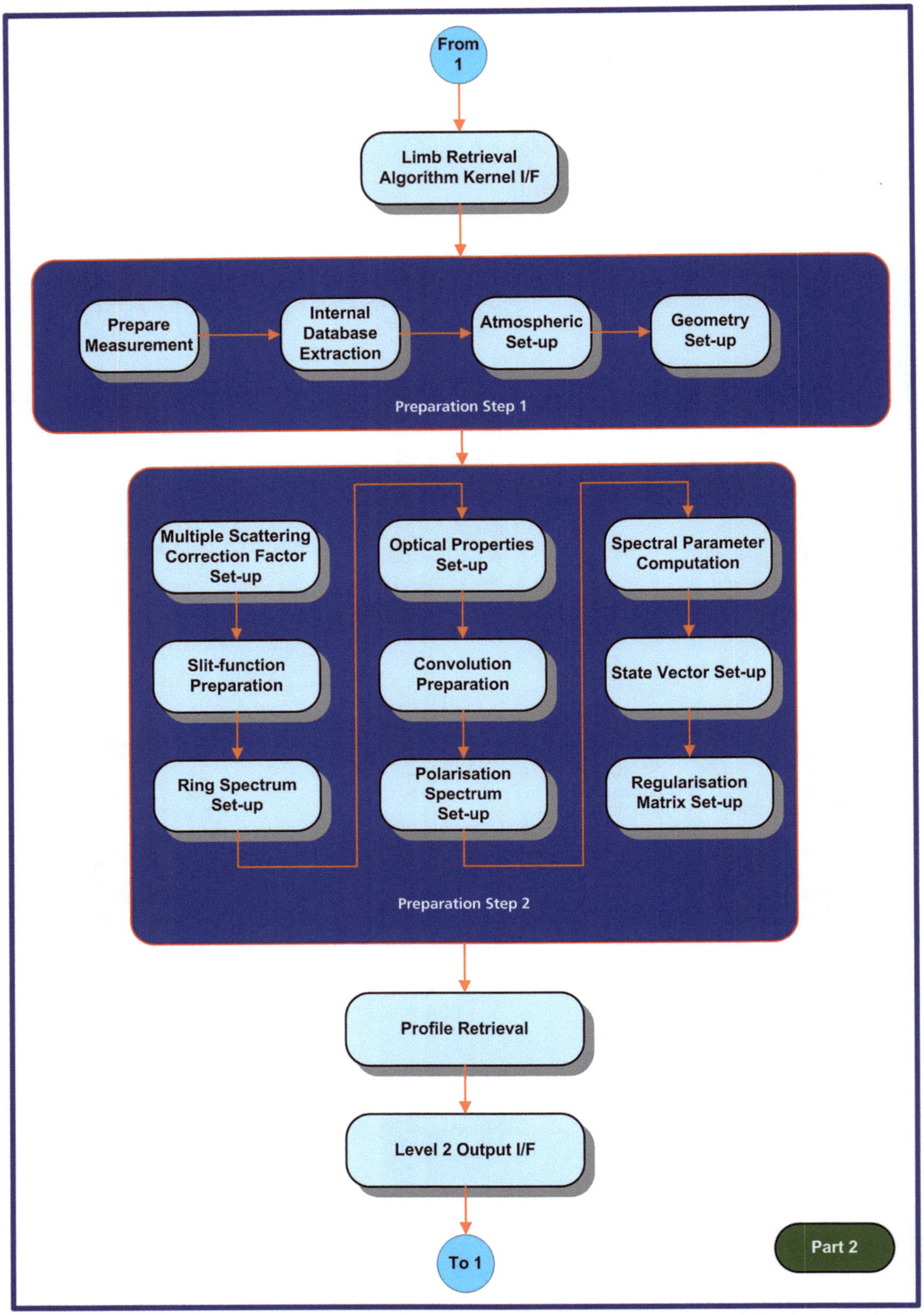

Fig. 8.9 Level 1b–2 data processing flow diagram for limb profile retrieval. At the end of the limb processing, the data flow returns to Fig. 8.8 for product output (Courtesy: DLR-IMF).

- Computation of the radiation field in the framework of the independent pixel approximation
- Calculation of the multiple scattering correction factors by means of a spherical radiative transfer model using Picard iteration with open boundaries

The profiles are calculated from a predefined minimum altitude up to 100 km. For each state a maximum of four profiles in East–West direction is determined depending on the integration time of the state. Figures 8.8 and 8.9 depict the s/w architecture of the limb retrieval algorithm with two preparative steps before the profile retrieval itself.

Limb Cloud Detection

In limb mode SCIAMACHY measures light scattered along the line-of-sight. If the line-of-sight intersects a cloud at a certain height, the spectrally resolved measurements differ from cloud-free measurements. To eliminate systematic uncertainties and to enhance the sensitivity of the cloud determination, differences and/or ratios of spectral measurements in a certain wavelength region are exploited. Actually radiance ratios at 750 and 1090 nm are used for the detection of clouds. In an ideal Rayleigh scattering atmosphere populated only by molecules, radiances at two wavelengths about 300 nm apart differ considerably while for larger particles like cloud droplets this difference is reduced. The cloud top height (CTH) above the limb tangent point is then derived by analysing light from the cloud layer and from above.

8.4 Scientific and Value-Added Products

Scientific Products

As outlined above, scientific products are a vital service of the mission. Although scientific products usually refer to level 2, scientific level 1c data are also generated from operational level 1b or level 0 data by applying calibration steps differing from what is done in the operational ground segment. Since scientific product generation does not necessarily follow the stringent requirements of the operational chains – which does not preclude them from being as precise as possible and even permitting the utilisation of the most accurate and evolved algorithms – they may be implemented in various ways. Therefore a generic description of scientific processor architectures similar to the operational ones is not applicable. Table 8.2 presents a listing of geophysical parameters available as scientific products. Details can be found on SCIAVALIG's website under http://www.sciamachy.org/products/ with successive species or parameter dependent links.

Value-Added Products

The value-added (VA) products are generated on a routine basis for various applications. Requirements for value-adding originate from very different user groups. Whereas scientific applications – if not based on level 2 data – usually need data sets on regular grids with complete information on pixelwise data quality, applications for governmental monitoring duties, public services or education purposes ask for synoptic maps, i.e. overviews at one point in time, on different scales without any gaps and for long time series of continuous observations. In some cases, NRT data provision is also requested to initialise forecast services.

VA products are based on the operational products generated within the PDS or on scientific products. In most cases value adding transforms level 2 products from their satellite projection and the time of the satellite overpasses to integrated datasets (level 3 and level 4) which contain a geophysical parameter as observed by one instrument. Averaging level 2 data over a certain period and grid cell generates level 3 products yielding daily, weekly, monthly or annual mean datasets. Level 4 products are based on the combination of satellite observations (level 2) and atmo-

Table 8.2 Scientific SCIAMACHY level 2 product suite (due to the nature of scientific products, the table is evolving and does not claim to be complete)

Geophysical parameters			
O_3	Vertical column, profile stratosphere and mesosphere	IO	Slant column
NO_2	Slant/vertical/tropospheric column, profile	H_2O	Vertical column
BrO	Slant/vertical column, profile	HDO/H_2O	
SO_2	Slant/vertical column	CO	Vertical column
HCHO	Vertical column	CH_4	Vertical column
CHOCHO	Vertical column	CO_2	Vertical column
OClO	Slant column, profile	$T_{mesopause}$	Mesopause temperature
Clouds	Cover, pressure, top height, optical thickness, liquid water path, phase index, top-of-atmosphere (TOA) reflectance, droplet effective radius, PSC index, NLC index		
Aerosol	AAI, AOT		
UV	Index, dose		

spheric models such as, e.g. meteorological circulation and chemistry. In addition, synergistic value-adding is conducted by either combining level 2 products of several sensors into one comprehensive level 3 and level 4 data set or by exploiting level 1 products (top-of-atmosphere radiances or spectra) from different sensors to retrieve a level 2 product with new information.

A wide set of requirements on atmosphere-related monitoring and forecasting are targeted within the ESA GMES service element project MACC, previously PROMOTE, where atmospheric satellite observations are exploited. Topics are, e.g. distribution and trends of the stratospheric ozone layer, UV irradiance at the surface, near-surface air quality, greenhouse gases and climate. Application fields range from global and European protocol monitoring (Montreal, Kyoto, CLRTAP – Convention on Long-range Transboundary Air Pollution) to public information services such as skin exposure sunburn time warning, street level air quality forecasts or flight route warning of dust/volcanic ash events. Further currently developed applications include services to support planning and management of solar power plants (ESA market development project ENVISOLAR – Environmental Information Services for Solar Energy Industries) or to improve the quality of satellite land images by providing an automatic atmospheric correction with actual information on the state of the atmosphere. In all these application areas, quality information is essential to improve confidence in the VA products. In the future the use of value-adding will most likely increase as new

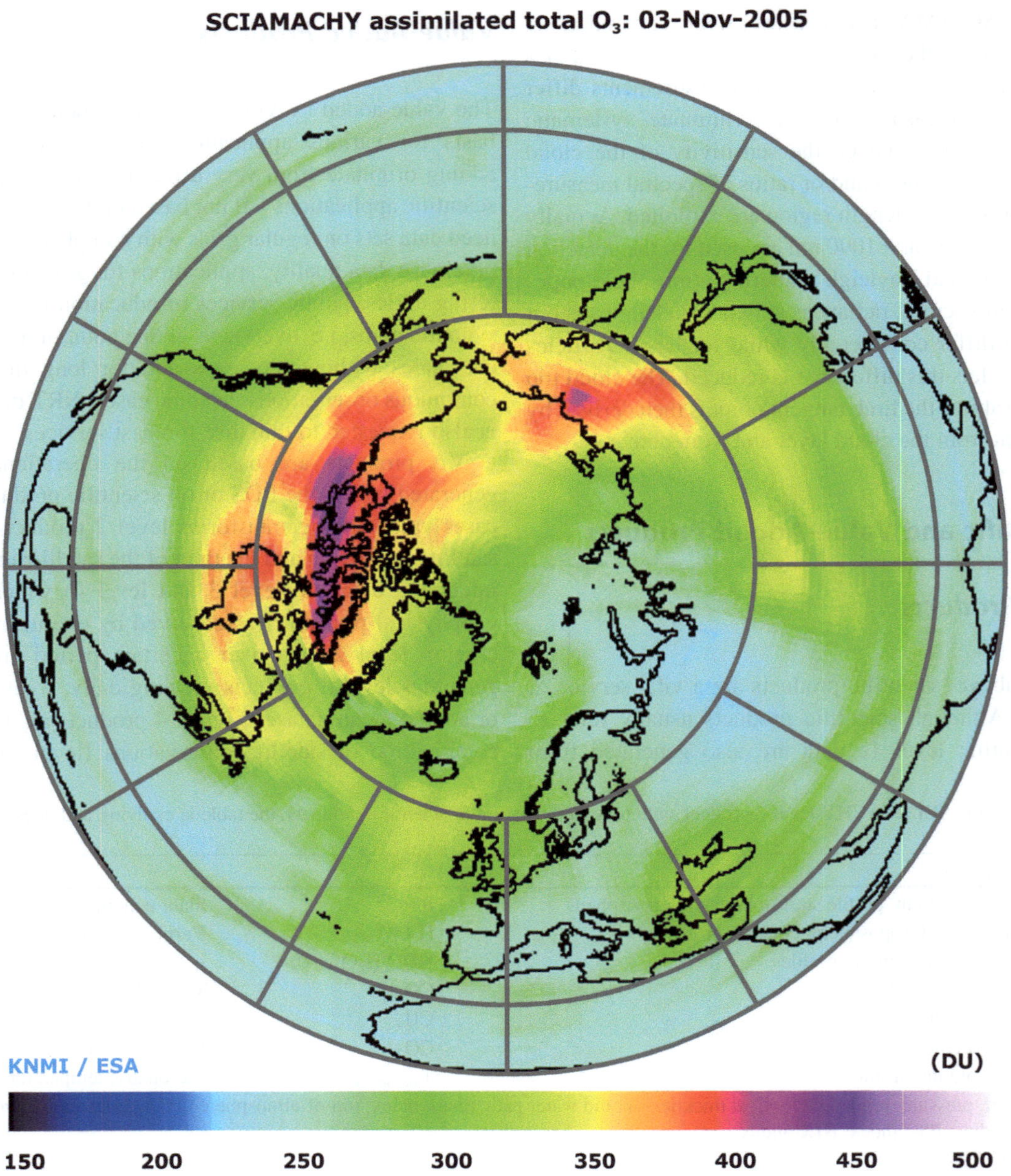

Fig. 8.10 A forecasted North Pole view of the assimilated total ozone column field for 3 November 2005 at 12:00 UTC based on SCIAMACHY data (Courtesy: KNMI/ESA).

requirements call for integration of all observations – satellite, aircraft, balloons, ground – together with models to extract the best possible and comprehensive knowledge on the state of the atmosphere. For more details the reader is referred to the IGACO (Integrated Global Atmospheric Chemistry Observations) theme strategy.

One key tool for achieving the goals described above is the use of data assimilation techniques which provide a mathematically consistent technology to integrate measurements and models together with their statistical errors. Data assimilation generates synoptic trace gas fields from asynoptic spaceborne measurements. It enables the derivation of interpolated concentration fields as well as information about transport mechanisms.

A major application of assimilation schemes to SCIAMACHY data concern the combination of stratospheric modelling and nadir column data. This is ideally suited for applications such as scientific studies of the evolution of the ozone layer and of special events (e.g. ozone hole or low ozone episodes) or inter-comparisons with models (e.g. study of dynamical and chemical processes). Since assimilated fields are globally available, comparison with independent observations can be performed without space/time mismatches. Stratospheric O_3 is also of particular interest as it can be assimilated into operational weather forecasts, thus improving the model representation of stratospheric wind fields and thereby the quality of the forecast. Since forecast services are provided on short timescales, the fast availability of O_3 data – via the operational fast delivery service or NRT scientific products – is essential (Eskes et al. 2005). Such ozone columns serve as an input for the assimilation analysis and a subsequent forecast of how the stratospheric ozone layer will develop in the upcoming nine days. The assimilation yields a complete picture of the global ozone distribution (Fig. 8.10). Within the error margins of both, model and observations, it is consistent with observations and our knowledge of atmospheric transport and chemistry (Eskes et al. 2002; Eskes et al. 2003). SCIAMACHY ozone columns are currently also assimilated operationally in the numerical weather prediction model of the ECMWF. Several centres use the ozone forecasts for UV radiation predictions.

References

Doicu A, Schreier F, Hess M (2002) Iteratively regularized Gauss–Newton method for atmospheric remote sensing. Comput Phys Commun 148:214–226

Doicu A (2005) SCIAMACHY Level 1b to 2 OFL Processing algorithm change description document limb profile retrieval algorithm (ENV-TN-DLR-SCIA-0062). Technical Document, DLR

Eskes HJ, van Velthoven PFJ, Kelder HM (2002) Global ozone forecasting based on ERS-2 GOME observations. Atmos Chem Phys 2:271–278

Eskes HJ, van Velthoven PFJ, Valks PJM, Kelder HM (2003) Assimilation of GOME total ozone satellite observations in a three-dimensional tracer transport model. Quart J R Meteorol Soc 129:1663–1681

Eskes HJ, van der A RJ, Brinksma EJ, Veefkind JP, de Haan JF, Valks PJM (2005) Retrieval and validation of ozone columns derived from measurements of SCIAMACHY on Envisat. Atmos Chem Phys Discuss 5:4429–4475

Falk WR (1984) Data reduction from experimental histograms. Nucl Instrum Methods Phys Res 220:473–478

Kleipool QL, Jongma RT, Gloudemans AMS, Schrijver H, Lichtenberg GF, van Hees RM, Maurellis AN, Hoogeveen RWM (2007) Inflight proton-induced radiation damage to SCIAMACHY's extended-wavelength InGaAs near-infrared detectors. Infrared Phys Technol 50:30–37. doi:10.1016/j.infrared.2006.08.001

Lerot C, Van Roozendael M, van Geffen J, van Gent J, Fayt C, Spurr R, Lichtenberg G, von Bargen A (2009) Six years of total ozone column measurements from SCIAMACHY nadir observations. Atmos Meas Tech 2:87–98

Lichtenberg G, Bovensmann H, Van Roozendael M, Doicu A, Eichmann K-U, Hess M, Hrechanyy S, Kokhanovsky A, Lerot C, Noël S, Richter A, Rozanov A, Schreier F, and Tilstra LG (2010) SCIAMACHY Offline level 1b-2 processor algorithm theoretical baseline document (ENV-ATB-QWG-SCIA-0085). Technical Document, SQWG/DLR-IMF. Available at http://atmos.caf.dlr.de/sciamachy/documentation.html

Loyola D (2000) Cloud retrieval for SCIAMACHY, Proceedings of ERS-ENVISAT symposium, Gothenburg, Sweden, ESA-SP-641

Noël S, Buchwitz M, Bovensmann H, Hoogen R, Burrows JP (1999) Atmospheric water vapor amounts retrieved from GOME satellite data. Geophys Res Lett 26(13):1841–1844

Noël S, Buchwitz M, Burrows JP (2004) First retrieval of global water vapour column amounts from SCIAMACHY measurements. Atmos Chem Phys 4:111–125

Rozanov VV, Kokhanovsky AA (2004) The semi-analytical cloud retrieval algorithm as applied to the cloud top altitude and the cloud geometrical thickness determination from the top of atmosphere reflectance measurements in the oxygen absorption bands. J Geophys Res 109:D05202. doi:10.1029/2003JD004104

Schreier F, Gimeno-Garcia S, Hess M, Doicu A, Lichtenberg G (2009) Carbon monoxide vertical column density retrieval from SCIAMACHY infrared nadir observations. In: Nakajiama T, Yamasoe MA (eds) IRS 2008: Current problems in atmospheric radiation. American Institute of Physics. doi:10.1063/1.3116983

Slijkhuis S (2004) ENVISAT-1 SCIAMACHY Level 0 to 1c processing, algorithm technical basis document (ENV-ATB-DLR-SCIA-0041). Technical Document, DLR. Available at http://atmos.caf.dlr.de/sciamachy/documentation.html

Spurr RJD, Van Roozendael M, Loyola DG (2004) Algorithm theoretical basis document for gome total column densities of ozone and nitrogen dioxide (ERSE-DTEX-EOPG-TN-04-0007). Technical Document. Available at http://atmos.caf.dlr.de/gome/documentation.html

Thomas W, Spurr RJD (1999) GOME Level 1 to 2 Algorithms description (ER-TN-DLR-GO-0025). Technical Document, DLR

Tilstra LG, De Graaf M, Noël S, Aben I, Stammes P (2007) Sciamachy's absorbing aerosol index and the consequences of instrument degradation. Proceedings of 3rd workshop on the atmospheric chemistry validation of ENVISAT (ACVE-3), Frascati, Italy, ESA-SP-642

Van Roozendael M, Spurr RJP, Lerot C, Loyola D, Schröder T, von Bargen A (2006) SCIAMACHY Level 1b-2 off-line data processing: Algorithm theoretical baseline document for trace gas retrieval from UV/VIS Nadir Spectra (ENV-ATB-BIR-SCIA-0074). Technical Document, BIRA-IASB

Chapter 9
Validation

J.-C. Lambert, A. Piters, A. Richter, S. Mieruch, H. Bovensmann, M. Buchwitz, and A. Friker

Abstract Satellite validation has to ensure that geophysical quantities derived from in-orbit radiometric measurements meet quality requirements for the intended scientific studies and applications. It is not a 'once a mission lifetime' task, but requires regular proof measurements throughout the in-orbit phase. An extensive SCIAMACHY validation programme has been developed jointly by Germany, The Netherlands and Belgium, with the support of ESA and a large number of international partners. Due to its status as an AO instrument SCIAMACHY is embedded in two collaborating validation structures, ESA's Atmospheric Chemistry Validation Team and the SCIAVALIG, a subgroup of the SCIAMACHY Science Advisory Group. In the first years of the mission intensive independent measurement campaigns were organised, setting the basis for a continuous validation programme over the instrument lifetime, harmonised with the steadily improving operational processors and the scientific algorithms. All validation activities have to obey well-established validation principles including a thorough analysis when comparing geophysical parameters derived from SCIAMACHY with correlative measurements. Such measurements use all available means – from ground, ships, aircraft and balloons. Additionally the SCIAMACHY measurements were compared with results from other space-borne atmospheric sensors. Meanwhile validation has confirmed the current accuracy for a wealth of retrieved parameters. The list includes O_3, NO_2, BrO, OClO, SO_2, HCHO, CHOCHO, H_2O, CO, CO_2, CH_4, cloud fraction and cloud top pressure, aerosol index and aerosol optical thickness.

Keywords Validation • Trace gases • Ground station networks • Shipborne campaigns • Airborne campaigns • Balloon-borne campaigns • Satellite intercomparisons

The geophysical validation of SCIAMACHY data has to face complex requirements in terms of measured species, altitude range, spatial and temporal scales, intended applications, and sustainability. Therefore an extensive SCIAMACHY validation programme has been developed jointly by Germany, The Netherlands and Belgium, with the support of ESA and a large number of international partners. The successful preparation of the validation included (Piters et al. 2006):

- An organisational structure to coordinate large-scale validation campaigns, to monitor continuously the validation results, and to foster exchanges between the different validation parties
- Numerous independent validation measurements of all planned SCIAMACHY products, performed intensively in the first 2 years of operation and continued on sustainable basis afterwards
- Adequate manpower to analyse the data in the first 2 years of operation, for a large part funded by the national space agencies of the three instrument-providing countries
- Sustained support from the national agencies and from ESA to ensure – after the two first years – continuity of the organisation, measurements and analysis manpower over the lifetime of the instrument

While the extensive SCIAMACHY validation setup was in place with the start of the mission, the analysis itself had to be adjusted to the actual availability of operational SCIAMACHY data products in the ENVISAT ground segment as

J.-C. Lambert (✉)
Belgian Institute for Space Aeronomy (BIRA-IASB), 3 Avenue Circulaire, 1180 Brussels, Belgium
e-mail: jean-christopher.lambert@aeronomie.be

A. Piters
Royal Netherlands Meteorological Institute (KNMI), Wilhelminalaan 10, 3732 GK De Bilt, The Netherlands

A. Richter, S. Mieruch, H. Bovensmann, and M. Buchwitz
Institute of Environmental Physics/Institute of Remote Sensing (IUP-IFE), University of Bremen, Otto-Hahn-Allee 1, 28359 Bremen, Germany

A. Friker
German Aerospace Center, Space Agency, Königswinterer Str. 522-524, 53227 Bonn, Germany

M. Gottwald and H. Bovensmann (eds.), *SCIAMACHY - Exploring the Changing Earth's Atmosphere*,
DOI 10.1007/978-90-481-9896-2_9, © Springer Science+Business Media B.V. 2011

well as data products derived by involved science teams. Since the first release of early SCIAMACHY data in summer 2002, the operational processors were upgraded regularly. More and more of the envisaged data products did achieve good quality (level 1b spectra, O_3, NO_2, BrO and cloud data). Also the science data products were successfully improved on a regular basis thus enabling meaningful validation of O_3, NO_2, BrO, SO_2, OClO, HCHO, CHOCHO, CH_4, CO, H_2O, CO_2, and aerosols.

The requirement to ensure maximum product quality throughout the mission and to achieve best possible consistency with follow-on missions made it necessary to continue SCIAMACHY validation throughout the instrument's lifetime – and even beyond, anticipating that algorithm updates, reprocessing of data and the development of new data products will continue after the in-orbit life of SCIAMACHY.

9.1 Validation Strategy

The rationale of satellite validation is to ensure that geophysical quantities derived from in-orbit radiometric measurements meet quality requirements for the intended scientific studies and applications. Starting from this perspective of scientific usability, considering the major scientific objectives of the mission, and based on the GOME validation experience, the SCIAMACHY Validation and Interpretation Group (SCIAVALIG) elaborated a list of validation requirements (SCIAVALIG 1998) and a detailed validation plan (SCIAVALIG 2002). These documents underlined the importance of

- Performing correlative studies with well-characterised data obtained by complementary measurement systems and modelling tools
- Validation as a diagnostic tool in the improvement of retrieval algorithms

The methods and practices developed and used for SCIAMACHY validation arise from the arguments and considerations in these documents. Over the years, they have been refined according to the developing SCIAMACHY validation experience, and enriched through exchanges of scientific and organisational nature occurring in international forums and projects.

Validation Principles

Satellite validation is often understood as a simple comparison exercise concluding to a once-and-for-all assessment of the difference between the satellite data being validated and a reference dataset of 'validated' quality. The reality is somewhat different. Data comparisons are indeed the basis for investigating the quality of the satellite data. Provided that the satellite and the correlative measurements offer the same perception of the atmospheric profile, its variability and its gradients – so that comparison errors remain small – the simplified approach outlined above is sufficient to determine whether the satellite and correlative data agree within their respective error bars, hence, whether the theoretical bias and precision estimates of the satellite data may be realistic. If the two measurements sample and smooth differently the atmospheric field and its variations, the verification of theoretical error bars through data comparisons is much more complex in the presence of atmospheric structures and variability. Furthermore, straightforward comparisons are by no means sufficient for assessing the usefulness of the data for their intended scientific applications. In addition obtaining an agreement within the estimated error bars offers no guarantee that the retrieved values do contain enough information coming from the measurement itself. For example a good agreement at altitudes where the satellite cannot measure at all for physical reasons, e.g. in the lower troposphere masked by thick clouds, might simply reflect the use of an excellent climatology as first guess data and of appropriate retrieval constraints. Therefore, beyond the calculation of differences between SCIAMACHY and correlative datasets, SCIAVALIG recommended the use of complementary validation methods, each with its specific contribution to the overall assessment of the usefulness of the data.

Throughout the validation it is important to investigate, both qualitatively and quantitatively, how well SCIAMACHY data capture known geophysical signals that are either observed by other measurement systems or deduced from our understanding of the atmosphere. Depending on the species and the type of data product, e.g. total or tropospheric column or vertical profile, these signals may include meridional and zonal structures, geographical structures linked to the distribution of emission sources and to the orography, vertical structures, long-term trends, temporal cycles on seasonal, day-to-day and diurnal scales, and special events of tropospheric pollution. Unpredictable events like the Antarctic vortex split of September 2002 (von Savigny et al. 2005), the solar proton events of October and November 2003 (Rohen et al. 2005), and a few volcanic eruptions (Afe et al. 2004), have been instrumental in testing the real capabilities of SCIAMACHY.

Level 1b spectra measured by SCIAMACHY contain information from the target species (usually absorption features) along the optical path. However, the optical path of light through the atmosphere is controlled by many factors, e.g. the solar elevation, the satellite viewing angles, and the presence of clouds, aerosols, and interfering species. Geophysical quantities (level 2 data) are retrieved from SCIAMACHY spectra (level 1b data) using auxiliary information such as absorption cross sections, output from radiative transfer models and climatologies. Consequently, the sensitivity of SCIAMACHY retrievals to the real atmospheric state depends on measurement, instrumental, and algorithmic parameters. Simplifications or misinterpretations therein can result in systematic errors in the retrieved quantities. It is necessary to study the influence of these parameter-dependent systematic errors on the intended scientific use. For example, polar ozone loss assessments relying on successive SCIAMACHY measurements along isentropic trajectories might be affected by any dependence of ozone-related products on the solar zenith angle and the latitude, and by altitude registration biases associated with pointing errors of the instrument. Global and regional chemical family budgets might be altered by fictitious spatial structures and temporal signals generated by the retrieval algorithms and superimposed on the actual geophysical signals. These retrieval dependent systematic errors need to be tracked down systematically and characterised in detail. To achieve this, a good communication is needed between retrieval and validation experts.

As a first stage, prior to performing the thorough geophysical validation of a mature data product, ad hoc algorithm verification and validation studies performed on a selected subset of data have often played and still play an important diagnostic role in the maturity of retrieval algorithms. In support, the use of data assimilation tools has been powerful in revealing internal inconsistencies in SCIAMACHY datasets, such as gaps, shifts, systematic biases between data acquired at different viewing angles, temporal drifts, abnormal cycles, etc. Intercomparison of SCIAMACHY data retrieved with independent algorithms, either directly or indirectly using correlative measurements as a standard transfer, has also given new insights for algorithm improvements.

Comparison Errors and Representativeness

From a metrological point of view, the comparison of remotely sensed geophysical quantities with correlative measurements is not straightforward. A major difficulty results from the convolution of atmospheric variability and structures with the smoothing/scanning properties inherent to the remote sensing approach. Different observation platforms, measurement techniques and retrieval methods yield different sampling of the atmosphere in time and in space, different averaging of its variations and structures, and different sensitivity to ancillary atmospheric and instrumental parameters. As a direct consequence of those differences in the perception of the atmospheric field, atmospheric structures and variability can critically corrupt the reliability of the comparison by introducing systematic biases and additional noise. Similar considerations apply to the comparison of remote sensing measurements with in situ measurements and with modelling results.

The first step of a comparison process consists in selecting the satellite and correlative data offering the best colocation in time and space. Ideally, data should not be selected for comparison if their time and space mismatches are associated with atmospheric gradients and variability exceeding the individual errors bars of the measurements. In practice, such an ideal coincidence of the data is rarely available due to the above considerations, and a compromise has to be found between – on one hand – representativeness and statistical significance of the datasets to compare – and on the other hand – accuracy of the comparison process. The most common selection practice is to compare satellite and correlative data within an arbitrary time/space coincidence window, spanning typically from 200 to 1000 km and from 1 h to 2 days. As expected, it works satisfactorily for long-lived species with negligible variability in space and in time, and for which the retrieval has a moderate sensitivity to the vertical structure. When atmospheric variability increases, differences in smoothing and sensitivity increase the comparison noise. For example using the same time/distance selection window, the 1σ standard deviation between an ozone column derived from SCIAMACHY and correlative data can increase from a few percent at mid-latitudes to several 10% near the polar vortex edge. Stronger effects, including systematic biases, have been observed for short-lived species, especially those exhibiting large meridian gradients and a diurnal cycle like NO_2 and BrO. In addition to time and space mismatches, differences in vertical and horizontal smoothing of the atmospheric variability can result in large discrepancies between satellite and correlative data. In particular, data products with a poor vertical resolution are often affected by the so-called 'vertical smoothing error', which must be considered appropriately when interpreting comparisons with data obtained at better vertical resolution. Altitude registration uncertainties due to pointing errors of the instrument and attitude uncertainties of the ENVISAT platform are a source of vertical mismatch for limb data validation.

According to the needs, more sophisticated methods have been developed to deal with differences in representation and representativeness. They comprise the use of

– Radiative transfer tools to better characterise the vertical and line-of-sight smoothing of both SCIAMACHY and

correlative data: modelling of apparent slant column densities, of weighting functions and of averaging kernels (see Chapter 7)
- Chemical-transport modelling and assimilation tools to deal with transport and photochemical effects (including diurnal cycles)
- Meteorological analyses to discriminate the effects of dynamic variability: e.g. the use of backward trajectories and the transformation of coordinates (latitude, longitude, time) to flow-tracking coordinates like equivalent latitude and potential temperature
- Complementary correlative data sources offering different smoothing and sampling properties, sensitivity and errors budgets in a synergistic way

The latter aspect is of prime importance for SCIAMACHY validation. The SCIAMACHY data products do support an assortment of scientific applications and thematic domains, covering regional to global scales, from the ground up to the mesosphere, from short-term to decadal time frames. The synergistic use of complementary validation sources can deal with this variety of products and scales. Local studies carried out at single stations constitute the preferred approach to detailed investigations. They benefit from local research and excellent understanding of local geophysical features leading to full control and accurate error budgets of the instrumentation and the availability of adequate ancillary data. Complementary studies exploiting pseudo-global sources like monitoring networks yield access to error patterns, sensitivity, and atmospheric structures on the global scale. Satellite-to-satellite comparisons, using MIPAS and GOMOS data as well as observations from other satellites, extend network-based validation ranges to higher altitudes and more regular geographical sampling. The differences in geographical coverage and in latitude/time sampling between different satellite instruments may introduce artefacts, but the massive amount of possible co-locations, at least for nadir viewing instruments, considerably improves the significance of statistical quantities and the representativeness in terms of atmospheric states.

9.2 Validation Organisation

The status as an announcement of opportunity (AO) instrument places the responsibility for SCIAMACHY validation with the AO provider. Since ESA takes care of the operational SCIAMACHY data processor, the validation of SCIAMACHY has also been included by ESA into their ENVISAT validation programme. Therefore two collaborating complementary validation structures co-exist. The overall organisation is sketched in Fig. 9.1.

ESA Validation Structure

In 1997 ESA raised an Announcement of Opportunity (AO) for the use of ENVISAT data. After review of proposals by representatives of the instrument science advisory groups, additional activities had been added to improve coverage of the validation programme. The principal investigators of the approved projects dealing with the validation of SCIAMACHY, GOMOS and MIPAS were gathered in the Atmospheric Chemistry Validation Team (ACVT). SCIAMACHY validation is performed in the following subgroups:

- ACVT/GBMCD: Ground-based measurements and campaign database
- ACVT/ESABC: ENVISAT stratospheric aircraft and balloon campaign
- ACVT/MASI: Models and data assimilation, satellite inter-comparisons
- SCCVT: SCIAMACHY Calibration and Verification Team (a subgroup of the overall ENVISAT Calibration and Validation team)

The ESABC subgroup is more than a working group. ESA, DLR, and the French Space Agency CNES together financed dedicated campaigns for the validation of SCIAMACHY, MIPAS, and GOMOS. These campaigns are referred to as ESABC campaigns and have been prepared and coordinated in the ESABC group. Preparation and results of other campaigns are only presented and discussed within the ESABC group. In addition, ESA developed a dedicated validation data centre, the ENVISAT Validation Data Centre (EVDC), hosting the validation database. This database is operated by the Norwegian Institute for Air Research (NILU) and integrated into its NILU Atmospheric Database for Interactive Retrieval (NADIR). The EVDC hosts all correlative results arising from the various validation campaigns (ground-based and ship-borne, balloon-borne, as well as from aircraft and satellites) generated for all ENVISAT instruments by the various calibration/validation teams and allows all groups involved access to data generated by the other teams.

Eight years after the launch of ENVISAT, this validation organisation still exists and subgroup participants work together episodically, not only for the preparation of Atmospheric Chemistry Validation of ENVISAT (ACVE) conferences but also to ensure sustainability of ENVISAT validation on the long-term. For this purpose, for delta validation of data processor upgrades and multi-mission consistency between ENVISAT and other atmospheric composition satellites, ESA supports a few long-term validation projects, which interact with algorithm maturation activities carried out in the framework of the SCIAMACHY Quality

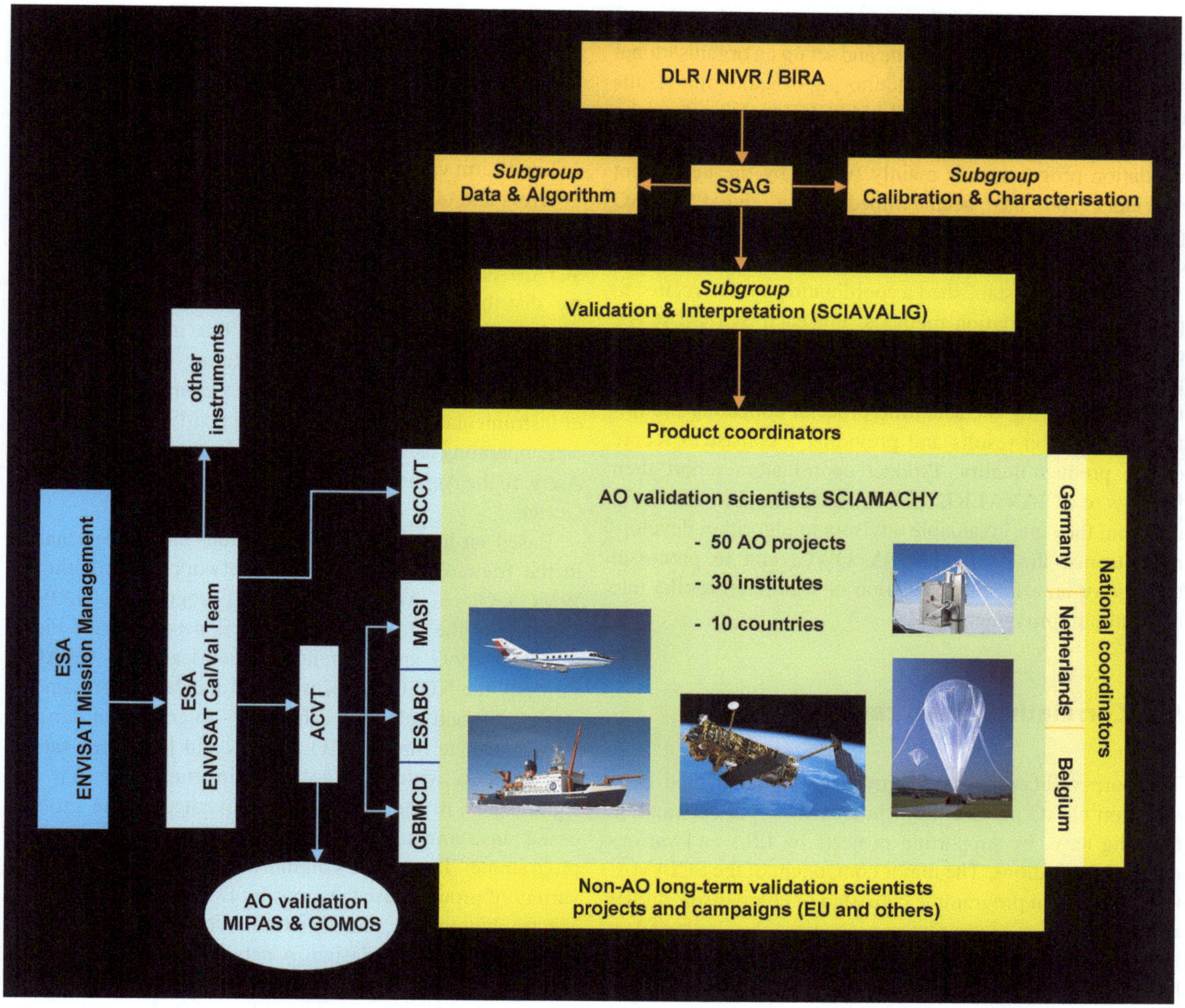

Fig. 9.1 A scheme of the SCIAMACHY validation organisational structures setup by SCIAVALIG (*light orange*) and by ESA (*pale blue*). The validation scientists actually doing the work are supported by both organisations, if they have an approved AO proposal for SCIAMACHY validation (*green*) (Courtesy: KNMI/DLR-IMF).

Working Group (SQWG). The ESA multi-mission oriented validation projects exploit major, continuously operated validation sources such as ground-based networks of the Network for the Detection of Atmospheric Composition Change (NDACC) and the Total Carbon Column Observing Network (TCCON) as well as independent satellite data. These ESA activities are complemented by national validation projects under the auspices of the SCIAMACHY AOP, e. g. balloon campaigns dedicated to improve the retrieval under specific atmospheric conditions.

SCIAVALIG Structure

The organisation of the AOP part of SCIAMACHY validation is delegated to SCIAVALIG, a subgroup of the SCIAMACHY Science Advisory Group (SSAG). SCIAVALIG, being co-chaired by KNMI, BIRA-IASB and IUP-IFE, University of Bremen, consists of an international scientific consortium of representatives from 12 institutes participating in the validation. Besides having established a list of validation requirements, SCIAVALIG also defined an essential 'core' validation programme and set up an organisational structure for the continuous monitoring of validation results throughout the lifetime of SCIAMACHY and for the delta validation, necessary in case of processor upgrades. The core validation programme is mainly funded by the instrument providers and has been embedded in the ESA AO programme via several AO projects involving 'national coordinators' of SCIAVALIG.

SCIAVALIG established a coordination system related to products and validation methods. For each SCIAMACHY data product a 'product coordinator' was selected who maintains an overview of the validation results. The coordinator's task is monitoring the scientific process, collecting the different validation results and providing a consistent record of the product quality. Product coordinators report their findings to SCIAVALIG, ACVT, and external scientists. In addition, they are invaluable advisors to algorithm development teams, the SCIAMACHY QWG, and to processor experts for an efficient translation of validation results into algorithm improvements.

9.3 Correlative Measurements

The core validation programme was complemented by a selection of AO projects from international partners and, in the long term, by supporting projects by ESA and various national institutions. The major component of the SCIAMACHY validation programme consisted of comparison studies with correlative measurements acquired by independent instrumentations from various platforms, namely, ground-based stations, ships, aircraft, stratospheric balloons, and satellites. The reference dataset collected within this initial effort is and remains the basis for every comparison. However, distinguishing between instrument ageing effects and real atmospheric trends requires continuous instrument monitoring and dedicated reference measurements throughout the mission.

Ground-Based Instruments

Ground-based instruments provide the appropriate correlative data to fulfil four main tasks of the SCIAMACHY validation programme:

- Quick validation before public release of a new product or just after the release of a near-realtime product
- Detailed geophysical validation from pole-to-pole and for a variety of geophysical states, including dependences on measurement, instrument and atmospheric parameters such as solar zenith angle, viewing angle and atmospheric temperature
- Verification of correctness of changes and preliminary quality assessment of the reprocessed data after a major improvement of a retrieval algorithm
- Long-term validation, including detection of drifts, cyclic errors and other time-varying features

The list of stations providing correlative measurements for SCIAMACHY validation is given in Fig. 9.2. These stations are distributed globally, but with a strong clustering in northern latitudes (see Fig. 9.3). The nationally funded core validation programme, constituting the backbone of the ground-based validation, includes complementary types of instrumentation, yielding together nearly all targeted species, operating at about forty stations distributed from the Arctic to the Antarctic and from South America to the Indian Ocean.

Based on long-lasting collaborations established mainly in the framework of monitoring networks contributing to WMO's Global Atmospheric Watch programme (GAW) – particularly the affiliated ozonometric networks (see Fioletov et al. 1999 and references therein) and the NDACC, formerly the NDSC (see Kurylo and Zander 2001; Lambert et al. 1999 and references therein) – international partners also contribute through AO projects and long-term validation projects with a long list of instruments which add significantly to the geographical coverage of the ground-based instrumentation included in the core validation programme. The ozone column amount is monitored at a variety of ground-based stations by Dobson and Brewer UV spectrophotometers and by Russian/NIS UV filter radiometers of the M-124 design. A network of about 30 DOAS instruments, all certified for the NDACC, monitor the column amount of species absorbing in the UV-VIS-NIR part of the spectrum such as O_3, NO_2, BrO, OClO, IO, HCHO, SO_2, and H_2O. Some of them have multi-axis observation capabilities yielding separation of the tropospheric and stratospheric columns. About ten Fourier Transform Infrared (FTIR) spectrometers, also NDACC certified, monitor the vertical column amount and, where possible, the vertical distribution of a series of species including O_3, NO_2, CO, CH_4, N_2O, CO_2, HCHO, and H_2O. Many of the NDACC FTIR teams are also expanding their measurement capabilities to the NIR, in an effort to join the network of TCCON stations. Six NDACC microwave radiometers measure the thermally induced rotational emission of selected species in the stratosphere and lower mesosphere, such as O_3, H_2O, and ClO. Differential Absorption Lidars (DIAL), certified for NDACC, and electro-chemical ozone sondes yield the vertical distribution of tropospheric and stratospheric ozone at high and moderate vertical resolution. Aerosol and cloud properties are recorded by lidar and aerosol instruments.

Fig. 9.2 Ground-based stations contributing to SCIAMACHY validation and associated SCIAMACHY data products. The last column includes UV, CF (cloud fraction), CTP (cloud top pressure) and AAI (absorbing aerosol index) (Courtesy: KNMI/DLR-IMF).

a

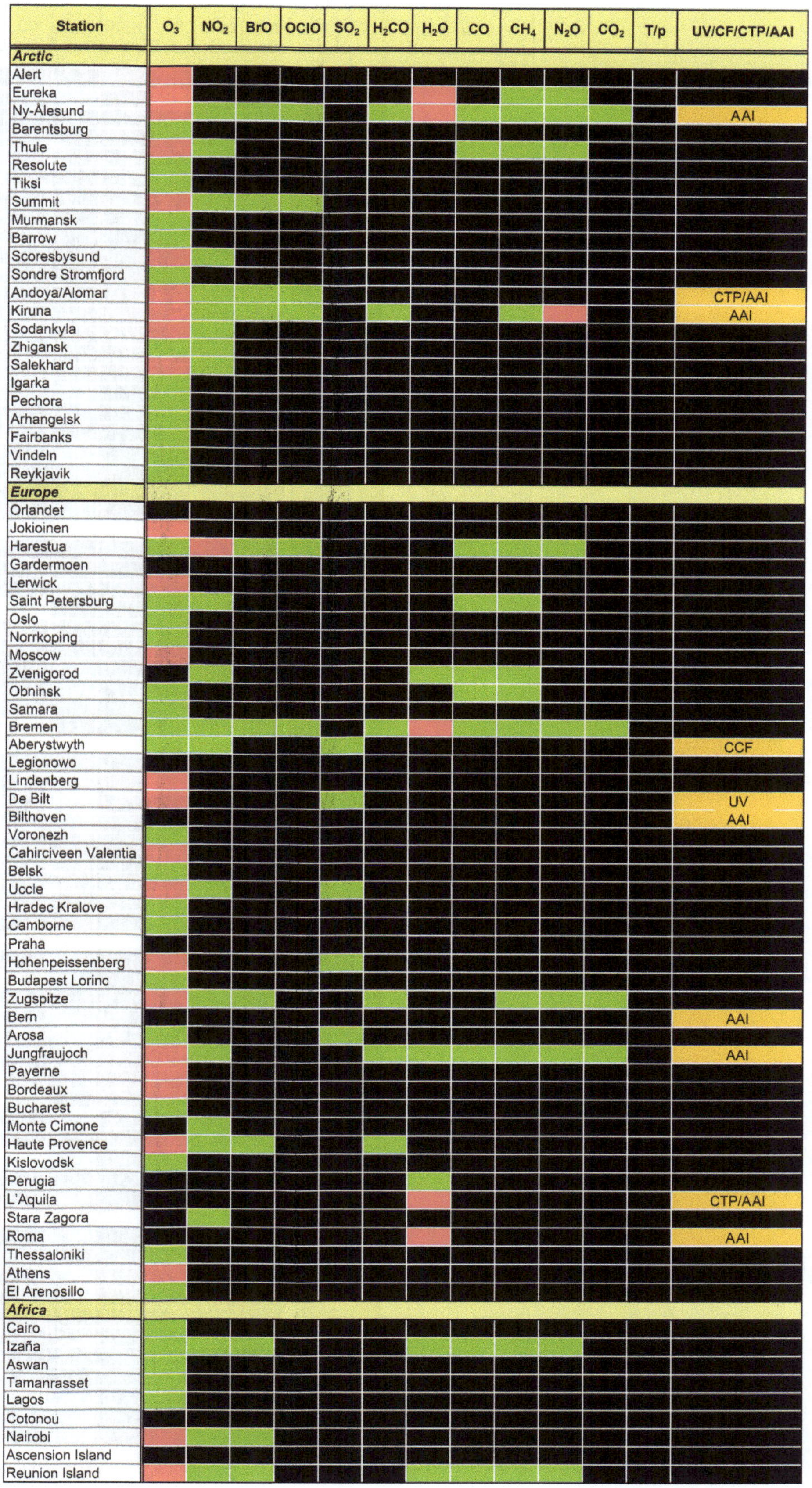

Fig. 9.2 (Continued)

b

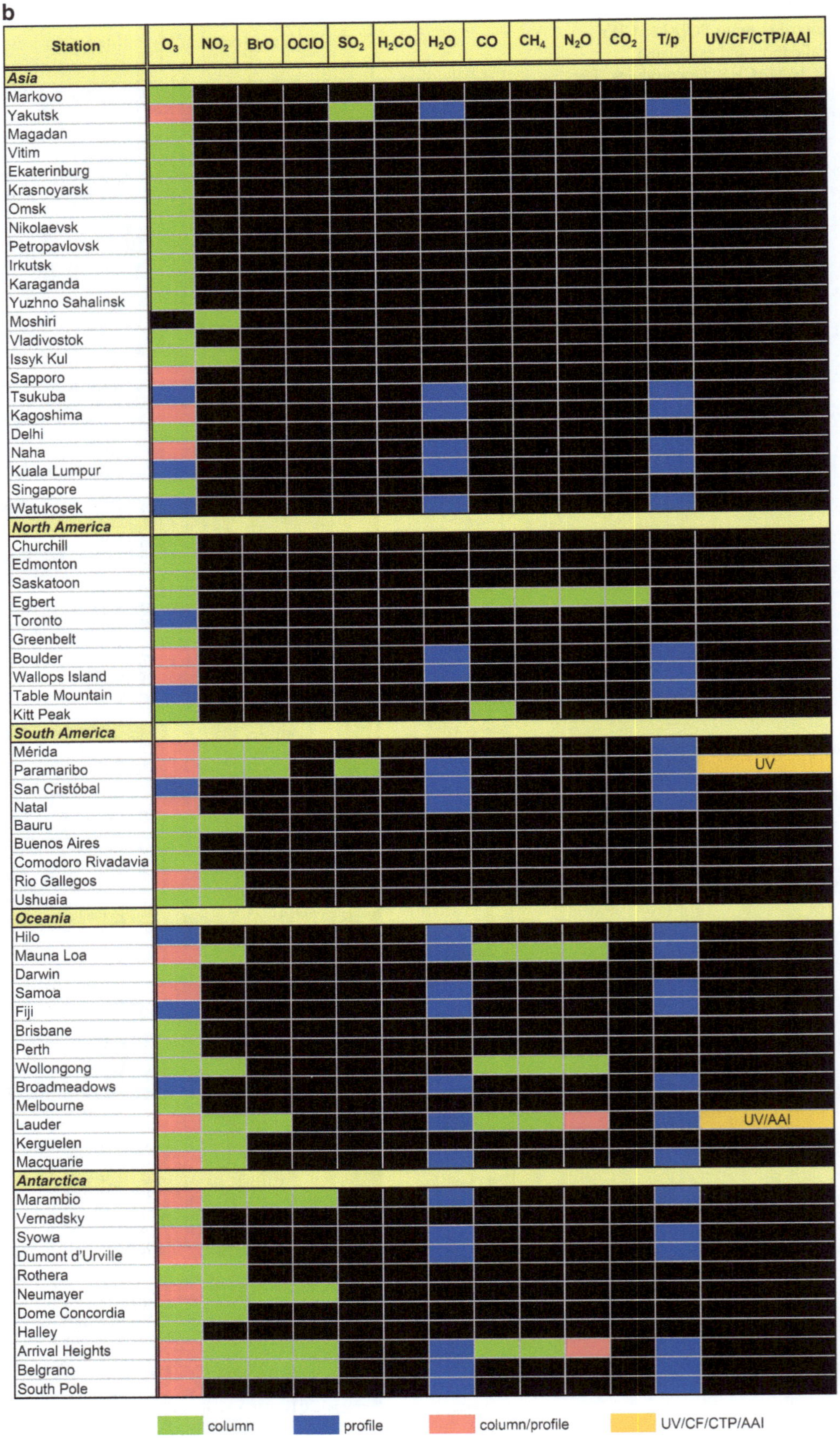

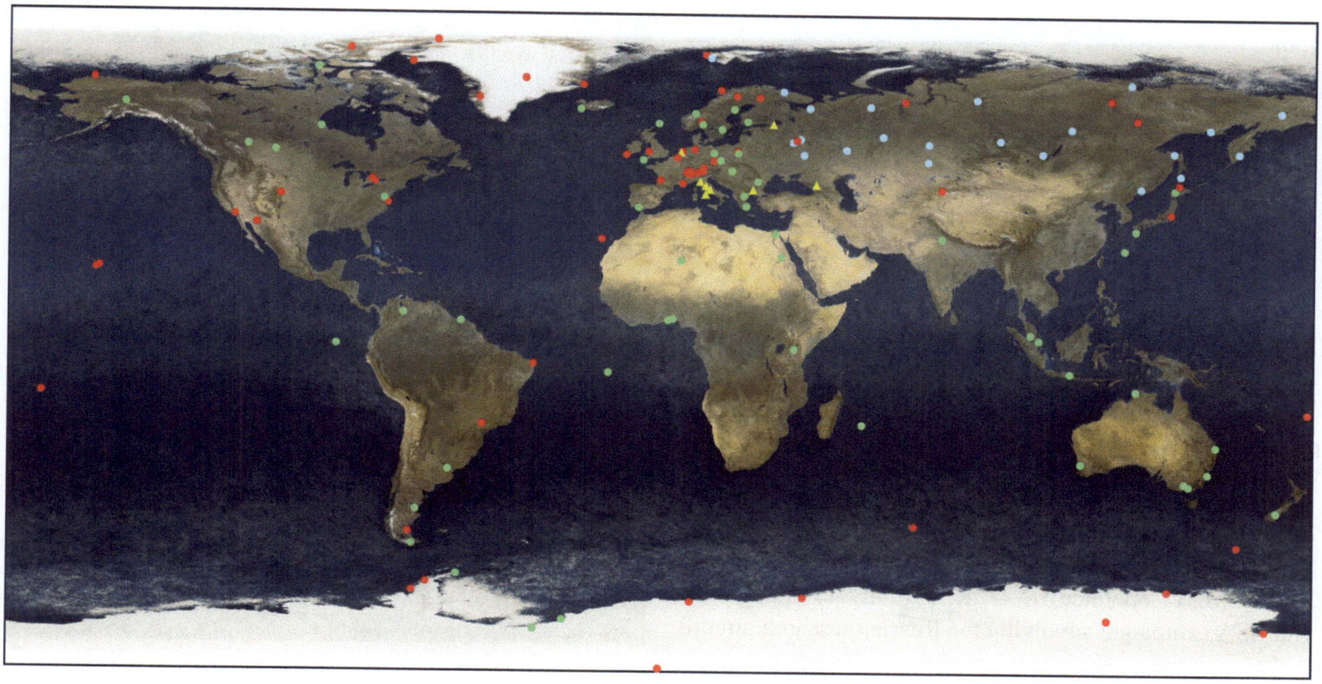

Fig. 9.3 Global distribution of validation sites. Symbols and colour codes indicate the prime network as follows: *red circles* – NDACC, *blue circles* – Russian/NIS M-124, *green circles* – GAW, *yellow triangles* – others (Courtesy: DLR-IMF/SRON/KNMI).

Triggered by new challenges in atmospheric research and the always higher accuracy requirements of satellite validation, several international measurement field campaigns took place in Europe in the last years with the specific objective to improve our understanding of the error bars of the respective types of measurement, of the intrinsic differences between these measurements, and of their potential synergies. At the same time they served identifying the best means to understand and validate observations from satellites. Some of the campaigns were supported by SCIAMACHY validation partners and by ESA. These comprised the SAUNA campaigns in Sodankylä (Finland) in 2006 and 2007 with the focus on ozone measurements at high latitudes, low solar elevations and large ozone column ranges. Other examples were the campaigns DANDELIONS in 2005 and 2006 (Brinksma et al. 2008) and CINDI in 2009, all taking place in Cabauw/The Netherlands. Their focus was on NO_2 and on other tropospheric measurements including species such as O_3, HCHO, CHOCHO, BrO and aerosols.

Shipborne Campaigns

In addition to the instruments operating continuously at ground-based sites, two instruments are operated on-board the German research vessel *Polarstern* to facilitate the validation of SCIAMACHY measurements in remote marine regions: a MAX-DOAS (Multi-Axis DOAS) and an FTIR instrument. The *Polarstern* made three cruises within the time period relevant for initial SCIAMACHY validation: the first between November 2001 and May 2002, the second between October 2002 and February 2003, the third between October 2003 and July 2004 (Figs. 9.4 and 9.5). The moveable MAX-DOAS experiment measured not only constantly during these initial cruises but also during further *Polarstern* expeditions from 2005–2009. The FTIR instrument was operating during the second and third campaign from Bremerhaven to Africa. The unique *Polarstern* dataset was and is most useful for all investigations concerning large scale latitudinal cross sections of atmospheric trace gases.

Besides dedicated validation measurements on-board the *Polarstern*, data acquired during primarily research-oriented ship campaigns were also made available for the validation of SCIAMACHY. An example is the TransBrom campaign in autumn 2009 with the German research vessel *Sonne*. During the TransBrom cruise through the Pacific Ocean from Tomakomai/Japan to Townsville/Australia the exchange of halogens between the ocean and the lower and upper atmosphere was investigated. Two MAX-DOAS and an FTIR instrument delivered important validation data in the rarely covered tropical Pacific area.

Airborne Campaigns

The German aircraft validation activities were concentrated on missions with the meteorological research aircraft Falcon

Fig. 9.4 The research vessel *Polarstern* (Photo: G. Chapelle/Alfred Wegener Institute for Polar and Marine Research).

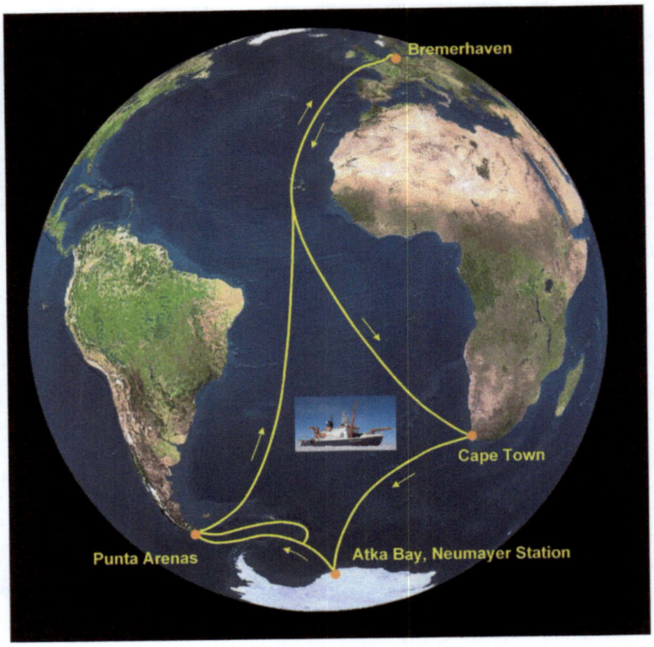

Fig. 9.5 Route of the *Polarstern* cruise during the *ANT XIX* campaign between November 2001 and May 2002 (Courtesy: DLR-IMF).

20 (D-CMET) operated by DLR, partially in cooperation with ESA campaigns involving the Russian research aircraft M55-Geophysika. Many features make the Falcon an excellent aircraft for validation. Three large optical windows, two in the floor and one in the roof enable operation of large lidar experiments for both tropospheric and stratospheric research. Specially manufactured polyethylene windows allow remote sensing in the microwave spectral region. The aircraft carries a data acquisition system and an extensive instrument package capable of measuring position, altitude, static pressure, and temperature. Within the SCIA-VALUE project (SCIAMACHY Validation and Utilization Experiment) two major campaigns with 28 flights were executed in September 2002 and February/March 2003 (Fig. 9.6). Both campaigns provided large-scale latitudinal cross sections from the polar regions to the tropics as well as longitudinal cross sections at polar latitudes. To validate SCIAMACHY, three different types of remote sensing instruments were installed on-board the Falcon 20 (Fig. 9.7). The AMAX-DOAS (Airborne Multi-Axis DOAS), which is an experiment developed jointly by the Universities of Heidelberg and Bremen, is capable of measuring tropospheric and stratospheric columns of key gases such as O_3, NO_2, BrO, and OClO, all absorbing in the UV-VIS wavelength range (see Fig. 9.8). ASUR (Airborne Submillimetre Radiometer), operated by the University of Bremen, is a passive microwave sensor. A broad range of molecular lines can be detected containing the molecules that play an important role in the catalytic destruction of ozone. The frequency band includes emission lines of O_3, ClO, HCl, HNO_3, N_2O, H_2O, HO_2, CH_3Cl, NO, HCN, and BrO. The Ozone Lidar Experiment (OLEX), developed and operated by DLR, completes the scientific payload of the Falcon. In the zenith viewing mode this instrument provides high resolution two-dimensional cross sections of ozone number densities, aerosol extinction and cirrus cloud cover information, from about 2 km above the aircraft flight level up to a height of 30 km (Fix et al. 2005).

The stratospheric research aircraft M55-Geophysika has also been involved in ENVISAT's validation. It performed two mid-latitude campaigns in July and October 2002 from a base in Forli/Italy and a high latitude campaign in January and March 2003 from Kiruna. For the ENVISAT validation flights, the M55-Geophysika was equipped with two sets of instruments. The so-called 'chemical flights' were performed with six in situ and one remote sensing instruments, capable of measuring, among others, concentrations of H_2O, O_3, NO, NO_y, N_2O, CH_4, BrO, and columns of O_3 and NO_2 (Kostadinov et al. 2003, Heland et al. 2003). For the so-called 'cloud/aerosol flights', the remote sensing instrument was replaced by six instruments dedicated to the characterisation of aerosol and cloud properties. Although the in situ instruments remained on-board, these flights were optimised for the cloud and aerosol investigations.

Within the MOZAIC programme (Measurements of Ozone and water vapour by Airbus In-service aircraft, Marenco et al. 1998) which started in 1994, five long-range Airbus A340 aircrafts were equipped with in situ instruments measuring O_3, H_2O, CO, and NO_y. They provide data from all over the world along their flight tracks in the upper troposphere/lower stratosphere at an altitude level from 9–12 km down to the ground in the vicinity of about 60 airports. These measurements are a unique dataset at the tropopause region and are especially useful for the development and validation of products distinguishing between troposphere and stratosphere. Other interesting datasets, which are not yet exploited for

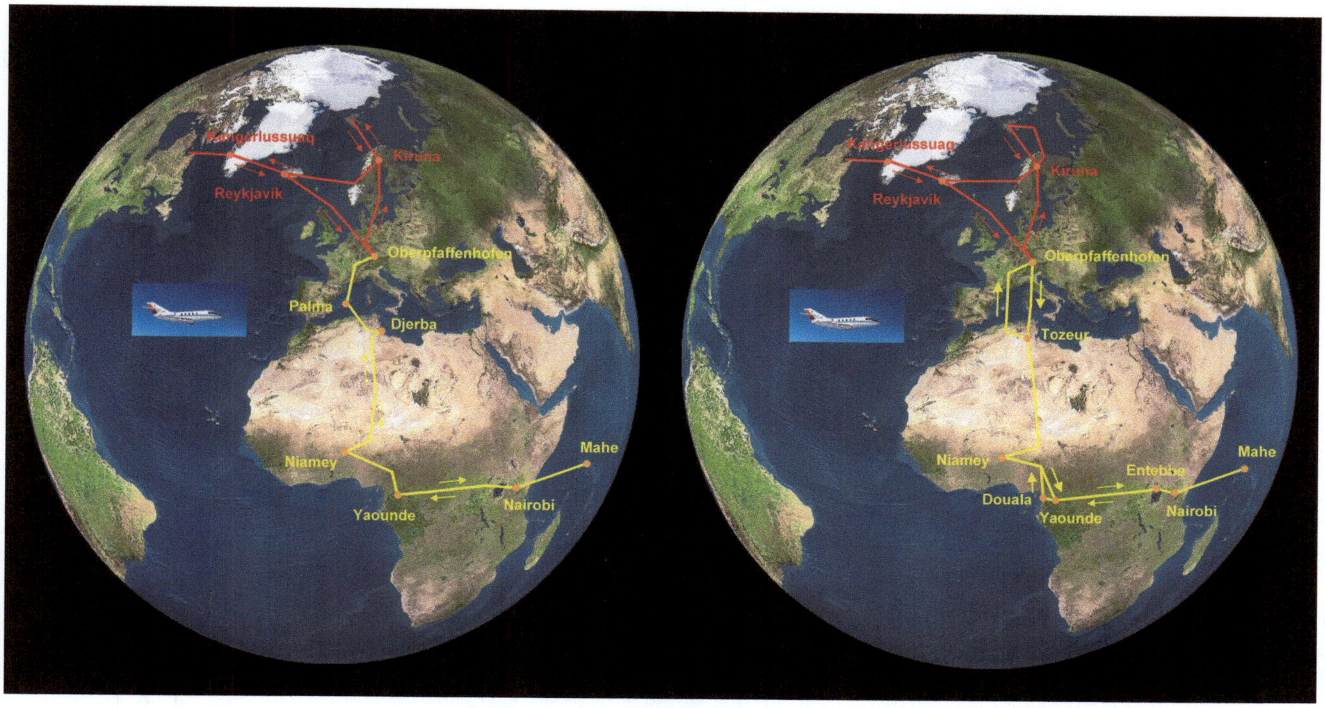

Fig. 9.6 Falcon flight tracks for the September 2002 (*left*) and February/March 2003 (*right*) SCIA-VALUE airborne campaigns. In red are the northern tracks (3–8 September 2002 and 19 February – 3 March 2003) while the southern tracks (15–28 September 2002 and 10–19 March 2003) are displayed in yellow (Courtesy: DLR-IMF).

Fig. 9.7 The Falcon aircraft with the viewing directions of the validation instruments (Courtesy: Fix et al. 2005).

SCIAMACHY validation, may come from the programmes CARIBIC (Civil Aircraft for the Regular Investigation of the Atmosphere Based on an Instrument Container, Brenninkmeijer et al. 2007) and NOXAR (Measurements of Nitrogen Oxides and Ozone along Air Routes).

Since 2005 data from several scientifically driven airborne campaigns also contributed to SCIAMACHY's validation and algorithm improvements (Martin et al. 2006). One of them is led by ICARTT (International Consortium of Atmospheric Research on Transport and Transformation)

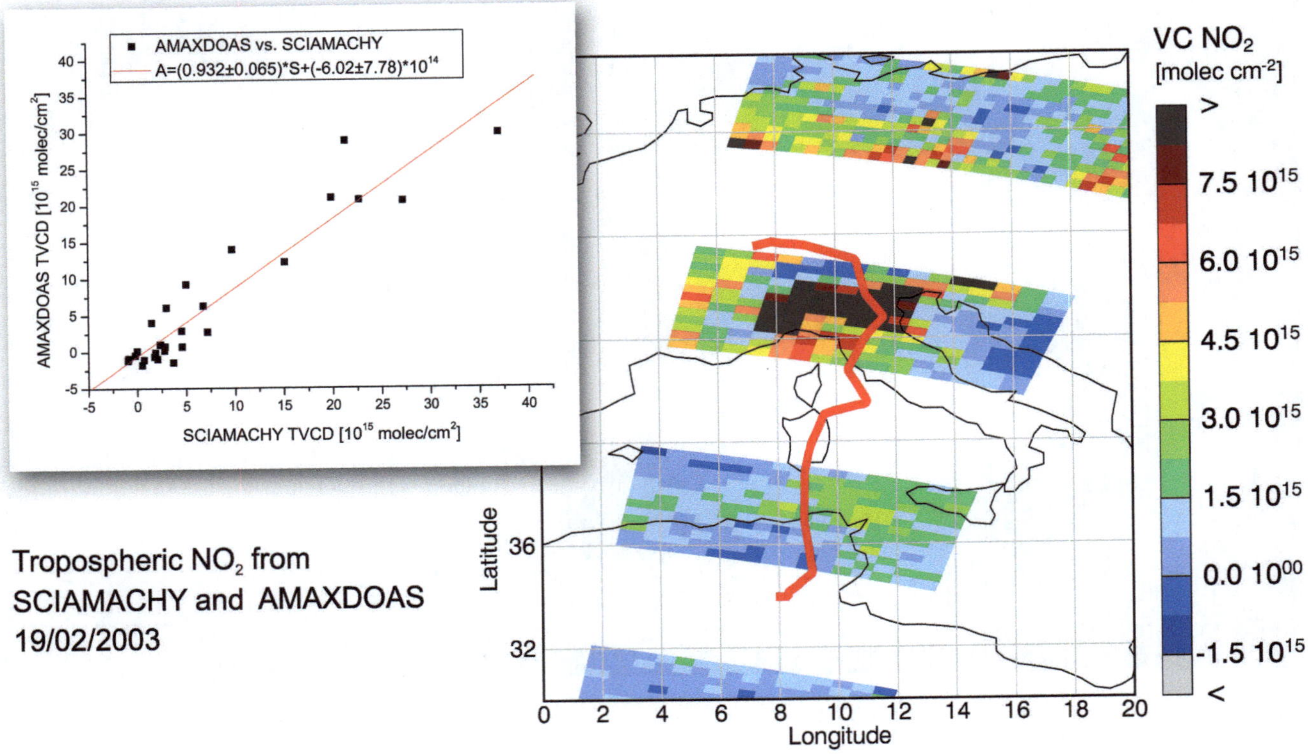

Fig. 9.8 Tropospheric NO_2 column obtained by SCIAMACHY together with the Falcon flight track *(in red)* showing where AMAX-DOAS measured almost simultaneously. In the inset tropospheric NO_2 columns from AMAX-DOAS are plotted versus those from SCIAMACHY. The SCIAMACHY tropospheric NO_2 columns are retrieved by IUP-IFE, University of Bremen (Courtesy: Heue et al. 2005).

combining several regional research projects independently developed by various international groups in the US and in Europe. The purpose of their cooperation is to derive a better understanding of the evolution of anthropogenic emission injected into the atmosphere. The major participants of the consortium are NOAA with their programme *New England Air Quality Study – Intercontinental Transport and Chemical Transformation* (NEAQS – ITCT 2004) program, NASA with their campaign *Intercontinental Chemical Transport Experiment – North America* (INTEX-NA) and, on European side, the project *Intercontinental Transport of Ozone and Precursors – North Atlantic Study* (ITOP) project. These regional undertakings studied local air quality by executing several aircraft flights in the New England area, the Northern Atlantic, and Western Europe. Another example is POLARCAT (Polar study using aircraft, remote sensing, surface measurements and modelling of climate, chemistry, aerosols and transport) which, from 2007–2008, investigated the transport of air pollution to the Arctic and its effect on atmospheric chemistry and climate. The measurements, taken with airborne sensors carried on several platforms, were closely coordinated with satellite overpasses including ENVISAT thus permitting validation of satellite observations of tropospheric species and aerosol parameters.

Balloon-Borne Campaigns

Balloon-borne measurements provide snapshot type vertical profile measurements of high precision. For the initial phase of intensive validation, the dedicated balloon campaigns for the atmospheric chemistry instruments GOMOS, MIPAS, and SCIAMACHY were funded by ESA, DLR, and CNES, with the costs and responsibilities shared according to an agreement between the three agencies. Part of this agreement was to use all balloon flights as far as possible for all three satellite instruments. CNES provided the facilities and staff for launching scientific payloads with large stratospheric balloons from dedicated stations. The availability of the CNES equipment was an important constraint for the implementation of campaigns. Within the ACVT subgroup ESABC, the involved scientists from the balloon teams and representatives of the agencies met frequently to organise the campaigns (Fig. 9.9). The launch sites and campaign times were selected to cover mid-latitudes, northern latitudes, and the tropics during several seasons, including the possibility for ozone depletion conditions in spring, as far as possible within the available resources.

Fig. 9.9 Launch of the TRIPLE payload in Aire sur l'Adour in September 2002. (Photo: W. Gurlit, IUP-IFE, University of Bremen).

Explicitly funded for the validation of SCIAMACHY were:

- LPMA-DOAS: Combining a Limb Profile Monitoring of the Atmosphere FTIR and a UV-VIS DOAS instrument
- TRIPLE/TWIN: Combining sensors such as a resonance fluorescence ClO/BrO instrument – HALOX-B, an in situ Stratospheric Hygrometer – FISH, a cryogenic total air sampler – BONBON, and, occasionally, a tunable diode laser measuring H_2O and CH_4
- MIPAS-B: MIPAS balloon version

These constitute a part of the German contribution to the balloon-borne validation of ENVISAT. All three balloon payloads measured atmospheric profiles of O_3, NO_2, OClO, BrO, CH_4, N_2O, H_2O, CO, CO_2, temperature and pressure which allow validation of corresponding parameters measured by SCIAMACHY during co-located overpasses of the satellite. Solar irradiances and limb radiances for level 1 validation were determined by radiometric calibration of the DOAS instruments on-board the LPMA-DOAS gondola.

In the initial phase of balloon-borne validation until mid 2005, many balloon campaigns (see Tables 9.1 and 9.2), dedicated to SCIAMACHY validation measurements, were performed from launch sites in Kiruna/Sweden, Aire sur l'Adour/France, Bauru/Brazil, Vanscoy/Canada, and Fort Sumner/New Mexico, US. Most of the individual flights lasted typically less than 1 day. However a few long duration trajectories were also included (Borchi and Pommereau 2007). These provided a large number of coincidences between spaceborne, including SCIAMACHY, and balloon-borne measurements as demonstrated by, e.g. the validation of water vapour profiles exploiting results from a 39 day long flight in the frame of the HIBISCUS campaign in February/March 2004 in Bauru/Brazil (Montoux et al. 2009).

Table 9.1 SCIAMACHY validation specific payloads used in balloon campaigns (after Piters et al. 2006 and Oelhaf et al. 2009)

Payload	Launch dates	Launch site	Target species
MIPAS-B	September 2002	Aire sur l'Adour	O_3, NO_2, N_2O, H_2O, CO, CO_2, T/p
	December 2002	Kiruna	
	March 2003	Kiruna	
	July 2003	Kiruna	
	June 2005	Teresina	
	June 2008	Teresina	
	March 2009	Kiruna	
TRIPLE +TWIN	September 2002	Aire sur l'Adour	CO_2, CH_4, N_2O, NO_2, H_2O, BrO
	March 2003	Kiruna	
	June 2003	Kiruna	
	June 2005	Teresina	
	June 2008	Teresina	
	March 2009	Kiruna	
LPMA-DOAS	August 2002	Kiruna	O_3, NO_2, OClO, BrO, CH_4, N_2O, H_2O, CO, T/p, irradiance
	March 2003	Kiruna	
	October 2003	Aire sur l'Adour	
	March 2004	Kiruna	
	June 2005	Teresina	
	June 2008	Teresina	
	September 2009	Kiruna	

When this phase was terminated, data suitable for SCIAMACHY validation were still acquired during additional balloon flights. These had either different atmospheric chemistry objectives or were intended for ENVISAT validation in general. Balloon launch sites included the locations listed above together with Teresina/Brazil and Niamey/Niger. One of them was performed in May/June 2008 within the SCOUT-O3 project in collaboration with an ENVISAT validation campaign, operated from Teresina, a site already visited by CNES and balloon scientists for an earlier ENVISAT validation campaign in June/July 2005. The balloon payload also comprised, in different combinations, the instruments funded for the dedicated SCIAMACHY validation such as MIPAS-B, TRIPLE, and LPMA/DOAS.

Satellite Intercomparisons

Correlative measurements by independent instruments on-board the same and other satellite platforms add significantly to the required pole-to-pole validation of

Table 9.2 Further balloon-borne experiments suitable for ENVISAT and/or SCIAMACHY validation

Payload	Launch dates	Launch site
SAOZ-MIR	February–March 2003	Bauru
	February–April 2004	Bauru
SAOZ	August 2002	Kiruna
+SAOZ-BrO	October 2002	Aire sur l'Adour
+SAOZ-H2O	February 2003	Bauru
	March 2003	Kiruna
	January 2004	Bauru
	February 2004	Bauru
	June 2004	Aire sur l'Adour
	August 2004	Vanscoy
	May 2005	Aire sur l'Adour
	October 2005	Aire sur l'Adour
	August 2006	Niamey
FIRS-2	October 2002	Ft. Sumner
	September 2003	Ft. Sumner
	September 2004	Ft. Sumner
MANTRA	September 2002	Vanscoy
	September 2004	Vanscoy
SALOMON	September 2002	Aire sur l'Adour
	March 2004	Kiruna
	June 2004	Aire sur l'Adour
	January 2006	Kiruna
	June 2007	Aire sur l'Adour
	August 2009	Kiruna
SPIRALE	October 2002	Aire sur l'Adour
	January 2003	Kiruna
	June 2005	Teresina
	January 2006	Kiruna
	June 2008	Teresina
	August 2009	Kiruna
SDLA-LAMA	August 2002	Kiruna
	September 2003	Aire sur l'Adour
ELHYSA	January 2003	Kiruna
	March 2004	Kiruna
	August 2009	Kiruna
AMON	March 2003	Kiruna
µRADIBAL	March 2004	Kiruna
	August 2009	Kiruna
LPMA-IASI	August 2002	Kiruna
	June 2005	Teresina
	March 2006	Kiruna
	June 2008	Teresina
	August 2009	Kiruna

SCIAMACHY. A few satellite instruments provide nearly global coverage at nearly daily frequency and, therefore, are well suited for global validation of SCIAMACHY data products in space and time. Complementarily, other satellites cover only a portion of the globe, often evolving with the season, but with a more regular geographical sampling and higher altitude range than the sampling and range achievable by ground-based instrumentation. Table 9.3 lists the satellite instruments used for the core validation of SCIAMACHY data products in the early phase of the mission. Some of these sensors are now no longer in operation (marked by a '*' below), others are still functioning. Additional instruments joined the fleet of spaceborne atmospheric instruments recently.

SCIAMACHY's precursor GOME on board ERS-2 follows ENVISAT with a delay of 30 min. Because the GOME channels are almost identical to the UV-VIS channels of SCIAMACHY, GOME was the first choice for validating UV-VIS nadir products. Over 15 years after launch, GOME is still a suitable validation source for SCIAMACHY. However, since a tape recorder anomaly in June 2003, GOME measurements are restricted to the North Atlantic sector and to the visibility sector of a few worldwide ground antennas. Launched in October 2006 on-board EUMETSAT's EPS/METOP-A, an improved version of GOME, GOME-2, precedes ENVISAT by 30 min and became the first choice for validating UV-VIS nadir measurements acquired after March 2007. Operational since 2004, OMI provides similar nadir UV-VIS data products, but with a relatively large time difference of about 5 h.

TOMS* and SBUV-2, other nadir looking instruments, deliver ozone total column and profile data. The solar occultation sensors HALOE*, SAGE II*/III*, POAM III* and ACE provide trace gas profiles at sunset and sunrise. SABER observes infrared emissions in limb, retrieving ozone and water vapour profiles at high altitudes. Also operating in limb mode, OSIRIS acquires vertical profiles of several UV-VIS absorbing species. In the infrared, MOPITT and TES generate nadir products. SUSIM and SOLSTICE results are used for comparison with solar irradiance measurements while MERIS and AATSR spectral reflectances are confronted with SCIAMACHY reflectance data. Both comparisons are required to check the radiometric calibration of SCIAMACHY. In addition, intercomparisons are performed between the three atmospheric chemistry instruments on board ENVISAT, i.e. MIPAS, GOMOS, and SCIAMACHY.

9.4 Validation Results

The goal of validation is to generate a clear description of the quality of all SCIAMACHY data to allow users to readily evaluate the fitness of the data for their purpose. Besides providing evidence of traceability to established standards in terms of bias, precision and uncertainties, quality assessment of a data product also involves specific criteria on data availability, product and algorithm description, as well as software version control. Given the evolving nature of algorithm development together with defining new products, this goal has to be pursued continuously, but likely is never completely achieved. Since the Commissioning Phase of the instrument, more than 50 products were created, in part retrieving identical

Table 9.3 Satellite instruments used for SCIAMACHY validation in the early phase of the mission (2002–2004)

ESA			
GOME	Global ozone monitoring experiment	ERS-2	Columns: O_3, NO_2, SO_2, BrO, HCHO. Profiles: O_3
AATSR	Advanced along track scanning radiometer	ENVISAT	Spectral reflectance (555, 659, 865 nm), cloud cover, cloud top height
MERIS	Medium resolution imaging spectrometer	ENVISAT	Spectral reflectance (390–1,040 nm), cloud cover, aerosol
NASA			
SUSIM	Solar ultraviolet irradiance monitor	UARS	Solar UV energy
HALOE*	Halogen occultation experiment	UARS	Profiles: O_3, NO_2, NO, CH_4, N_2O, CO_2, H_2O
MLS*	Microwave limb sounder	UARS	Profiles: O_3
TOMS	Total ozone mapping spectrometer	Earth Probe	Columns: O_3, SO_2 AAI
SAGE II* and III*	Stratospheric aerosol and gas experiment II and III	ERBS and METEOR-3M	Profiles: O_3, NO_2, H_2O, aerosols
SABER	Sounding of the atmosphere using broadband emission radiometer	TIMED	Profiles: O_3, H_2O
SOLSTICE*	Solar stellar irradiance comparison experiment	UARS	Solar UV spectral irradiance
MOPITT	Measurements of pollution in the troposphere	EOS-Terra	Columns/profiles: CO
MODIS	Moderate resolution imaging spectroradiometer	EOS-Terra	Cloud cover, cloud top pressure, aerosol
NRL			
POAM III*	Polar ozone and aerosol measurement III	SPOT-4	Profiles: O_3, H_2O, NO_2, aerosols
CSA/SNSB			
OSIRIS	Optical spectrograph and infrared imaging system	ODIN	Profiles: O_3, NO_2
NOAA			
SBUV/2	Solar backscatter ultraviolet ozone experiment II	NOAA 14 and NOAA 16	Profiles: O_3

Instruments marked '*' stopped operations around 2005

parameters with different algorithms. Most of them are science products, i.e. produced on a non-routine basis by the involved scientific institutes. Twelve products are currently provided and distributed operationally by the ESA ENVISAT ground segment. Upcoming new algorithms, changing scientific attention with respect to regional atmospheric particularities or trends as well as the monitoring of instrument changes require continuous validation throughout the instrument's lifetime and even beyond, e.g. when data is expected to be used in long-term climatological studies.

The first years of the SCIAMACHY mission have proven the overall validation concept and provided initial results. A detailed overview of the validation results for the years 2002–2004 was given by Piters et al. (2006) and references therein. Since then, data products and their validation have been updated and upgraded on many occasions. As a result of the intensive work of the SQWG the new version 5 of the SCIAMACHY Ground Processor (SGP) has become operational in 2010. It improves data products delivered and operationally provides new data products. This version expands on the previous processor SGP V3.01 whose products have been validated in the recent years as described below. In addition, as a consequence of the creative work of the SCIAMACHY scientific retrieval teams, improved and new scientific products became available and were validated by mission participating institutions. A regularly updated summary of the validation status of operational and science data products can be found on http://www.sciamachy.org/products/.

Level 1 Irradiance and Reflectance UV-VIS-NIR-SWIR

The spectral solar irradiance is the most important extraterrestrial energy input into the Earth-Atmosphere system. SCIAMACHY measures the spectral solar irradiance from 214 to 2380 nm, with small gaps around 2000 and 2200 nm, on a daily basis. Skupin et al. (2005) demonstrated that after careful radiometric calibration, the solar spectral irradiance measured by SCIAMACHY agrees with other independently acquired datasets such as SIM, SOLSPEC, SOLSTICE, and SUSIM typically within 2–3%.

The solar radiation leaves the Earth's atmosphere after reflection at the surface and/or scattering in the atmosphere. Normalising this radiation by the incoming solar irradiance leads to the Sun-normalised spectral reflectance, an important parameter characterising the Earth's energy budget and being also the basis for some level 2 data products. Kokhanovsky et al. (2008) compared the top-of-atmosphere sun-normalised spectral reflectance measurements from SCIAMACHY, based on the up-to-date level 1b calibration, to those from MERIS and AATSR, all three operating on ENVISAT. They agree typically within 2–3% in the visible wavelength range, even taking into account the large spread in different reflectivity regimes from close to 1.0 over snow to close to 0.05 over the oceans. This agreement is within the calibration errors for the individual instruments and demonstrates the substantial progress made – due to SCIAMACHY's improved calibration – when compared with earlier intercomparisons (von Hoyningen-Huene et al. 2007; Tilstra and Stammes 2007).

Level 2 Products from Nadir UV-VIS-NIR

Ozone (O_3): The agreement of SCIAMACHY SGP V3.01 O_3 columns with correlative data from ground-based networks and satellites is within 0–2%, with a small underestimation of up to 5% at high solar zenith angles. The bias depends on solar zenith angle, season, viewing angle, and cloud fraction. Although dominated by comparison errors, the 3–10% scatter of the discrepancies seems to vary with the cloud optical depth derived from SCIAMACHY O_2 A-band spectra. When comparing on the long-term to ground-based stations, the difference in O_3 total column may show a small temporal drift, around 0.5% per year (Lerot et al. 2009). The drift will become smaller in the processor version 5, as instrument degradation is taken into account in the related level 1b product. The sign and amplitude of this trend seems to be variable from station to station (Fig. 9.10). SCIAMACHY O_3 columns generated by scientific algorithms also show a good agreement compared to ground-based data, with a bias of 1–1.5% (usually a slight underestimation) and a RMS of about 5% and the same trend as the operational product.

Nitrogen dioxide (NO_2): Validation of the SCIAMACHY SGP V3.01 NO_2 columns indicates good agreement to within 1–5 × 10^{14} molec/cm^2 with correlative data (from ground-based UV-VIS and FTIR, and from the GOME and GOME-2 satellite sensors) over areas free of tropospheric NO_2, although slightly low biased in the Southern Hemisphere, by about 5 × 10^{14} molec/cm^2 on average. The low bias exhibits a seasonal cycle and varies smoothly with latitude. Larger deviations are observed in cases of tropospheric pollution and have a clear correlation with cloud fraction and Air Mass Factor. The scientific NO_2 stratospheric columns also show good quality (Gil et al. 2008) as illustrated in Fig. 9.11. Large differences exist between the different tropospheric NO_2 column algorithms (Brinksma et al. 2008). Results of the CINDI 2009 intercomparison field campaign, still under investigation, are expected to bring more quantitative validation statements on tropospheric NO_2. Also first attempts were made to compare tropospheric NO_2 from SCIAMACHY with airborne (Martin et al. 2006) and boundary layer (Boersma et al. 2009) in situ measurements. In the latter case satellite data were linked to concentrations relevant for air quality applications.

Bromine oxide (BrO): While the previous operational processor SGP V3.01 did not include BrO as a deliverable trace gas, BrO columns are provided with the current level 2 processor version 5. BrO columns retrieved with scientific algorithms agree well with those obtained from GOME, GOME-2 and from ground-based measurements. This indicates that also the operationally generated results will be of good quality. For example, comparisons between ground-based BrO vertical columns measured over Reunion-Island (20.9°S, 55.5°E) and total BrO columns derived from SCIAMACHY nadir observations in a latitudinal band centred around 21°S present a good level of consistency (Theys et al. 2007).

Chlorine dioxide (OClO): Qualitatively, SCIAMACHY scientific OClO slant columns reproduce well OClO features measured by ground-based UV-VIS and airborne AMAX-DOAS spectrometers. They also agree well with GOME scientific retrievals. Quantitatively, the SCIAMACHY data show a relatively large offset and scatter compared to the GOME data. The accuracy of the SCIAMACHY OClO column retrieval appears to be much better at low column amounts of less than 0.5 × 10^{14} molec/cm^2, close to the detection limit. Oetjen et al. (2009) demonstrated excellent agreement between SCIAMACHY OClO slant columns and MAX-DOAS ground based data around 1 × 10^{13} molec/cm^2 or even lower (see Fig. 9.12).

Sulphur dioxide (SO_2): Quantitative validation of SCIAMACHY columns of SO_2 is hampered by the lack of independent measurements. Routine measurements near emission sources are currently being developed. A few international airborne campaigns were organised, e.g. INTEX and a campaign over Eastern China, during which SO_2 columns from SCIAMACHY and from OMI were validated with co-located in situ measurements. Validations conclude yearly average errors ranging from 10^{15} molec/cm^2 over clean ocean areas to 25 × 10^{15} molec/cm^2 over polluted regions (e.g. eastern China). This corresponds to errors of less than 10% in the first case and up to 35–70% in the second case (Lee et al. 2009).

Formaldehyde (HCHO): As in the case of SO_2, the quantitative validation of SCIAMACHY columns of HCHO requires independent measurements which are only rarely available. Comparisons to ground based DOAS measurements over Cabauw/The Netherlands and Nairobi/Kenya agree within their error bars with the tropospheric column derived from SCIAMACHY data (Wittrock et al. 2006). De Smedt et al (2008) compiled an error assessment of the HCHO product, indicating total errors of 20–40% for monthly and zonally averaged HCHO vertical columns.

Glyoxal (CHOCHO): Since glyoxal was for the first time detected from space with SCIAMACHY, quantitative vali-

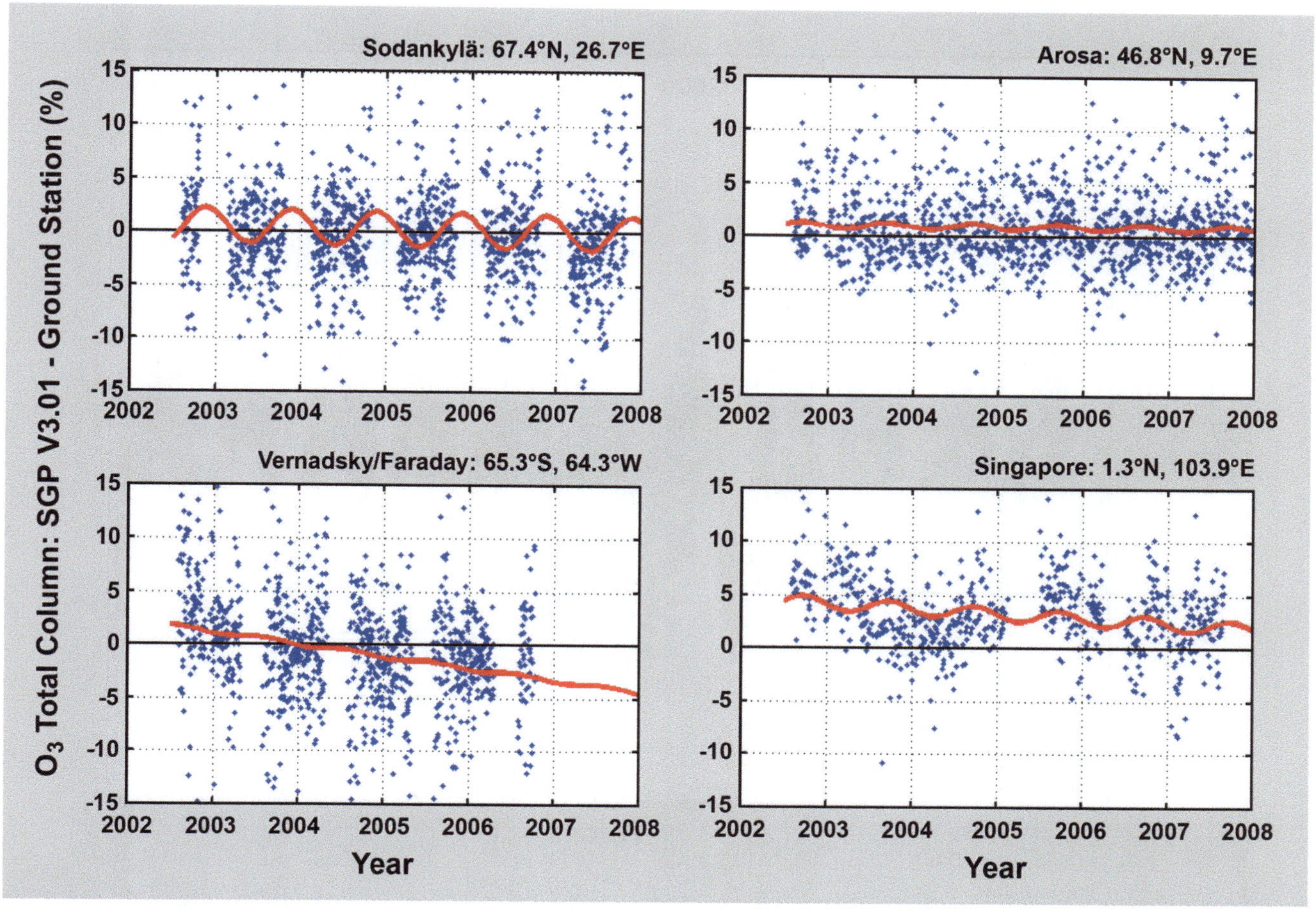

Fig. 9.10 The difference between total O_3 columns retrieved from SCIAMACHY and selected ground stations for the years 2002–2008. The red curve indicates seasonal variability and long-term trends obvious in the comparison. The trends range between −0.15% and −1.1% per year (Courtesy: adapted from Lerot et al. 2009).

dation of columns is even harder to accomplish due to the lack of independent measurements. Ground based DOAS measurements again from Cabauw/The Netherlands and Nairobi/Kenya indicate an agreement within the error bars for the tropospheric SCIAMACHY column (Wittrock et al. 2006). How the quality of such data relates to various chemical environments (urban, rural, ocean) is discussed by Vrekoussis et al. (2009) when comparing SCIAMACHY derived CHOCHO with in situ ground-based observations.

Water vapour (H_2O): Validation of the scientific SCIAMACHY AMC-DOAS water vapour columns V1.0 shows a systematic bias of −0.1 to −0.5 g/cm^2 with respect to co-located balloon-sonde observations (cryogenic frost point hygrometer) and SSM/I satellite observations, as well as ECMWF analyses (Fig. 9.13). This bias is attributed to the fact that SCIAMACHY retrievals are essentially cloud-cleared (Noël et al. 2005; Noël et al. 2007). The scatter on such comparisons, about 0.5 g/cm^2, is quite large due to the significant atmospheric variability of water vapour in the troposphere. Somewhat larger deviations are present between SCIAMACHY and SSM/I water vapour data over oceans.

Cloud Fraction (CF), Cloud Top Pressure/Cloud Top Height (CTP/CTH): The SCIAMACHY processor estimates the fractional cloud coverage (CF) using the OCRA algorithm, and then the cloud top height (CTH) or pressure (CTP) together with the cloud optical thickness (COT) using SACURA. The CF correlates well with scientific retrievals and with MODIS observations. From intercomparisons with ground-based cloud radar for single-layer cloud fields Kokhanovsky et al. (2009) concluded, that the uncertainty of the SACURA CTH retrieval is less than 0.34 km for low-level clouds and 2.22 km for high-level clouds with an underestimate in CTH on average for all clouds.

Absorbing Aerosol Index (AAI), Aerosol Optical Thickness (AOT): The SCIAMACHY scientific AAI compares well with TOMS results. Also the SCIAMACHY scientific AOT agrees reasonably well with MERIS data.

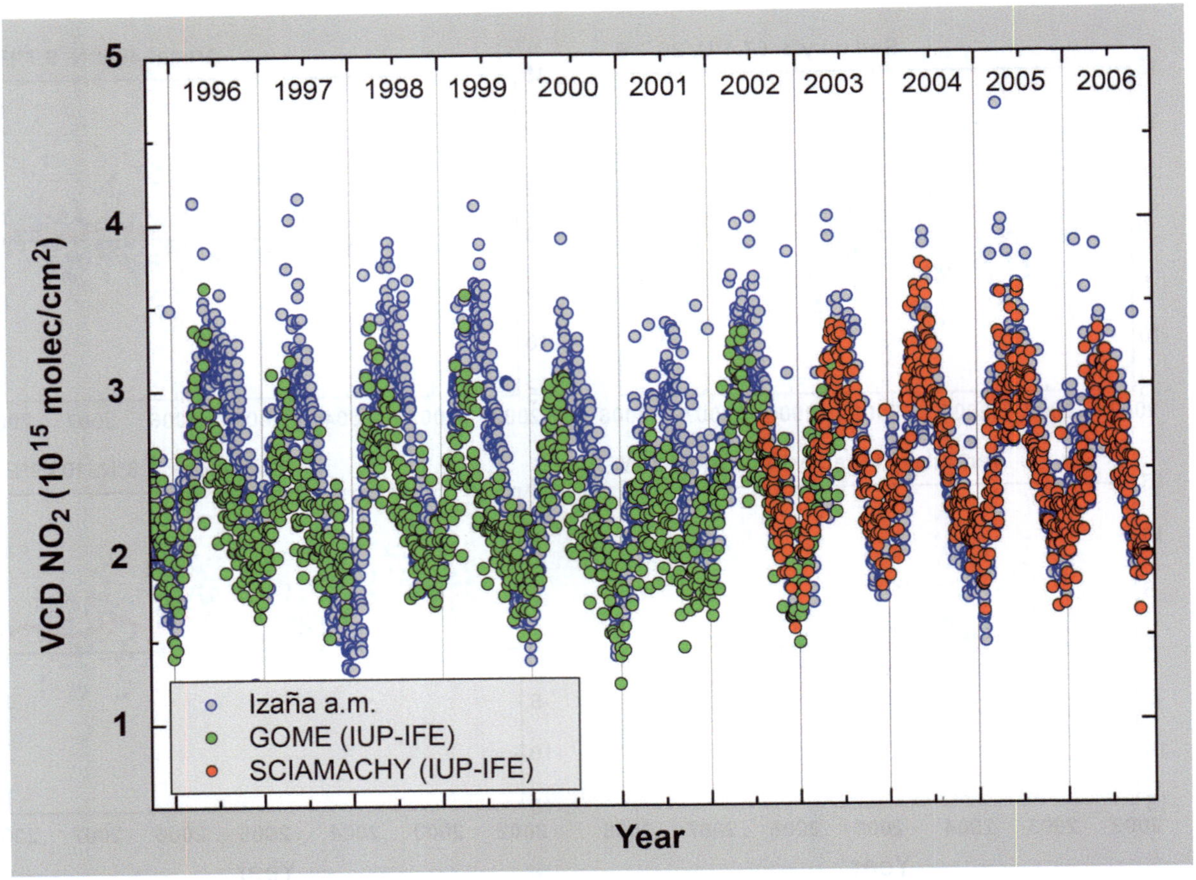

Fig. 9.11 NO_2 vertical column densities from SCIAMACHY nadir measurements (*red*) compared with ground-based morning data from Izaña (*grey*) and VCD from GOME (*green*). The GOME and SCIAMACHY VCD have been retrieved with a scientific algorithm from IUP-IFE, University of Bremen (Courtesy: Gil et al. 2008).

Level 2 Products from Nadir SWIR

The general potential of SCIAMACHY SWIR products had already been demonstrated, in particular its capabilities to detect areas of sources of CO, CH_4 and CO_2 and to track their long-range transport. Various algorithm versions have been validated by comparing their output with results from:

- Ground-based CO, CH_4, and CO_2 data from Fourier Transform Spectroscopy (FTS) via the pole-to-pole network of 12 NDACC-certified FTIR instruments (see e.g. Dils et al. 2006, 2007).
- Ship-based FTIR data taken during the bi-yearly cruises of the *Polarstern* vessel from Bremerhaven to Africa
- CO column data from the EOS-Terra MOPITT satellite
- CO and CH_4 data from the TM4 (KNMI), TM5 (IMAU), INCA (LSCE), and CTM2 (Oslo University) models
- Ancillary data such as fire maps produced by ERS-2 ATSR and EOS-AQUA MODIS

Observations from the Japanese TANSO/GOSAT satellite mission, launched in January 2009, are expected to contribute significantly to further characterise SCIAMACHY and GOSAT data quality.

Carbon monoxide (CO): Pre-launch estimates for the precision of scientific retrievals of SCIAMACHY nadir CO vertical columns amounted to 10% over land. The scientific CO data product derived with the WFM-DOAS algorithm has been compared with MOPITT (Buchwitz et al. 2007) and ground-based FTS measurements (Dils et al. 2006) finding a scatter of the CO product relative to the FTS retrievals of typically 20% for daily averages around the FTS sites, i.e. close to what was expected. Main reason for the somewhat larger scatter is the fact that not all CO lines can be used in the retrieval due to detector degradation. The mean bias is typically approximately 10%. Recently de Laat et al. (2010) analysed the quality of the SCIAMACHY CO data set for the years 2003 to 2007. Their main finding was that for stations not affected by local emissions or altitude effects, differences between SCIAMACHY and ground based FTS measurements are close to or within the SCIAMACHY CO total column precision of 0.1×10^{18}molecules/cm^2 (~5–10%) of the SCIAMACHY CO columns (de Laat et al. 2010).

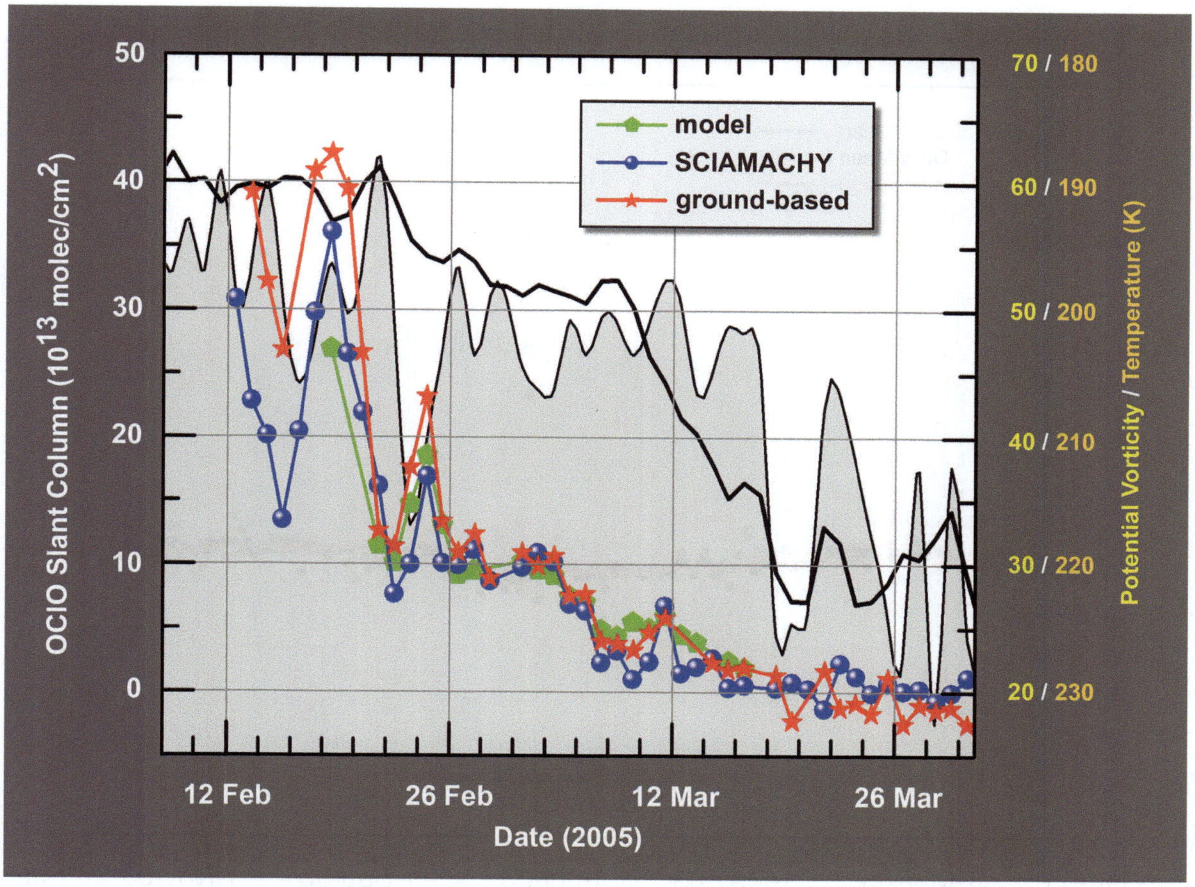

Fig. 9.12 OClO slant columns retrieved from SCIAMACHY data (*blue*) compared with ground-based measurements (*red*), sampled at the time of ENVISAT overpass, over Ny-Ålesund (79°N, 12°E) in spring 2005 and model data (*green*). Also shown are potential vorticity (*shaded area*) and temperature (*black solid line*) at the 475 K isentropic surface (Courtesy: adapted from Oetjen et al. 2009).

Initial results, illustrated in Fig. 9.14, demonstrate good agreement with CO columns from MOPITT. The data was already used in a variety of applications, like inverse modelling of emissions (Kopacz et al. 2010; Tangborn et al. 2009). CO is included in the operational processor version 5 released in 2010.

Methane (CH_4): High precision and accuracy of greenhouse gas measurements are a firm precondition for the potential improvement of existing emission catalogues by inverse modelling. With about 1–2% the pre-launch precision estimates for scientific retrievals of SCIAMACHY nadir CH_4 vertical columns were not far off the nominal requirements for such applications. A huge effort was spent in the last years to improve the precision and accuracy of real retrievals to finally reach the goal that SCIAMACHY CH_4 data is used for emission estimates (see Chapter 10). From the comparison with FTS and model (TM5) data and supported by an error analysis of the WFM-DOAS retrieval algorithm, the SCIAMACHY XCH_4 data set (column averaged mixing ratio) can be characterised globally by a single ground pixel retrieval precision of about 1.7% and a systematic low bias of about 1% (Schneising et al. 2009). Methane FTS observations, acquired over Paramaribo/Suriname, exhibit good agreement for the measured ratio of CH_4/CO_2 with SCIAMACHY results and represent the first validation of SCIAMACHY retrievals in the tropics using ground-based remote sensing techniques (Petersen et al. 2010).

Carbon dioxide (CO_2): As in the case of CH_4, the precision of scientific retrievals of SCIAMACHY nadir CO_2 vertical columns were estimated before launch to be 1–2%. Schneising et al. (2008) reported indeed that the XCO_2 dataset can be characterised globally by a single measurement retrieval precision (random error) of 1–2% (see Fig. 9.15), a systematic low bias of about 1.5%, and by a relative accuracy of about 1–2% for monthly averages at a spatial resolution of about 7° × 7°. In areas of enhanced atmospheric scattering due to aerosol or cirrus clouds the systematic errors could be higher. It is expected that in the near future the ongoing expansion of ground-based FTIR spectrometers to the near-infrared will improve capabilities for the validation of satellite greenhouse gas data products such as CO_2.

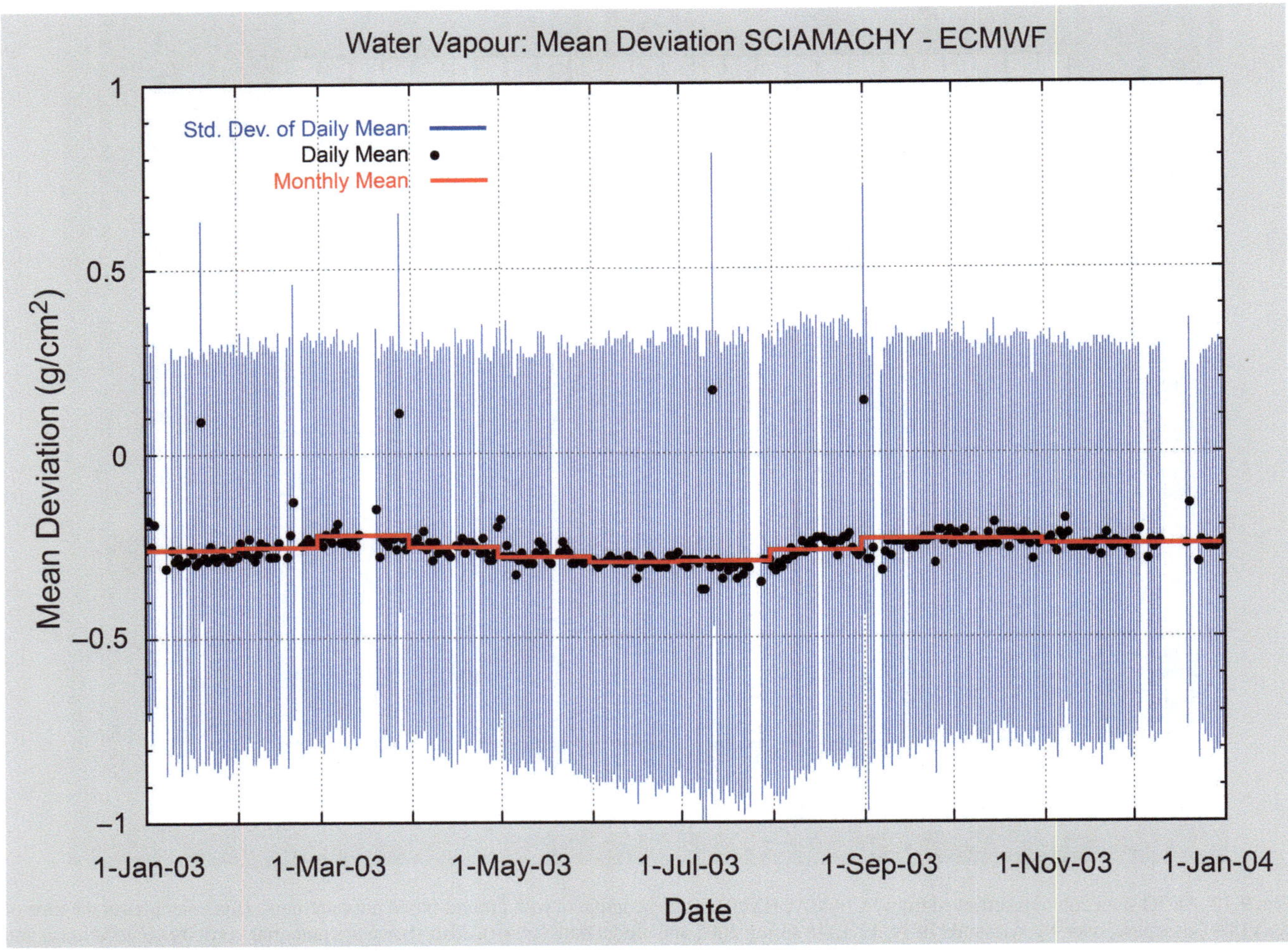

Fig. 9.13 AMC-DOAS V1.0 total column water vapour compared with model data provided by the European Centre for Medium-Range Weather Forecasts (ECMWF) over land and ocean for 2003. The comparison has been performed with co-located daily global averages on a 0.5° x 0.5° grid (Courtesy: S. Noël, IUP-IFE, University of Bremen).

Level 2 Products from Limb UV-VIS

Ozone (O_3): Vertical profiles of the ozone concentration retrieved from SCIAMACHY limb UV-VIS measurements with the SGP version V3.01 have been validated extensively against observations from ground-based networks (of ozonesondes, lidars, and microwave radiometers) and satellites, e.g. HALOE, SAGE-II, MLS (Fig. 9.16), ODIN, ACE, GOMOS, and MIPAS. Validations indicate that SCIAMACHY has a low bias (up to 10 %) around the ozone maximum and below, even after implementation of the limb pointing corrections. Due to the limited sensitivity of the retrievals below 20 km and above 40 km, the retrieval errors increase considerably past these altitudes. O_3 profiles retrieved with a scientific limb retrieval algorithm (Sonkaew et al. 2009) were additionally validated against ground-based and satellite data. Between 16 and 40 km the systematic bias of the retrieved profiles ranges within 5–10% compared to lidars and SAGE-II and is even smaller when MLS data are taken into account. Jones et al. (2009) assessed the quality of several long-term stratospheric O_3 satellite data records. They examined monthly average ozone values from various satellites for nine latitude and altitude bins covering 60°S–60°N and 20–45 km from the time period 1979–2008. The analysed data were from SAGE I/II, HALOE, SBUV/2, SMR (Submillimeter Radiometer), OSIRIS and SCIAMACHY. This investigation identified that instrumental drifts are not relevant for SCIAMCHY data in the mid northern latitudes, but become relevant for long-term trend assessment in the tropics and southern mid-latitudes. When comparing the SCIAMACHY O_3 profiles with observations from the ACE FTS solar occultation sensor the agreement is on average within 4% (Dupuy et al. 2009).

Nitrogen dioxide (NO_2): After photochemical correction for the time difference between SCIAMACHY limb data acquired in the morning and correlative solar occultation

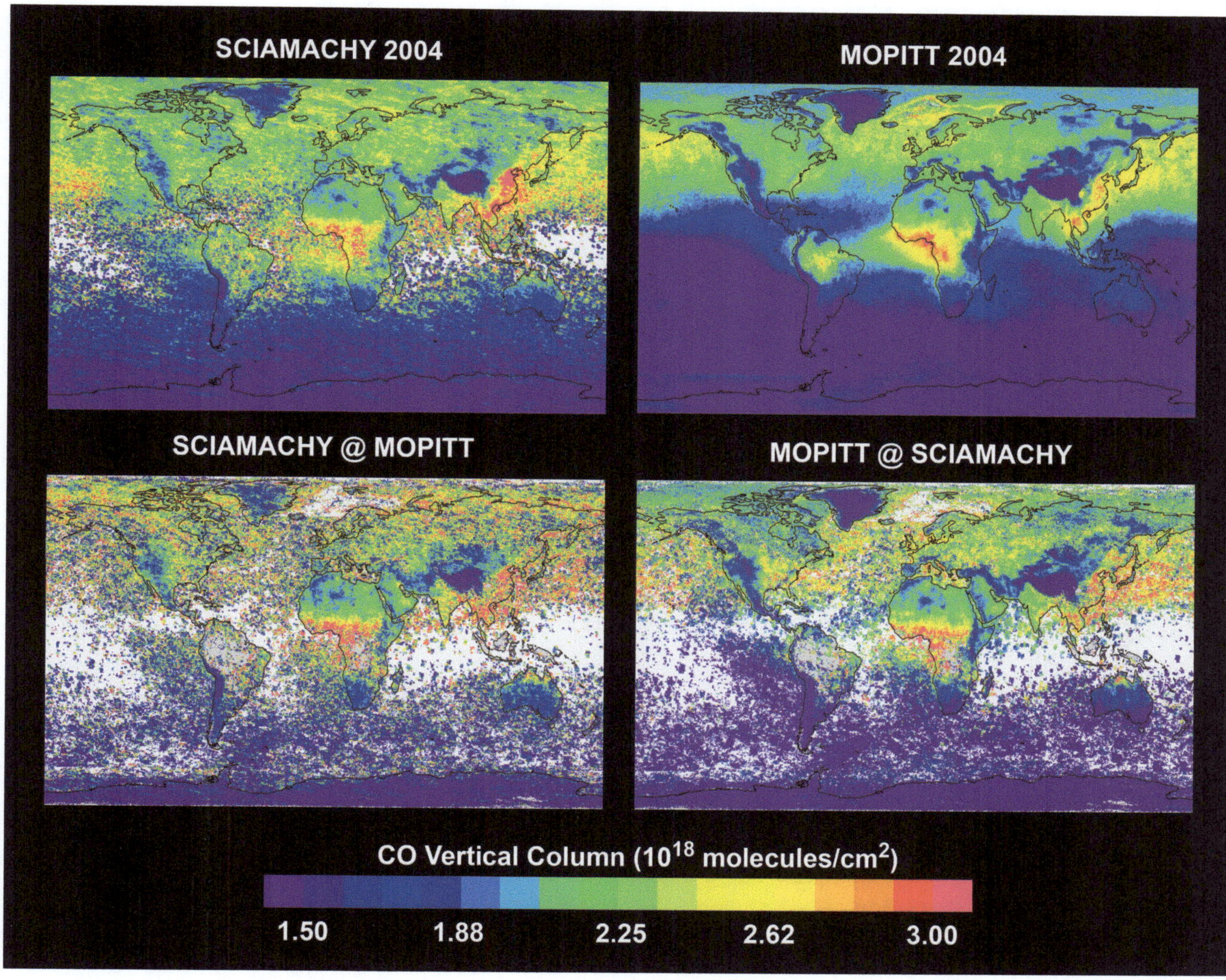

Fig. 9.14 Comparison of CO vertical columns for the year 2004 derived with SCIAMACHY (*top left*) and MOPITT (*top right*). The bottom row shows spatial coincidences between both sensors (Courtesy: Buchwitz et al. 2007).

profile measurements performed at twilight by other missions, SCIAMACHY limb NO_2 profiles retrieved with SGP V3.01 coincide well – on average at about 10–15% – with profiles derived by other spaceborne sensors. Comparisons with NO_2 profiles measured by ACE FTS (Kerzenmacher et al. 2008) show an even better coincidence (4% average). Further improved NO_2 retrieval algorithms are in preparation and are aiming at better results in the lower stratosphere towards the tropopause. Figure 9.17 illustrates, as an example, the results of such an algorithm when applied to SCIAMACHY limb data with an agreement of better then 10% for even a single balloon-borne measurement.

Bromine oxide (BrO): Figure 9.18 displays 5 years of SCIAMACHY BrO profiles, retrieved with a scientific algorithm (Rozanov et al. 2005), and how they match with ground-based UV-VIS profile measurements in Harestua/Norway (60°N, 11°E). SCIAMACHY and ground-based UV-VIS columns integrated over the 15–27 km range are in good agreement, with SCIAMACHY columns being lower than ground-based columns by 2% ± 20%. Both datasets reflect markedly the seasonal variability of the BrO column (Hendrick et al. 2009). Dorf et al. (2006) showed – using photochemical corrections for balloon observations along calculated air mass trajectories – that the SCIAMACHY BrO profiles coincide on average within approximately 20% with the balloon data.

Chlorine dioxide (OClO): Experimental OClO profile measurements were compared to non-co-located balloon data from earlier campaigns. The values of the SCIAMACHY OClO data and their variability are similar to those from the balloon-borne measurements, indicating that SCIAMACHY is providing geophysical meaningful OClO profiles. In addition, comparisons of SCIAMACHY OClO with ClO from ODIN SMR – both species are photo-chemically related – are supporting these findings (Kühl et al. 2008).

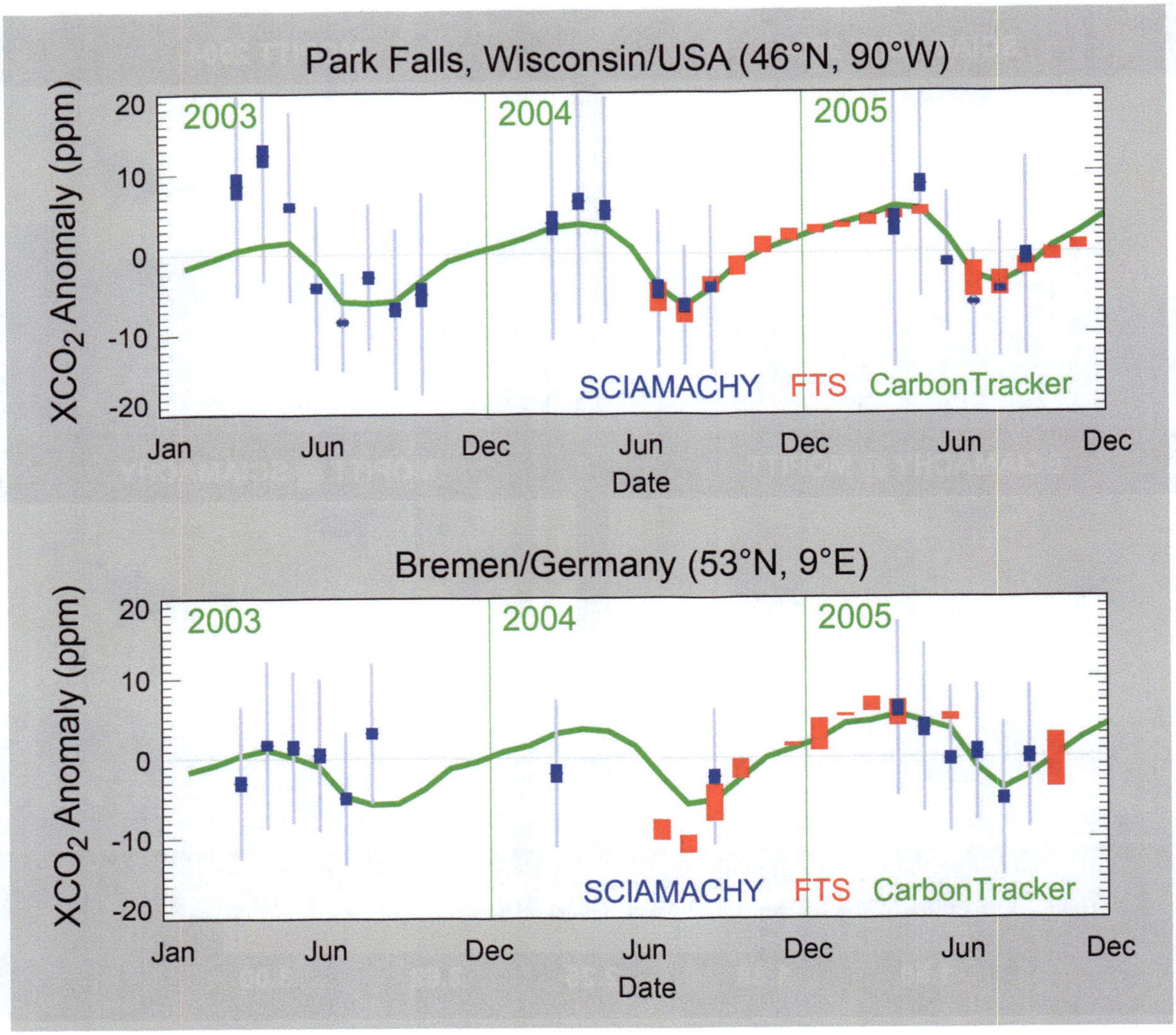

Fig. 9.15 Comparison of SCIAMACHY XCO_2 (*blue*) with ground based Fourier Transform Spectroscopy (FTS) measurements (*red*) for Park Falls (*top*) and Bremen (*bottom*). Also included are corresponding CarbonTracker results (*green*). Shown are comparisons of XCO_2 anomalies, i.e. the corresponding mean values have been subtracted. All quality filtered SCIAMACHY measurements within a radius of 350 km around the ground station are considered for the comparison (Courtesy: Schneising et al. 2008).

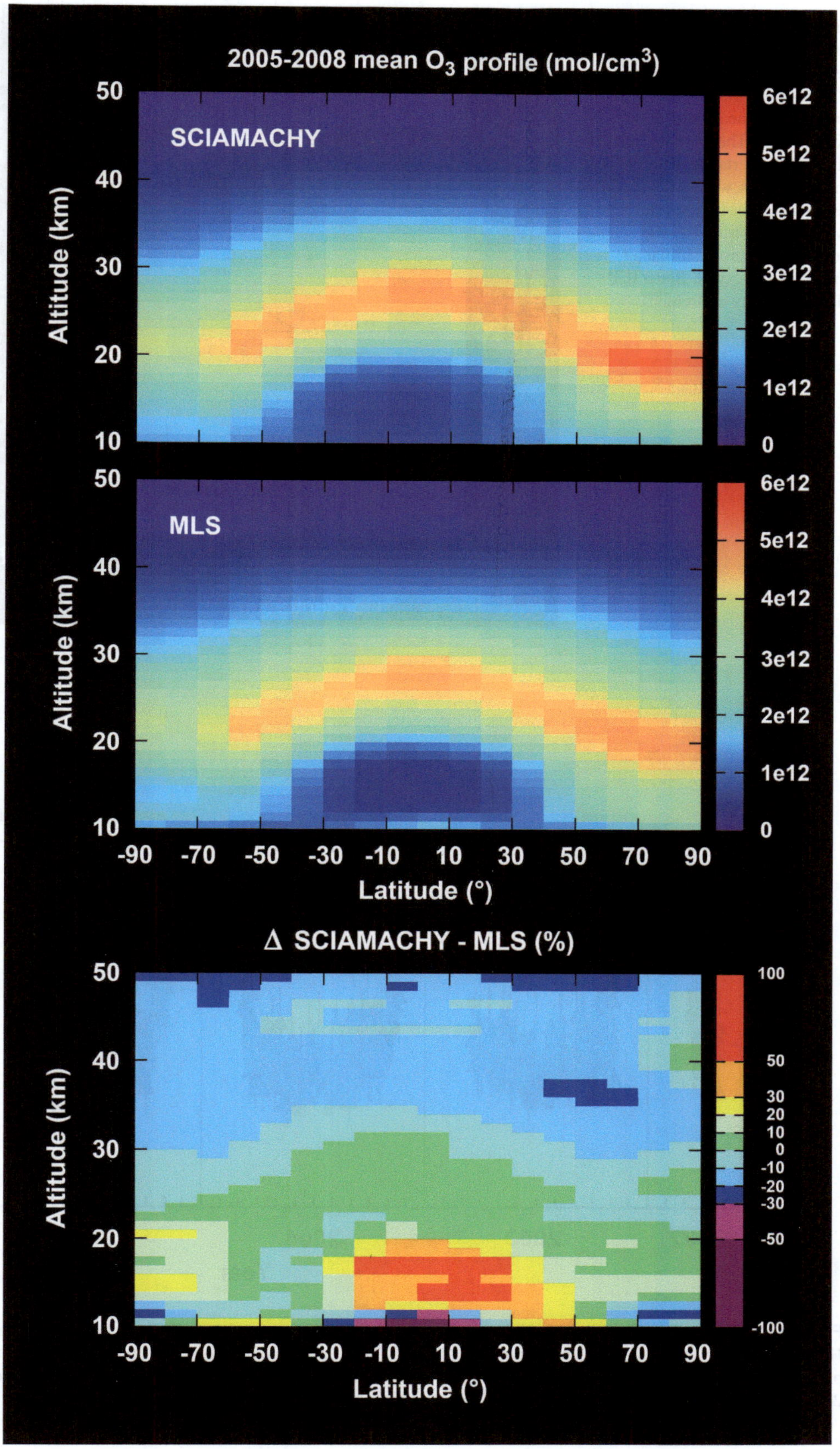

Fig. 9.16 Zonal mean O_3 from SCIAMACHY vertical profiles averaged from 2005 to 2008. Only co-located daily measurements with MLS have been used, where the criteria for co-locations are 100 km spatial and 10 h temporal difference. Co-locations with MLS measurements during night with solar zenith angles >90° have been excluded (Courtesy: S. Mieruch, IUP-IFE, University of Bremen).

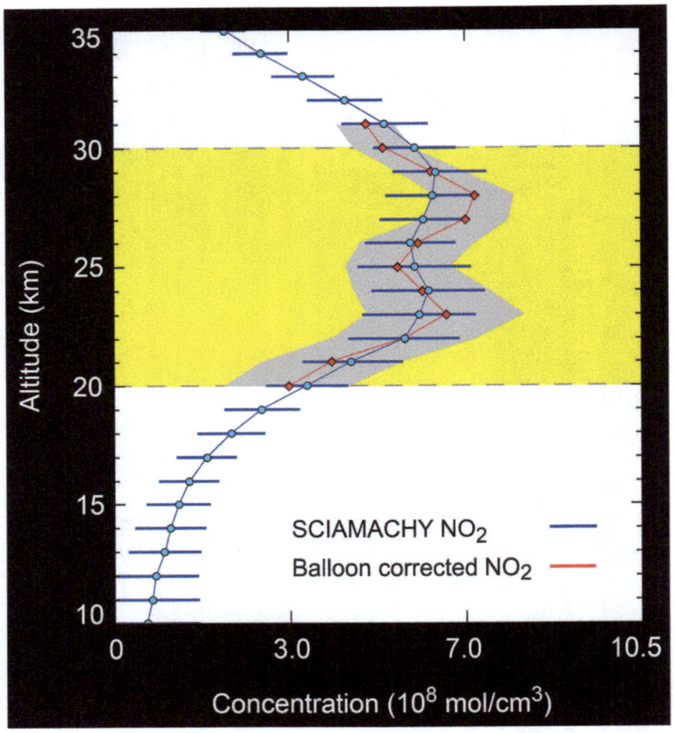

Fig. 9.17 Comparison between a SCIAMACHY NO_2 profile (*blue*) retrieved with the SCIATRAN V3.1 algorithm and co-located balloon NO_2 measurements at Kiruna (*red*). The grey shaded area shows the uncertainty of the balloon measurements, whereas the yellow region indicates the altitude range where both instruments are considered to probe similar air masses (Courtesy: R. Bauer IUP-IFE, University of Bremen).

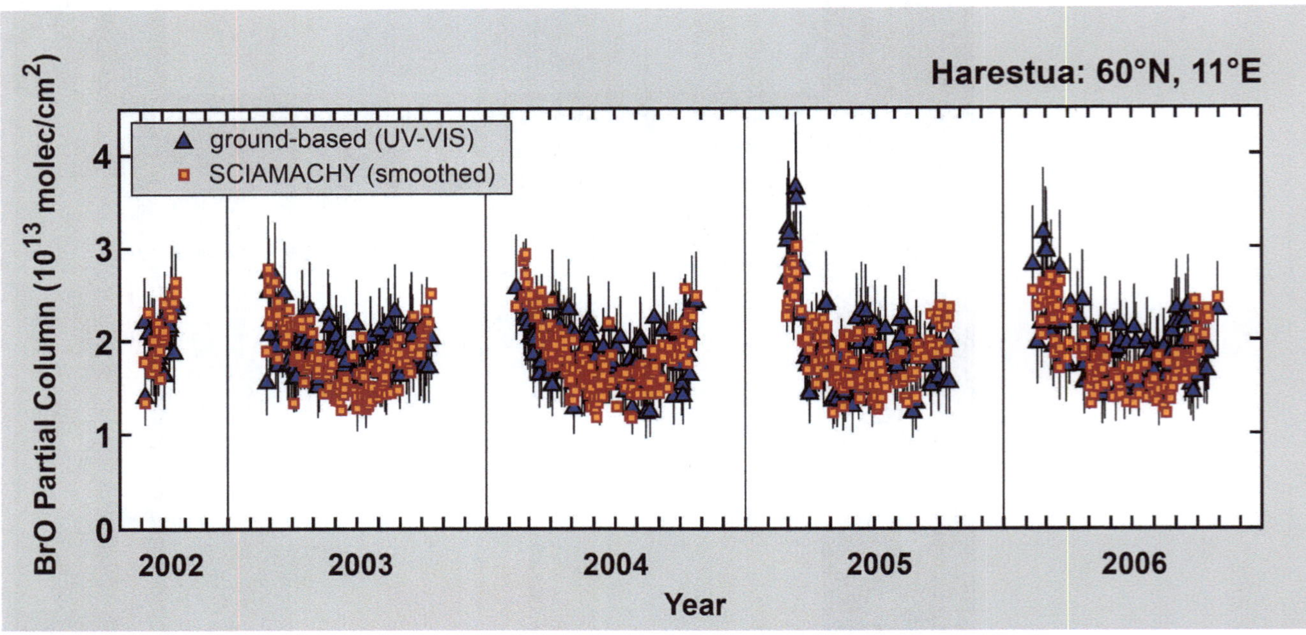

Fig. 9.18 Intercomparison of the 15–27 km BrO partial columns calculated from SCIAMACHY limb and ground-based UV-VIS profiles at Harestua/Norway for 2002–2006 (morning coincidences). To reduce comparison errors due to differences in vertical smoothing of the vertical profile, SCIAMACHY profiles were smoothed using coincident ground-based UV-VIS averaging kernels (Courtesy: adapted from Hendrick et al. 2009).

Water vapour (H_2O): Recently it was successfully demonstrated to derive H_2O concentrations in the upper troposphere and lower stratosphere region from SCIAMACHY limb measurements. Montoux et al. (2009) used data from the HIBISCUS campaign in Bauru/Brazil in the February/March 2004 period for validating satellite water vapour measurements in the southern tropical and subtropical UTLS region. Preliminary analysis suggests that the SCIAMACHY results reproduce the atmospheric variability of water vapour around the tropopause and indicate a dryer atmosphere by 20% when compared to HALOE.

A common issue to all limb viewing instruments is the accurate determination of the limb pointing and in particular of the tangent height. New corrections in tangent height registration and of misalignment parameters, already implemented in SGP V3.01 (Gottwald et al. 2007), yielded a reduction in the east-west offset and in the bias of the altitude registration of the limb profile products, being now smaller than about 200 m (see Section 6.5).

References

Afe OT, Richter A, Sierk B, Wittrock F, Burrows JP (2004) BrO emission from volcanoes: a survey using GOME and SCIAMACHY measurements. Geophys Res Lett 31:L24113. doi:10.1029/2004GL020994

Boersma KF, Jacob DJ, Trainic M, Rudich Y, DeSmedt I, Dirksen R, Eskes HJ (2009) Validation of urban NO_2 concentrations and their diurnal and seasonal variations observed from space (SCIAMACHY and OMI sensors) using in situ measurements in Israeli cities. Atmos Chem Phys 9:3867–3879

Borchi F, Pommereau J-P (2007) Evaluation of ozonesondes, HALOE, SAGE II and III, ODIN-OSIRIS and SMR, and ENVISAT-GOMOS, -SCIAMACHY and -MIPAS ozone profiles in the tropics from SAOZ long duration balloon measurements in 2003 and 2004. Atmos Chem Phys 7:2671–2690

Brenninkmeijer CAM, Crutzen P, Boumard F, Dauer T, Dix B, Ebinghaus R, Filippi D, Fischer H, Franke H, Frieß U, Heintzenberg J, Helleis F, Hermann M, Kock HH, Koeppel C, Lelieveld J, Leuenberger M, Martinsson BG, Miemczyk S, Moret HP, Nguyen HN, Nyfeler P, Oram D, O'Sullivan D, Penkett S, Platt U, Pupek M, Ramonet M, Randa B, Reichelt M, Rhee TS, Rohwer J, Rosenfeld K, Scharffe D, Schlager H, Schumann U, Slemr F, Sprung D, Stock P, Thaler R, Valentino F, van Velthoven P, Waibel A, Wandel A, Waschitschek K, Wiedensohler A, Xueref-Remy I, Zahn A, Zech U, Ziereis H (2007) Civil aircraft for the regular investigation of the atmosphere based on an instrumented container: the new CARIBIC system. Atmos Chem Phys 7:4953–4976

Brinksma EJ, Pinardi G, Braak R, Volten H, Richter A, Schönhardt A, van Roozendael M, Fayt C, Hermans C, Dirksen RJ, Vlemmix T, Berkhout AJC, Swart DPJ, Oetjen H, Wittrock F, Wagner T, Ibrahim OW, de Leeuw G, Moerman M, Curier RL, Celarier EA, Knap WH, Veefkind JP, Eskes HJ, Allaart M, Rothe R, Piters AJM, Levelt PF (2008) The 2005 and 2006 DANDELIONS NO_2 and aerosol validation campaigns. J Geophys Res 113, D16S46. doi:10.1029/2007JD008808.

Buchwitz M, Khlystova I, Bovensmann H, Burrows JP (2007) Three years of global carbon monoxide from SCIAMACHY: comparison with MOPITT and first results related to the detection of enhanced CO over cities. Atmos Chem Phys 7:2399–2411

de Laat ATJ, Gloudemans AMS, Schrijver H, Aben I, Nagahama Y, Suzuki K, Mahieu E, Jones NB, Paton-Walsh C, Deutscher NM, Griffith DW T, De Mazière M, Mittelmeier R, Fast H, Notholt J, Palm M, Hawat T, Blumenstock T, Rinsland C, Dzhola AV, Grechko EI, Poberovskii AM, Makarova MV, Mellqvist J, Strandberg A, Sussmann R, Borsdorff T, Rettinger M (2010) Validation of five years (2003–2007) of SCIAMACHY CO total column measurements using ground-based spectrometer observations, Atmos Meas Tech 3:1457–1471

De Smedt I, Müller J -F, Stavrakou T, van der AR, Eskes H, Van Roozendael M (2008) Twelve years of global observations of formaldehyde in the troposphere using GOME and SCIAMACHY sensors. Atmos Chem Phys 8:4947–4963

Dils B, De Mazière M, Müller JF, Blumenstock T, Buchwitz M, de Beek R, Demoulin P, Duchatelet P, Fast H, Frankenberg C, Gloudemans A, Griffith D, Jones N, Kerzenmacher T, Kramer I, Mahieu E, Mellqvist J, Mittermeier RL, Notholt J, Rinsland CP, Schrijver H, Smale D, Strandberg A, Straume AG, Stremme W, Strong K, Sussmann R, Taylor J, van den Broek M, Velazco V, Wagner T, Warneke T, Wiacek A, Wood S (2006) Comparisons between SCIAMACHY scientific products and ground-based FTIR data for total columns of CO, CH_4, CO_2 and N_2O. Atmos Chem Phys 6:1953–1976

Dils B, De Mazière M, Müller J -F, Buchwitz M, de Beek R, Frankenberg C, Gloudemans A, Schrijver H, Van den Broek M, Contributing NDSC FTIR teams (2007) The evaluation of SCIAMACHY CO, CH_4, CO_2 and N_2O scientific data, using ground-based FTIR measurements. In: Burrows JP, Borrell P (eds) Measuring tropospheric trace constituents from space, ACCENT-TROPOSAT-2 in 2005–6, pp 258–262

Dorf M, Bösch H, Butz A, Camy-Peyret C, Chipperfield MP, Engel A, Goutail F, Grunow K, Hendrick F, Hrechanyy S, Naujokat B, Pommereau J-P, Van Roozendael M, Sioris C, Stroh F, Weidner F, Pfeilsticker K (2006) Balloon-borne stratospheric BrO measurements: comparison with Envisat/SCIAMACHY BrO limb profiles. Atmos Chem Phys 6:2483–2501

Dupuy E, Walker KA, Kar J, Boone CD, McElroy CT, Bernath PF, Drummond JR, Skelton R, McLeod SD, Hughes RC, Nowlan CR, Dufour DG, Zou J, Nichitiu F, Strong K, Baron P, Bevilacqua RM, Blumenstock T, Bodeker GE, Borsdorff T, Bourassa AE, Bovensmann H, Boyd IS, Bracher A, Brogniez C, Burrows JP, Catoire V, Ceccherini S, Chabrillat S, Christensen T, Coffey MT, Cortesi U, Davies J, De Clercq C, Degenstein DA, De Mazière M, Demoulin P, Dodion J, Firanski B, Fischer H, Forbes G, Froidevaux L, Fussen D, Gerard P, Godin-Beekmann S, Goutail F, Granville J, Griffith D, Haley CS, Hannigan JW, Höpfner M, Jin JJ, Jones A, Jones NB, Jucks K, Kagawa A, Kasai Y, Kerzenmacher TE, Kleinböhl A, Klekociuk AR, Kramer I, Küllmann H, Kuttippurath J, Kyrölä E, Lambert J-C, Livesey NJ, Llewellyn EJ, Lloyd ND, Mahieu E, Manney GL, Marshall BT, McConnell JC, McCormick MP, McDermid IS, McHugh M, McLinden CA, Mellqvist J, Mizutani K, Murayama Y, Murtagh DP, Oelhaf H, Parrish A, Petelina SV, Piccolo C, Pommereau J-P, Randall CE, Robert C, Roth C, Schneider M, Senten C, Steck T, Strandberg A, Strawbridge KB, Sussmann R, Swart DPJ, Tarasick DW, Taylor JR, Tétard C, Thomason LW, Thompson AM, Tully MB, Urban J, Vanhellemont F, Vigouroux C, von Clarmann T, von der Gathen P, von Savigny C, Waters JW, Witte JC, Wolff M, Zawodny JM (2009) Validation of ozone measurements from the Atmospheric chemistry experiment (ACE). Atmos Chem Phys 9:287–343

Fioletov E, Kerr JB, Hare EW, Labow GJ, McPeters RD (1999) An assessment of the world ground-based total ozone network performance from the comparison with satellite data. J Geophys Res 104:1737–1747

Fix A, Ehret G, Flentje H, Poberaj G, Gottwald M, Finkenzeller H, Bremer H, Bruns M, Burrows JP, Kleinböhl A, Küllmann H, Kuttippurath J, Richter A, Wang P, Heue K-P, Platt U, Pundt I, Wagner T (2005) SCIAMACHY validation by aircraft remote measurements: design, execution, and first results of the SCIA-VALUE mission. Atmos Chem Phys 5:1273–1289

Gil M, Yela M, Gunn LN, Richter A, Alonso I, Chipperfield MP, Cuevas E, Iglesias J, Navarro M, Puentedura O, Rodrguez S (2008) NO_2 climatology in the northern subtropical region: diurnal, seasonal and interannual variability. Atmos Chem Phys 8:1635–1648

Gottwald M, Krieg E, von Savigny C, Noël S, Bramstedt K (2007) SCIAMACHY extra misalignment model (PO-TN-DLR-SH-0016). Technical Document, DLR

Heland J, Schlager H, Schiller C, Sitnikov N, Ulanovsky A, Ravegnani F, Volk CM, Werner A, Petritoli A, Kostadinov I, Giovanelli G, Bortoli D, Stroh F, von Hobe M, The Geophysika Team (2003) Validation of MIPAS on Envisat by in situ instruments on the M55-Geophysika. Proceedings on Envisat validation workshop, Frascati, Italy, ESA SP-531

Hendrick F, Rozanov A, Johnston PV, Bovensmann H, De Mazière M, Fayt C, Hermans C, Kreher K, Lotz W, Sinnhuber B-M, Theys N, Thomas A, Burrows JP, Van Roozendael M (2009) Multi-year comparison of stratospheric BrO vertical profiles retrieved from SCIAMACHY limb and ground-based UV-visible measurements. Atmos Meas Tech 2:273–285

Heue K -P, Richter A, Bruns M, Burrows JP, v. Friedeburg C, Platt U, Pundt I, Wang P, Wagner T (2005) Validation of SCIAMACHY tropospheric NO_2 columns with AMAXDOAS measurements. Atmos Chem Phys 5:1039–1051

Jones A, Urban J, Murtagh DP, Eriksson P, Brohedel S, Haley C, Degenstein D, Bourassa A, von Savigny C, Sonkaew T, Rozanov A, Bovensmann H, Burrows J (2009) Evolution of stratospheric ozone and water vapour time series studied with satellite measurements. Atmos Chem Phys 9:6055–6075

Kerzenmacher T, Wolff MA, Strong K, Dupuy E, Walker KA, Amekudzi LK, Batchelor RL, Bernath PF, Berthet G, Blumenstock T, Boone CD, Bramstedt K, Brogniez C, Brohede S, Burrows JP, Catoire V, Dodion J, Drummond JR, Dufour DG, Funke B, Fussen D, Goutail F, Griffith DWT, Haley CS, Hendrick F, Höpfner M, Huret N, Jones N, Kar J, Kramer I, Llewellyn EJ, López-Puertas M, Manney G, McElroy CT, McLinden CA, Melo S, Mikuteit S, Murtagh D, Nichitiu F, Notholt J, Nowlan C, Piccolo C, Pommereau J-P, Randall C, Raspollini P, Ridolfi M, Richter A, Schneider M, Schrems O, Silicani M, Stiller GP, Taylor J, Tétard C, Toohey M, Vanhellemont F, Warneke T, Zawodny JM, Zou J (2008) Validation of NO_2 and NO from the Atmospheric chemistry experiment (ACE). Atmos Chem Phys 8:5801–5841

Kokhanovsky AA, Schreier M, von Hoyningen-Huene W (2008) The comparison of spectral top-of-atmosphere reflectances measured by AATSR, MERIS, and SCIAMACHY Onboard ENVISAT. IEEE Geosci Remote Sens Lett 5(1):53

Kokhanovsky AA, Naud CM, Devasthale A (2009) Intercomparison of ground-based radar and satellite cloud-top height retrievals for overcast single-layered cloud fields. IEEE Trans Geosci Remote Sens 47(1):1901

Kopacz M, Jacob DJ, Fisher JA, Logan JA, Zhang L, Megretskaia IA, Yantosca RM, Singh K, Henze DK, Burrows JP, Buchwitz M, Khlystova I, McMillan WW, Gille JC, Edwards DP, Eldering A, Thouret V, Nedele P (2010) Global estimates of CO sources with high resolution by adjoint inversion of multiple satellite datasets (MOPITT, AIRS, SCIAMACHY, TES). Atmos Chem Phys 10:855–876

Kostadinov I, Giovanelli G, Petritoli A, Bartoli D, Ravegnani F, Radaelli G, Ulanovsky A, Yuzhkov V (2003) Combined in-situ and quasi in-situ measurements aboard the M55 Geophysika stratospheric aircraft dedicated for ENVISAT satellite data validation. Proceedings on ENVISAT validation workshop, Frascati, Italy, ESA SP-531

Kühl S, Pukite J, Deutschmann T, Platt U, Wagner T (2008) SCIAMACHY limb measurements of NO_2, BrO and OClO. Retrieval of vertical profiles: algorithm, first results, sensitivity and comparison studies. Adv Space Res 42(10):1747–1764

Kurylo MJ, Zander RJ (2001) The NDSC – Its status after ten years of operation. In: NASDA (ed) Proceedings of Quadrennial ozone symposium 2000, Sapporo, Japan, pp 167–168

Lambert J-C, Van Roozendael M, De Mazière M, Simon PC, Pommereau J-P, Goutail F, Sarkissian A, Gleason JF (1999) Investigation of pole-to-pole performances of spaceborne atmospheric chemistry sensors with the NDSC. J Atmos Sci 56:176–193

Lee C, Martin RV, van Donkelaar A, O'Byrne G, Krotkov N, Richter A, Huey LG, Holloway JS (2009) Retrieval of vertical columns of sulfur dioxide from SCIAMACHY and OMI: air mass factor algorithm development, validation, and error analysis. J Geophys Res 114:D22303. doi:10.1029/2009JD012123

Lerot C, Van Roozendael M, van Geffen J, van Gent J, Fayt C, Spurr R, Lichtenberg G, von Bargen A (2009) Six years of total ozone column measurements from SCIAMACHY nadir observations. Atmos Meas Tech 2:87–98

Marenco A, Thouret V, Nédélec P, Smit H, Helten M, Kley D, Karcher F, Simon P, Law K, Pyle J, Poschmann G, von Wrede R, Hume C, Cook T (1998) Measurement of ozone and water vapor by Airbus in-service aircraft: the MOZAIC airborne program, an overview. J Geophys Res 103:25631–25642

Martin RV, Sioris CE, Chance K, Ryerson TB, Bertram TH, Wooldridge PJ, Cohen RC, Neuman JA, Swanson A, Flocke FM (2006) Evaluation of space-based constraints on global nitrogen oxide emissions with regional aircraft measurements over and downwind of eastern North America. J Geophys Res 111:D15308. doi:10.1029/2005JD006680

Montoux N, Hauchecorne A, Pommereau J-P, Lefèvre F, Durry G, Jones RL, Rozanov A, Dhomse S, Burrows JP, Morel B, Bencherif H (2009) Evaluation of balloon and satellite water vapour measurements in the Southern tropical and subtropical UTLS during the HIBISCUS campaign. Atmos Chem Phys 9:5299–5319

Noël S, Buchwitz M, Bovensmann H, Burrows JP (2005) Validation of SCIAMACHY AMC-DOAS water vapour columns. Atmos Chem Phys 5:1835–1841

Noël S, Mieruch S, Bovensmann H, Burrows JP (2007) A combined GOME and SCIAMACHY global water vapour data set. Proceedings of ENVISAT symposium, Montreux, Switzerland, ESA SP-636

Oelhaf H, Wetzel G, Kleinert A, Friedl-Vallon F, Maucher G (2009) Long-term validation of MIPAS products based on balloon measurements. Proceedings of ESA Atmospheric science conference, Barcelona, Spain, ESA SP-676

Oetjen H, Wittrock F, Richter A, Chipperfield MP, Medeke T, Sheode N, Sinnhuber B-M, Sinnhuber M, Burrows JP (2009) Evaluation of stratospheric chlorine chemistry for the Arctic spring 2005 using modelled and measured OClO column densities. Atmos Chem Phys Discuss 9:26539–26575

Petersen AK, Warneke T, Frankenberg C, Bergamaschi P, Gerbig C, Notholt J, Buchwitz M, Schneising O, Schrems O (2010) First ground-based FTIR-observations of methane in the tropics. Atmos Chem Phys 10:7231–7239

Piters AJM, Bramstedt K, Lambert J-C, Kirchhoff B (2006) Overview of SCIAMACHY validation: 2002–2004. Atmos Chem Phys 6:127–148

Rohen GJ, v. Savigny C, Sinnhuber M, Llewellyn EJ, Kaiser JW, Jackman CH, Kallenrode M -B, Schröter J, Eichmann K -U, Bovensmann H, Burrows JP (2005) Ozone depletion during the solar proton events of October/November 2003 as seen by SCIAMACHY. J Geophys Res 110, A09S39. doi:10.1029/2004JA010984

Rozanov A, Bovensmann H, Bracher A, Hrechany S, Rozanov V, Sinnhuber M, Stroh F, Burrows JP (2005) NO_2 and BrO vertical profile retrieval from SCIAMACHY limb measurements: sensitivity studies. Adv Space Res 36:846–854

Schneising O, Buchwitz M, Burrows JP, Bovensmann H, Reuter M, Notholt J, Macatangay R, Warneke T (2008) Three years of greenhouse gas column-averaged dry air mole fractions retrieved from satellite - Part 1: Carbon dioxide. Atmos Chem Phys 8:3827–3853

Schneising O, Buchwitz M, Burrows JP, Bovensmann H, Bergamaschi P, Peters W (2009) Three years of greenhouse gas column-averaged dry air mole fractions retrieved from satellite – Part 2: Methane. Atmos Chem Phys 9:443–465

SCIAVALIG (1998) SCIAMACHY Validation Requirements (SVDS-01). Technical Document, KNMI and NIVR. http://www.sciamachy.org/validation/

SCIAVALIG (2002) SCIAMACHY Detailed Validation Plan (SVDS-04). Technical Document, KNMI and NIVR. http://www.sciamachy.org/validation/.

Skupin J, Weber M, Noël S, Bovensmann H, Burrows JP (2005) GOME and SCIAMACHY solar spectral irradiance and Mg II solar activity proxy indicator. Memorie della Societa Astronomica Italiana 76:1038–1041

Sonkaew T, Rozanov VV, von Savigny C, Rozanov A, Bovensmann H, Burrows JP (2009) Cloud sensitivity studies for stratospheric and lower mesospheric ozone profile retrievals from measurements of limb-scattered solar radiation. Atmos Meas Tech 2:653–678

Tangborn A, Stajner I, Buchwitz M, Khlystova I, Pawson S, Burrows J, Hudman R, Nedelec P (2009) Assimilation of SCIAMACHY CO observations: global and regional analysis of data impact. J Geophys Res 114, D07307, 1–11. doi:10.1029/2008JD010781

Theys N, Van Roozendael M, Hendrick F, Fayt C, Hermans C, Baray J-L, Goutail F, Pommereau J-P, De Mazière M (2007) Retrieval of stratospheric and tropospheric BrO columns from multi-axis DOAS measurements at Reunion Island (21 S, 56 E). Atmos Chem Phys 7:4733–4739

Tilstra LG, Stammes P (2007) Earth reflectance and polarization intercomparison between SCIAMACHY onboard, Envisat and POLDER onboard ADEOS-2. J Geophys Res 112:D11304. doi:10.1029/2006JD007713

von Hoyningen-Huene W, Kokhanovsky AA, Wuttke MW, Buchwitz M, Noël S, Gerilowski K, Burrows JP, Latter B, Siddans R, Kerridge BJ (2007) Validation of SCIAMACHY top-of-atmosphere reflectance for aerosol remote sensing using MERIS L1 data. Atmos Chem Phys 7:97–106

von Savigny C, Rozanov A, Bovensmann H, Eichmann K-U, Noël S, Rozanov VV, Sinnhuber B-M, Weber M, Burrows JP (2005) The ozone hole break-up in September 2002 as seen by SCIAMACHY on ENVISAT. J Atmosph Sci 62:721–734

Vrekoussis M, Wittrock F, Richter A, Burrows JP (2009) Temporal and spatial variability of glyoxal as observed from space. Atmos Chem Phys 9:4485–4504

Wittrock F, Richter A, Oetjen H, Burrows JP, Kanakidou M, Myriokefalitakis S, Volkamer R, Beirle S, Platt U, Wagner T (2006) Simultaneous global observations of glyoxal and formaldehyde from space. Geophys Res Lett 33:L16804. doi:10.1029/2006GL026310

Chapter 10
SCIAMACHY's View of the Changing Earth's Environment

H. Bovensmann, I. Aben, M. Van Roozendael, S. Kühl, M. Gottwald, C. von Savigny, M. Buchwitz, A. Richter, C. Frankenberg, P. Stammes, M. de Graaf, F. Wittrock, M. Sinnhuber, B.M. Sinnhuber, A. Schönhardt, S. Beirle, A. Gloudemans, H. Schrijver, A. Bracher, A.V. Rozanov, M. Weber, and J.P. Burrows

Abstract Since August 2002 SCIAMACHY delivers a wealth of high-quality data permitting to study the status of the Earth's atmosphere. Enhanced concentrations of greenhouse gases are identified as the major source of global warming and their atmospheric concentrations are increasing. SCIAMACHY monitors the most prominent species such as CO_2, CH_4 and water vapour, the latter including isotope variants. Further anthropogenic impacts on the troposphere occur by emission of reactive trace gases contributing to pollution and affecting air quality. With SCIAMACHY their global, regional and even local signatures can be detected. Long-term analyses document how the emissions of NO_2, SO_2, HCHO, CHOCHO and CO evolve with time. In addition, the halogen cycle of polar BrO and IO, both of natural origin, is studied. The stratosphere is the layer where public interest in the Earth's atmosphere has begun to grow with the detection of the ozone hole in the mid-1980s. Until the mid-1990s a steady decrease has been observed in the ozone abundance. The most striking feature is the massive loss of stratospheric ozone over Antarctica during each southern spring. In order to detect possible signs of recovery, SCIAMACHY contributes to the continuous monitoring of the ozone layer, the ozone hole, Polar Stratospheric Clouds (PSC) and species impacting the ozone chemistry such as NO_2, OClO and BrO. A much more poorly explored region is the mesosphere and lower thermosphere, which forms the transition between interplanetary space and the terrestrial atmosphere. This region is dominated by extraterrestrial impacts as well as couplings to the lower atmosphere. With SCIAMACHY's limb viewing capabilities Noctilucent Clouds (NLC) are studied providing insight into generation and depletion mechanisms. At times of strong solar activity, SCIAMACHY measurements reveal how the chemistry of the upper atmosphere is disturbed. By analysis of emission lines in SCIAMACHY spectra the composition of the thermosphere above 100 km can be studied. SCIAMACHY is the first instrument to globally observe the metal layers in the upper mesosphere/lower thermosphere (MLT) region. When applying appropriate retrieval techniques it is meanwhile possible to derive vegetation information over land and phytoplankton characteristics in the oceans from SCIAMACHY data. Finally SCIAMACHY even has proven useful in planetary science by measuring spectra of our solar system neighbour Venus.

Keywords • Earth's atmosphere • Greenhouse gases • Air quality • Reactive gases • Ozone layer • Solar-terrestrial interactions • Troposphere • Stratosphere • Mesosphere • Lower thermosphere

H. Bovensmann (✉), C. von Savigny, A. Richter, F. Wittrock, M. Sinnhuber, B.M. Sinnhuber, A. Schönhardt, A.V. Rozanov, M. Weber, J.P. Burrows, M. Buchwitz
Institute of Environmental Physics/Institute of Remote Sensing (IUP-IFE), University of Bremen, Otto-Hahn-Allee 1, 28359 Bremen, Germany
e-mail: heinrich.bovensmann@iup.physik.uni-bremen.de

I. Aben, A. Gloudemans, and H. Schrijver
SRON, Netherlands Institute for Space Research, Sorbonnelaan 2, 3584 CA Utrecht, The Netherlands

M. Van Roozendael
Belgian Institute for Space Aeronomie (BIRA-IASB), 3 Avenue Circulaire, 1180 Brussels, Belgium

S. Kühl and S. Beirle
Max Planck Institute for Chemistry, Johann-Joachim-Becher-Weg 27, 55128 Mainz, Germany

M. Gottwald
Remote Sensing Technology Institute, German Aerospace Center (DLR-IMF), Oberpfaffenhofen, 82234 Wessling, Germany

C. Frankenberg
Jet Propulsion Laboratory, California Institute of Technology, 4800 Oak Grove Drive, 91109 Pasadena, CA, USA

P. Stammes and M. de Graaf
Royal Netherlands Meteorological Institute (KNMI), Wilhelminalaan 10, 3732 GK De Bilt, The Netherlands

A. Bracher
Alfred Wegener Institute for Polar and Marine Research, Bussestraße 24, 27570 Bremerhaven, Germany
and
Institute of Environmental Physics/Institute of Remote Sensing (IUP-IFE), University of Bremen, Otto-Hahn-Allee 1, 28359 Bremen, Germany

M. Gottwald and H. Bovensmann (eds.), *SCIAMACHY - Exploring the Changing Earth's Atmosphere*,
DOI 10.1007/978-90-481-9896-2_10, © Springer Science+Business Media B.V. 2011

Expectations were high when SCIAMACHY was launched in March 2002. Since August 2002 a continuous stream of high quality data has been acquired at the ground stations and disseminated for further processing, scientifically and operationally. With more than 8 years in orbit, SCIAMACHY's data 'harvest' considerably exceeded the original objective – and even more can be expected for the next years provided the ENVISAT mission extension operates as planned. During the past years numerous scientific results from the SCIAMACHY mission have been derived, demonstrating clearly the high 'return on investment' of this enterprise. New and exciting insights into the Earth-atmosphere system are obtained. They contribute significantly to atmospheric physics and chemistry as well as global climate change research. Many scientific groups at various institutes in Europe and abroad were and are actively involved in the analysis of the data. Their efforts made these excellent results possible. In the following tour through the atmosphere we will highlight some of the most spectacular findings. This summary is by no means complete but gives a good impression how SCIAMACHY explores our changing environment.

10.1 Tropospheric Composition – Greenhouse Gases

SCIAMACHY measurements provide information on tropospheric constituents as solar radiation penetrates the atmosphere down to the surface. Our civilisation generates significant stress upon the troposphere. Concentrations of greenhouse gases are increasing and have been identified as the source of global warming. SCIAMACHY permits not only the monitoring of the global status of the major greenhouse gases but also retrieves knowledge about the distribution of their sources such, as e.g. densely populated regions or wetlands.

Carbon Dioxide – CO_2

CO_2, the most important anthropogenic greenhouse gas, is regulated by the Kyoto Protocol and can be considered as a synonym for the impact of industrialisation on our environment. In pre-industrial times CO_2 mixing ratios dating back several thousands of years were about 300 ppm at maximum. Present values are around 390 ppm, i.e., 30% higher, with the increase mainly attributed to the past 50 years as a clear indication of an anthropogenic effect. Carbon, dumped into natural sinks over millions of years, is now released into the atmosphere by fossil fuel burning (oil, coal, gas). In addition other anthropogenic activities such as deforestation destroy important CO_2 sinks and reduce the ability of nature to efficiently recycle atmospheric CO_2. A thorough study of carbon dioxide is thus required to understand the global carbon cycle and to predict how greenhouse gas concentrations evolve with time. Currently, about 50% of the emitted CO_2 remains in the atmosphere, the other half is taken up by the oceans and in the biosphere. Photosynthesis extracts carbon dioxide from the troposphere over land. Thus large forest areas act as a CO_2 sink. The North American and Siberian boreal forests in summer are examples for such extended CO_2 sinks. These sink regions can be observed by SCIAMACHY, as illustrated in Fig. 10.1 displaying atmospheric CO_2 levels from April to June compared to July to September, where CO_2 concentrations are lower due to uptake by the terrestrial biosphere. This seasonal 'CO_2 breathing' is superimposed on the steady increase of atmospheric CO_2 with much higher concentrations in 2009 than in 2003. Both phenomena can be clearly observed by SCIAMACHY. It is even possible to detect localised elevated CO_2 over anthropogenic source regions such as Central Europe's highly populated and industrialised Rhine-Main area (Fig. 10.2). SCIAMACHY nadir observations in the SWIR spectral region formed the basis for the retrieved CO_2 information (Buchwitz et al. 2005a, 2007a; Bösch et al. 2006; Barkley et al. 2007; Reuter et al. 2010; Schneising et al. 2008). The CO_2 mixing ratio is obtained by normalising the CO_2 column with the simultaneously retrieved airmass from oxygen measurements (Schneising et al. 2008) or using meteorological surface pressure (Barkley et al. 2007).

Methane – CH_4

CH_4 is the second most important anthropogenic greenhouse gas after CO_2. It is regulated by the Kyoto Protocol as well. Compared to pre-industrial times CH_4 concentrations have more than doubled due to anthropogenic activities. Although the total sum of all CH_4 sources, about 550 Tg/year, is relatively well known, the distribution among different source categories is highly uncertain and impedes our capability to reliably predict CH_4 source strengths in a warming climate.

First results from SCIAMACHY showed substantially higher tropical CH_4 abundances than previously estimated (Frankenberg et al. 2005). Even though these results were recently partially revised (Frankenberg et al. 2008), it remains true that tropical emissions are very high, constituting about a third of all CH_4 emissions. In general, SCIAMACHY CH_4 retrievals have substantially matured and results from several independent studies (see also Buchwitz et al. 2005b; Schneising et al. 2009) draw a consistent picture of the global distribution of this greenhouse gas.

For CH_4 already improved emission estimates have been obtained on the basis of SCIAMACHY data. An atmospheric general circulation model, in which the current knowledge of global sources is implemented, is used to model the

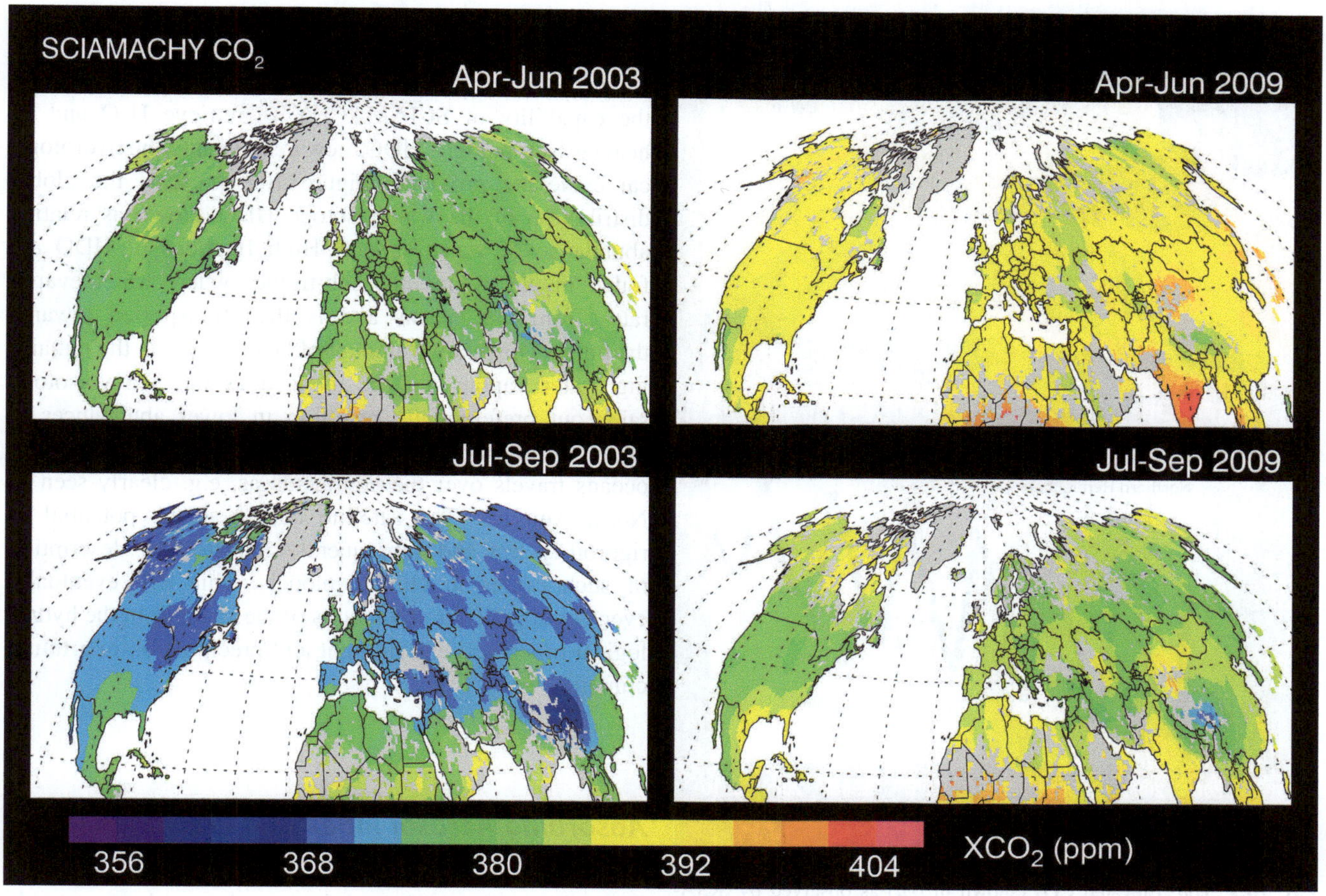

Fig. 10.1 Northern hemispheric CO_2 distribution as observed by SCIAMACHY. The differences between spring and summer are due to the CO_2 'breathing' of the vegetation (Courtesy: M. Buchwitz, IUP-IFE, University of Bremen).

worldwide CH_4 distribution. The source terms in this model can be adjusted in magnitude, region and timing until the modelled distribution provides the best match with SCIAMACHY observations, thus obtaining *inverted* source estimates using satellite data (Meirink et al. 2008). Recent inversion studies (Bergamaschi et al. 2009) result in significant changes in the spatial patterns of emissions and their seasonality compared to the bottom-up inventories. Large CH_4 emissions are attributed to various wetland regions in tropical South America and Africa, seasonally varying and opposite in phase with CH_4 emissions from biomass burning. As obvious in Fig. 10.3, India, China and South East Asia are characterised by pronounced emissions from rice paddies peaking in the third quarter of the year, in addition to further anthropogenic emissions throughout the year.

Water Vapour – H_2O

Water is the key to the Earth's climate system. As vapour it is the strongest greenhouse gas and as precipitation it is the essential ingredient for making our planet habitable. Water vapour is a highly variable component of the atmosphere with direct anthropogenic impact on its amount being usually negligible. Its contribution may reach up to 4% of the atmospheric volume in the tropics and amounts to less than 1% in dry air conditions. Due to the relation between temperature and humidity, water vapour acts as a positive feedback to anthropogenic radiative forcing and is thereby indirectly affected by human activity.

In contrast to microwave instruments, SCIAMACHY water vapour data is available over both land and ocean down to the surface for at least partly cloud-free scenes. Because of their independence from other in situ or remote sensing measurements, SCIAMACHY water vapour columns provide a new important global dataset (Noël et al. 2004; Schrijver et al. 2009). A combination of SCIAMACHY water vapour with corresponding data derived from GOME and follow-on instruments allows the study of water vapour long-term trends spanning now already more than 15 years with the potential of extension up to 2020 when GOME-2 data on METOP is considered. Using linear and non-linear methods from time series analysis and standard statistics the trends of H_2O columns and their errors have been derived from GOME and SCIAMACHY for the years 1996–2007 (Mieruch et al. 2008). The trends clearly show elevated water vapour levels in years of strong

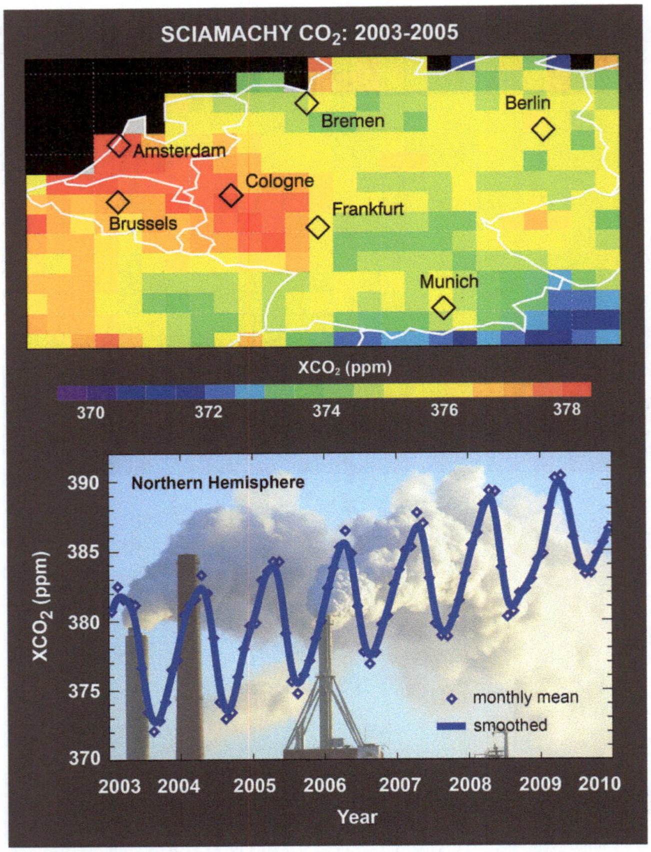

Fig. 10.2 *Top*: Elevated CO_2 (shown in *red*) as observed by SCIAMACHY over Central Europe's most populated Rhine-Main area. *Bottom*: Increase of Northern hemispheric CO_2 derived from SCIAMACHY measurements (Courtesy: M. Buchwitz, IUP-IFE, University of Bremen).

El-Niño activity. How these trends are distributed on a global scale is further demonstrated in Fig. 10.4. Increasing long-term trends in water vapour have been observed for Greenland, Eastern Europe, Siberia and Oceania, whereas decreasing trends occur for the northwest US, Central America, Amazonia, Central Africa and the Arabian Peninsula.

Heavy Water – HDO

When water evaporates from the Earth's oceans and surface, moves through the atmosphere and falls back as rain, evaporation and condensation processes change the content of heavy water (HDO). Therefore, the isotopic composition contains information about the history of water. SCIAMACHY's measurements permit obtaining a global view on the water vapour isotopic composition in the atmosphere (Frankenberg et al. 2009). These are the first global isotope measurements with high sensitivity towards the lowest layers of the atmosphere down to the surface, where most of the water vapour resides. By exploiting the capability of SCIAMACHY to retrieve H_2O and its heavier isotopologue HDO, new insights into the hydrological cycle are provided. Figure 10.5 presents the global distribution of the water isotope HDO shown as relative abundance of water vapour. High fractions of HDO are found in the tropics and sub-tropics where water evaporates from the oceans and is then transported towards the poles. The relative amount of heavy water in the remaining water vapour will be reduced as the heavy isotope rains out preferentially resulting in lower abundances at higher latitudes. The same occurs when moist air from the oceans travels over the continents as, e.g. clearly seen in North America. The satellite data bear the potential to rigorously test and subsequently improve the description of such cycles in climate models. This will eventually even result in better predictions of the changes in the hydrological processes, e.g. drought and precipitation in a future climate.

Absorbing Aerosol Index and Precipitation

Water in the form of precipitation plays a dominant role in local climate and weather. This is particularly the case in Africa where the monsoon is the driving mechanism for the climate and therefore also for the social and economical development. The northern part of Africa hosts large dry areas such as the Sahara and the Sahel. Dust storms arise frequently from the dry areas and have a profound impact on the weather conditions and lives of the local people. A linkage between the African monsoon systems and aerosol loading in Africa is suggested by the analysis of GOME and SCIAMACHY measurements. De Graaf et al. (2010) investigated multi-year satellite observations of UV-absorbing aerosols and compared these with precipitation data. The main UV-absorbing aerosol types occurring over Africa are desert dust and biomass burning aerosols. Their abundances can be characterised using Absorbing Aerosol Index (AAI) data from GOME and SCIAMACHY. Time series of regionally averaged AAI from 1995 to 2008 show the seasonal variations of aerosols in Africa. When relating the zonally averaged daily AAI to monthly mean precipitation data they indicate monsoon-controlled atmospheric aerosol loadings, which are different for the West African and East African monsoons due to their different dynamics caused by the asymmetric distribution of land masses around the equator. Figure 10.6

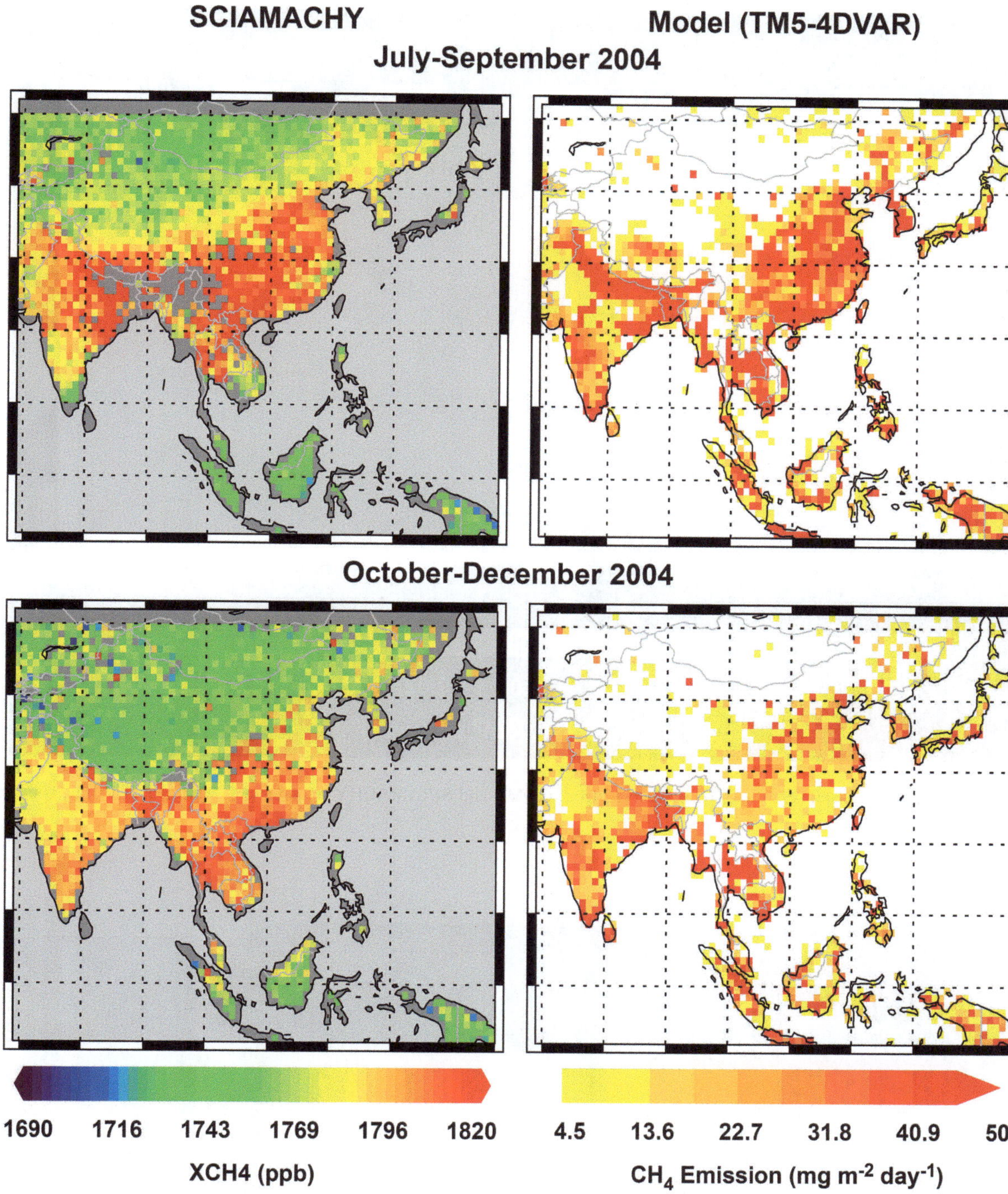

Fig. 10.3 Methane emission derived from SCIAMACHY data. *Left column*: Column-averaged CH_4 mixing ratios (XCH_4) over South-East Asia from SCIAMACHY for summer and autumn 2004. *Right column*: Modelled emissions per 1° × 1° grid cell (Courtesy: adapted from Bergamaschi et al. 2009. Reproduced by permission of American Geophysical Union).

clearly shows that the seasonal variation of the aerosol distribution is linked to the seasonal cycle of the monsoonal wet and dry periods in both areas. During dry periods the AAI varies freely, driven by emissions from deserts and biomass burning events. During wet periods the AAI depends linearly on the amount of precipitation, due to scavenging of aerosols and the prevention of aerosol emissions from wet surfaces.

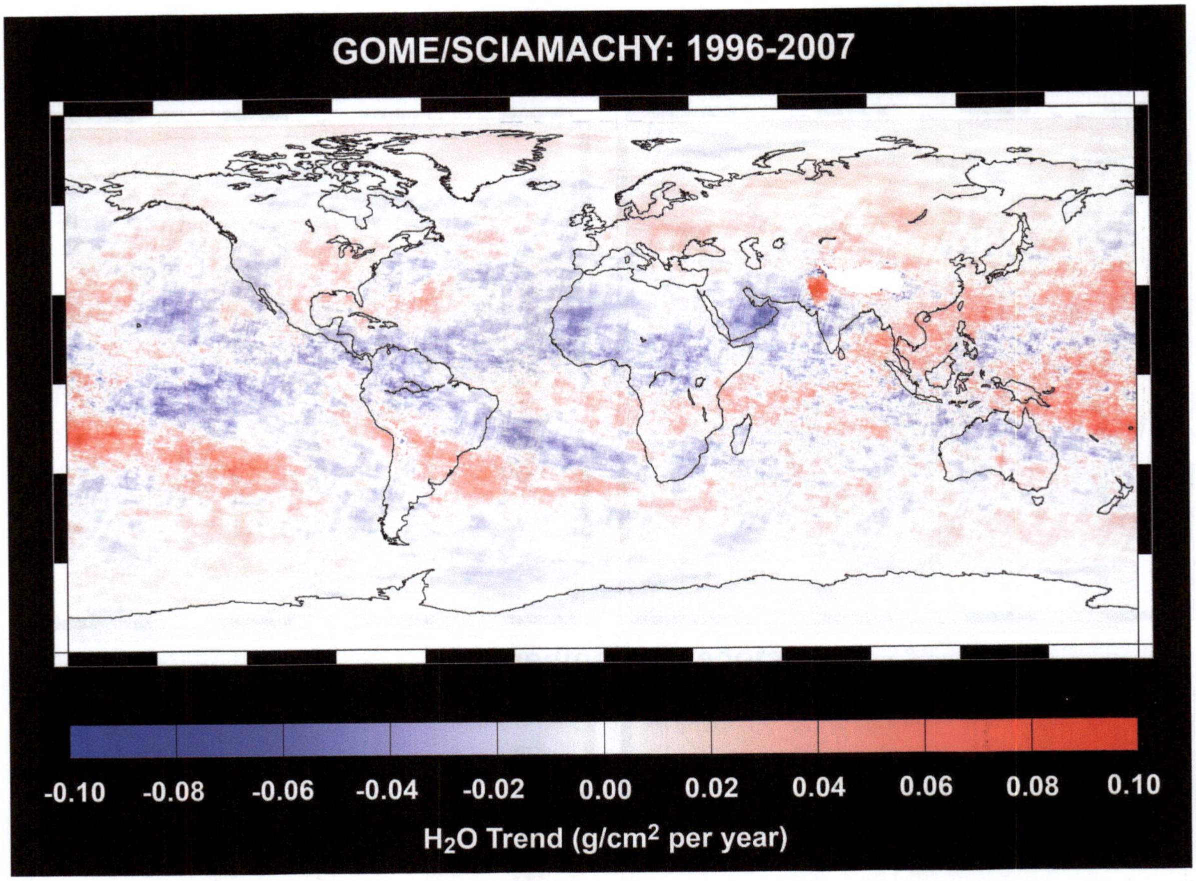

Fig. 10.4 Water vapour trends for 1996 to 2007 as derived from GOME and SCIAMACHY (Courtesy: S. Mieruch, IUP-IFE, University of Bremen).

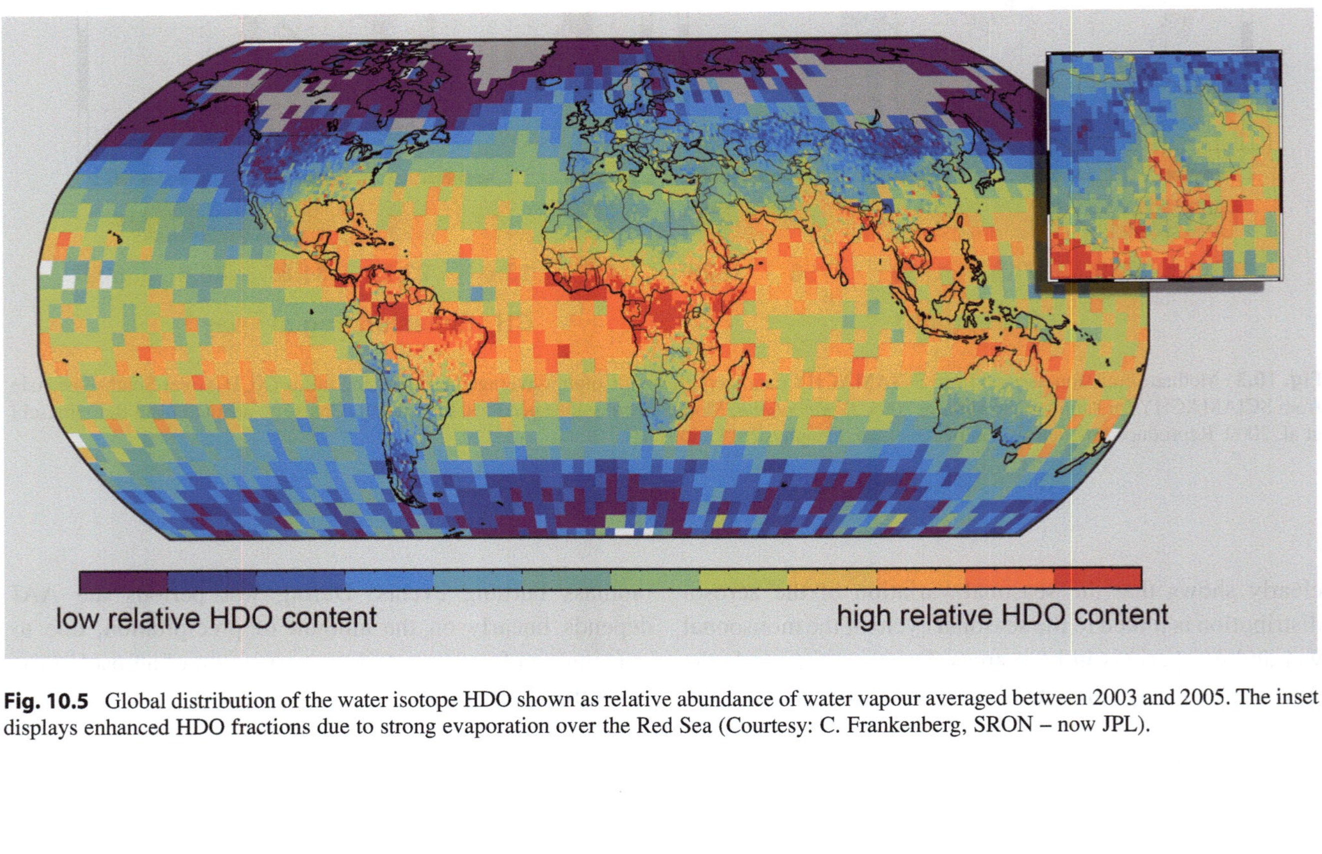

Fig. 10.5 Global distribution of the water isotope HDO shown as relative abundance of water vapour averaged between 2003 and 2005. The inset displays enhanced HDO fractions due to strong evaporation over the Red Sea (Courtesy: C. Frankenberg, SRON – now JPL).

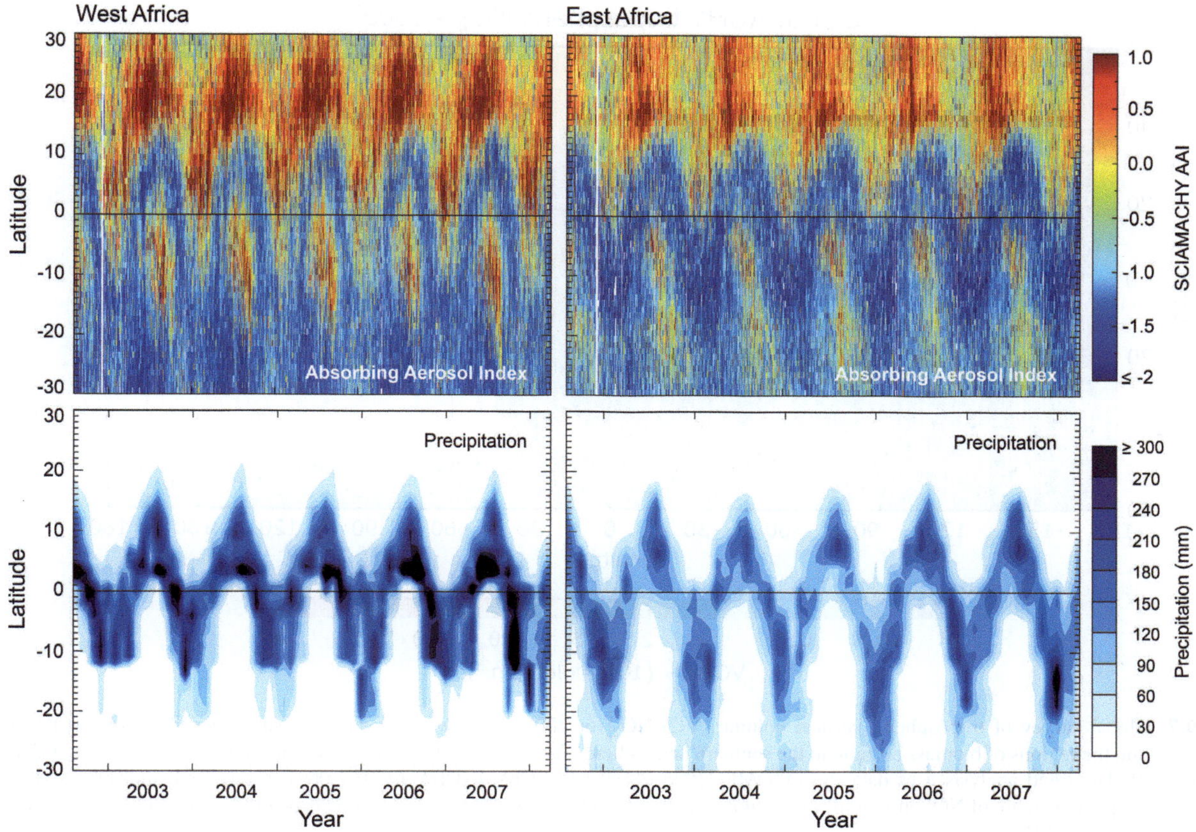

Fig. 10.6 Zonally averaged AAI from SCIAMACHY for each day between August 2002 and April 2008 as a function of latitude in Western and Eastern Africa (*upper panels*). The bottom panels display the monthly and zonally averaged precipitation for the same areas and the same period (Courtesy: M. de Graaf, KNMI).

10.2 Tropospheric Composition – Reactive Gases

Emissions of greenhouse gases are not the only anthropogenic impact onto the lowest layer of the Earth's atmosphere. Pollution and air quality have become a major concern in an ever increasing industrialised world. SCIAMACHY is able to detect and monitor the global, regional and local signatures of trace gases contributing to air pollution and to follow how emissions evolve with time.

Nitrogen Dioxide – NO_2

NO_2 is an important indicator of air pollution and a cause of summer smog. NO_2 catalyses ozone production, contributes to acidification and also adds to radiative forcing. The main sources of NO_2 are anthropogenic in origin, e.g. power plants, traffic, forced biomass burning and both heavy and agricultural industry. Other but slightly less important sources comprise natural biomass burning, lightning and microbiological soil activity. NO_2 emissions have increased by more than a factor of 6 since pre-industrial times with concentrations being highest in large urban areas.

Global monitoring of tropospheric NO_2 emissions is a crucial task. SCIAMACHY's predecessor GOME has already demonstrated the unique ability to monitor tropospheric air pollution. Figure 10.7 shows a global survey of tropospheric NO_2 as seen by SCIAMACHY. The inset in Fig. 10.7 presents a time series of these concentrations over China. The periodic trend in NO_2 columns each year can mainly be explained by seasonal variations in energy consumption while the overall increase in tropospheric NO_2 over China is a result of the increase in industrial activity (Richter et al. 2005a). The inset also demonstrates how well SCIAMACHY matches with GOME and GOME-2. NO_2 vertical columns of both instruments perfectly overlap around the turn of the year 2002 and from 2007 onwards.

Due to its higher spatial resolution (60 × 30 km as compared to 320 × 40 km for GOME), SCIAMACHY enables very detailed observations of polluted regions. This is obvious in Fig. 10.8 where industrialised regions of North America, Europe and China are displayed. As a result of these new

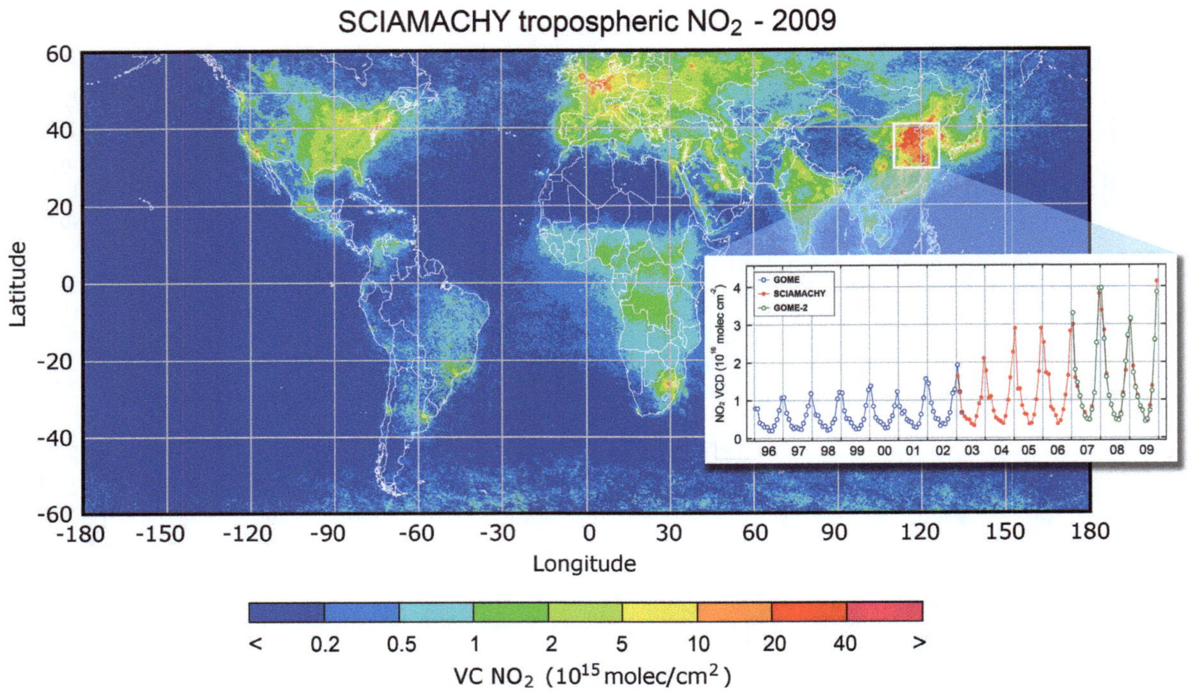

Fig. 10.7 Global survey of tropospheric vertical column (VC) NO_2 for 2009. Clearly visible are the industrialised regions in the northern hemisphere and the regions of biomass burning in the southern hemisphere. The inset illustrates how NO_2 concentrations have risen in China from 1996 to 2009. The trend analysis uses data from GOME (1996–2002) and SCIAMACHY (2003–2009). While 'old' industrialised countries were able to stop the increase of NO_2 emissions, the economical growth in China turns out to be a strong driver for pollution (Courtesy: A. Richter, IUP-IFE, University of Bremen).

datasets, individual cities (Beirle et al. 2004) and even large power plants (Kim et al. 2009) can be identified. Similar small scale structure in NO_2 emissions can also be detected over the oceans. The high sensitivity and spatial resolution of the SCIAMACHY measurements permits localising frequently used ship routes (Richter et al. 2004), as illustrated in Fig. 10.9. Using data from GOME, SCIAMACHY and GOME-2, Franke et al. (2009) could even derive temporal changes in these 'tiny' signatures of anthropogenic activity.

Particularly interesting are studies concerning the global trend in NO_2 concentrations (Richter et al. 2005a; Stavrakou et al. 2008; van der A et al. 2008). By combining SCIAMACHY results with those of previous missions, e.g. GOME, it is possible to investigate how the tropospheric NO_2 load has changed over the past decade. A strong increase in nitrogen dioxide is observed by SCIAMACHY in countries and areas with a booming economy, in particular in China (see inset Fig. 10.7) while in Europe SCIAMACHY has observed a stabilisation of NO_2 levels, attributed partly to slightly increased traffic emissions after a period of reducing nitrogen dioxide levels in the 1990s resulting from EU regulations. For the US SCIAMACHY results indicate a decrease in NO_2 emissions, related to the recent implementation of pollution controls for power plants (Kim et al. 2006).

The spatial and temporal characteristics of a multi-year dataset provides much improved constraints for attempts to identify main emission sources and to quantify emission strengths by inverse modelling. It also facilitates the derivation of independent top-down estimates of emissions not only on a country-by-country basis but even on regional scales. Konovalov et al. (2008, 2010) investigated such trends within Europe for a 10 year time period, showing that for many countries emissions are decreasing, but especially in the Mediterranean area, along the coastlines as well as in Eastern Europe emissions have increased. Van der A et al. (2008) analysed the spatial and temporal patterns in a multi-year dataset of GOME and SCIAMACHY tropospheric NO_2 and identified from this data the most dominating emission sources (Fig. 10.10). Whereas in the northern hemisphere the NO_2 mainly stems from anthropogenic sources and from soils, in the southern hemisphere biomass burning is the dominating origin of tropospheric NO_2.

By combining SCIAMACHY NO_2 observations at 10:00 local time with NO_2 observations from OMI at 13:30 local time it becomes even possible to get a first glimpse on the diurnal variations of tropospheric chemistry and emissions from space. These measurements suggest a decrease in tropospheric NO_2 between 10:00 and 13:30 over fossil fuel source regions due to photochemical loss. Over tropical biomass burning regions the opposite effect is obvious – an increase due to a midday peak in emissions (Boersma et al. 2008).

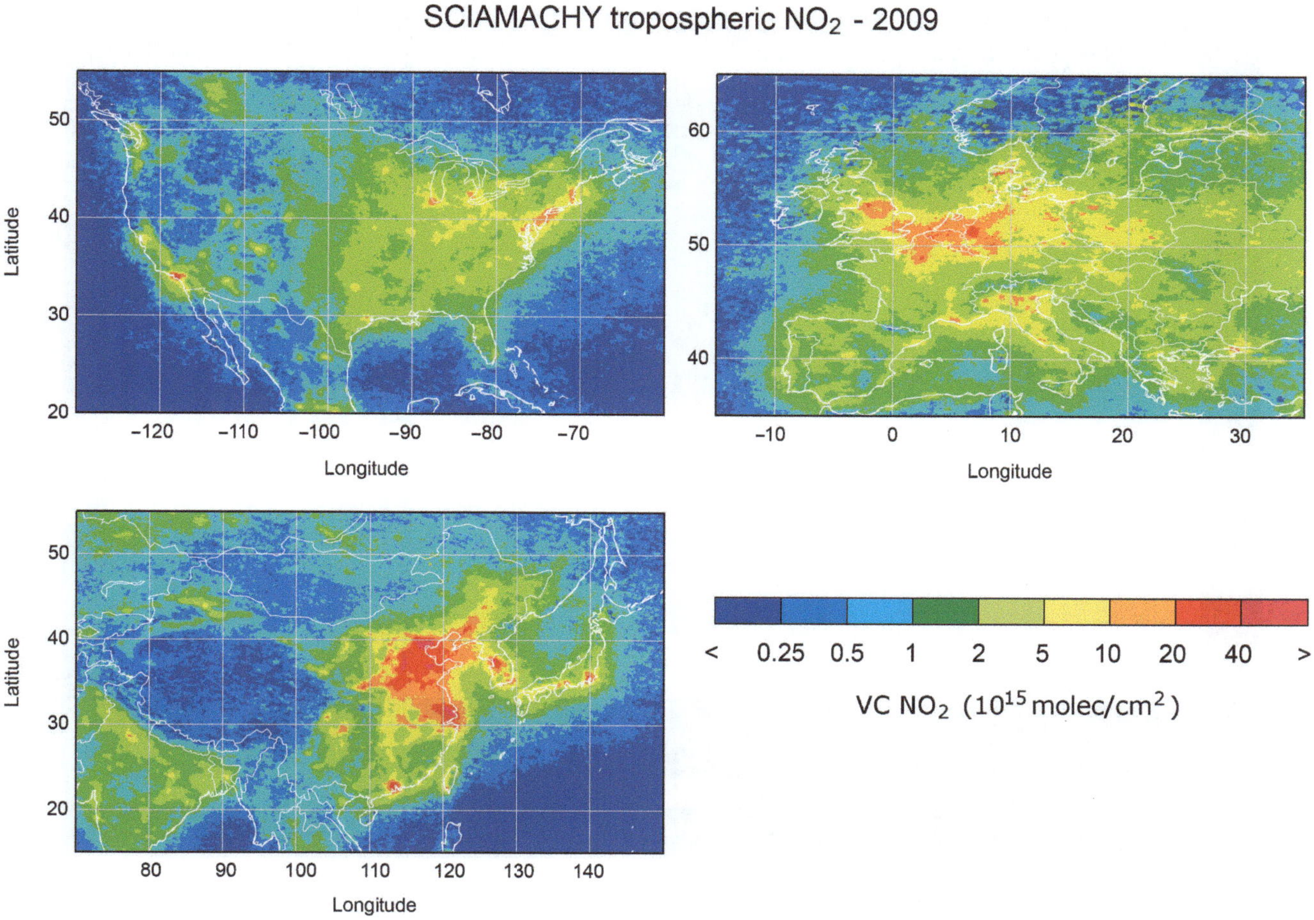

Fig. 10.8 Mean tropospheric vertical column (VC) NO_2 densities over Europe, the United States and China in 2009 (Courtesy: A. Richter, IUP-IFE, University of Bremen).

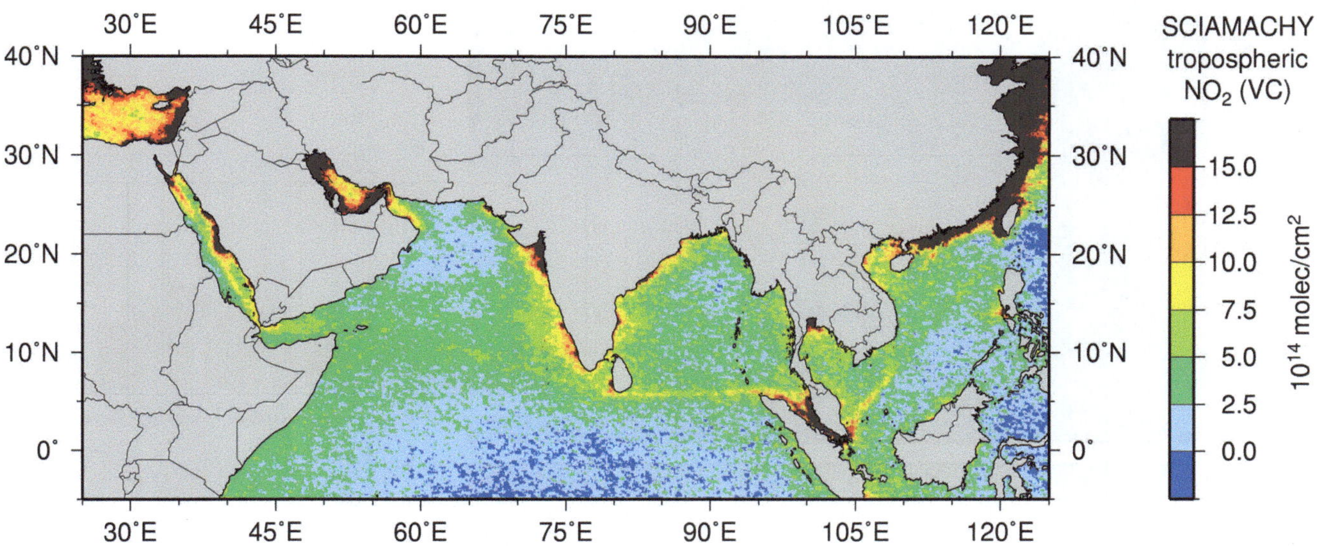

Fig. 10.9 NO_2 densities over the Indian Ocean and the Red Sea as derived from measurements between August 2002 and April 2004. Ship routes from Eastern Asia to the Suez Canal can be clearly seen. Along the industrialised coast lines strong NO_2 concentrations are related to urban areas (Courtesy: A. Richter, IUP-IFE, University of Bremen).

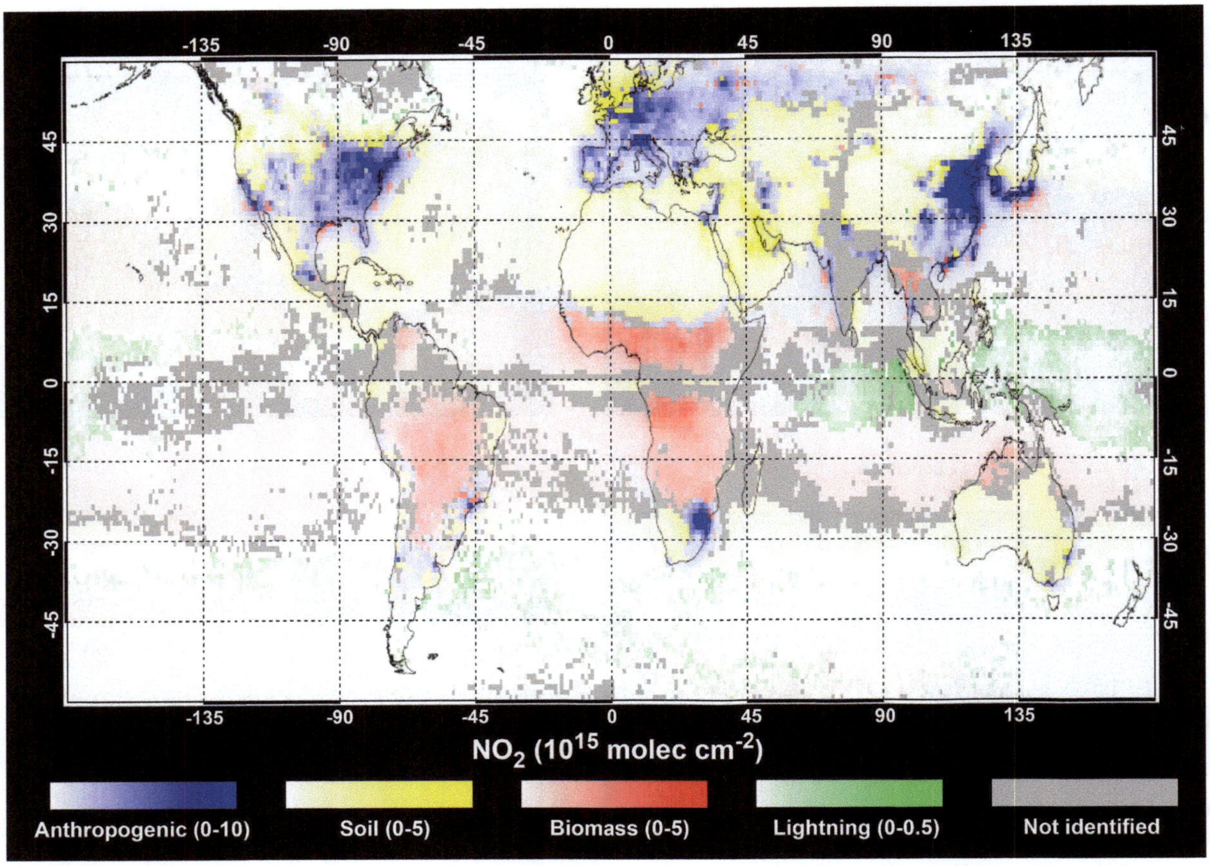

Fig. 10.10 Dominant NO_x source identification based on analyses of the time series of measured tropospheric NO_2 from GOME and SCIAMACHY satellite observations (1996–2006) (Courtesy: van der A et al. 2008. Reproduced by permission of the American Geophysical Union).

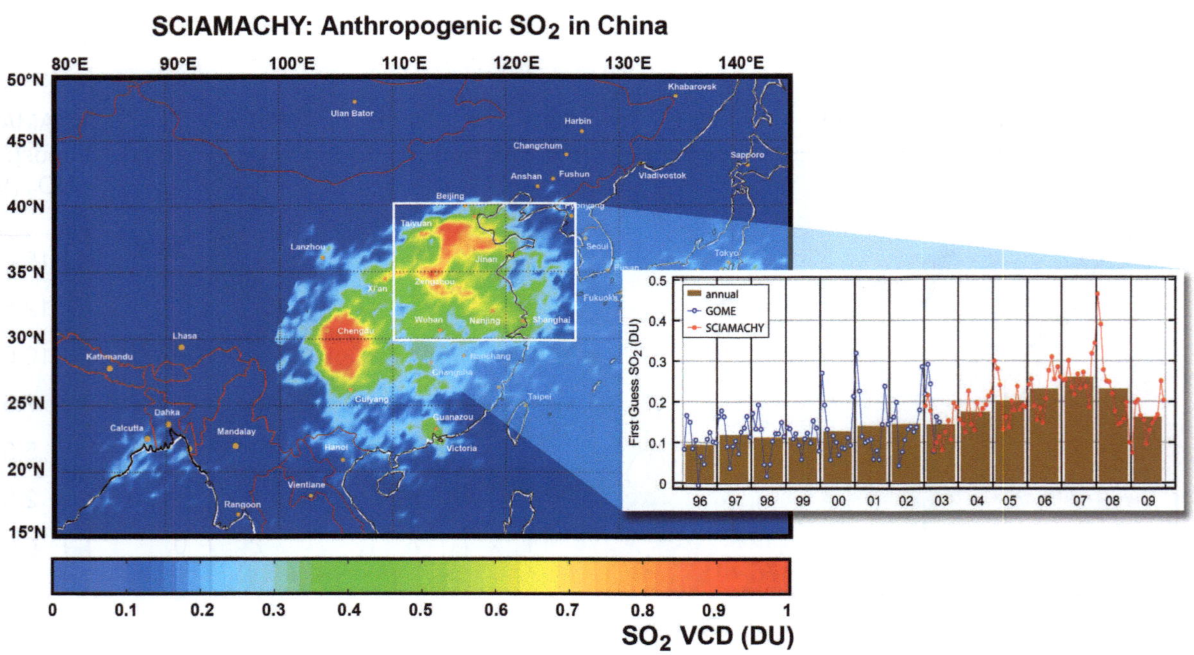

Fig. 10.11 Average SO_2 vertical column densities (VCD) over eastern China during the year 2003. SCIAMACHY's improved spatial resolution permits to identify localised emissions due to anthropogenic activities. The trend analysis uses data from GOME (1996–2002) and SCIAMACHY (2003–2009) (Courtesy: map – M. Van Roozendael, BIRA-IASB; trend – A. Richter, IUP-IFE, University of Bremen).

Fig. 10.12 The Mt. Etna volcanic eruption in 2002 with obvious SO_2 emissions (redish plume). The SCIAMACHY nadir measurement is overlayed on a MERIS image showing that the plume of SO_2 and the visible ash cloud match well (Courtesy: ESA and Brockmann Consult).

Sulphur Dioxide - SO_2

SO_2 is another pollutant leaving a clear absorption signature in SCIAMACHY spectra. Sources of SO_2 are combustion of sulphur rich coal and other fossil fuels or volcanic eruptions including degassing. Although SO_2 emissions have been reduced significantly over the last decades, clear signals can be detected over the Eastern US and in particular the polluted areas of China (Fig. 10.11). As in the case of NO_2, the high spatial resolution facilitates source identification and makes the dataset an interesting new data source for air pollution research. By combining GOME and SCIAMACHY results, the long-term trend of SO_2 emissions in heavily polluted areas can be inferred and compared with similar analyses for NO_2 (see inset of Fig. 10.11).

Volcanoes are a natural source of SO_2. Since the start of SCIAMACHY's routine observations a few of the several hundred existing active volcanoes have been erupting and were overpassed by ENVISAT. In October 2002, Mt. Etna on the island of Sicily entered an explosive phase. The rectangular overlay in Fig. 10.12 represents a SCIAMACHY nadir measurement displaying SO_2 emissions which match well with the ash plume visible in a simultaneously obtained image by MERIS on ENVISAT. The SO_2 emissions of volcanic eruptions are usually associated with such volcanic ash plumes. These could be a major threat for air traffic. Therefore SCIAMACHY SO_2 retrievals are used in early warning services in support of aviation control. Whenever elevated SO_2 densities are obtained in regions with known volcanoes, an alert is issued indicating a potential volcanic ash cloud.

Formaldehyde (HCHO) and Glyoxal (CHOCHO)

In the presence of nitrogen oxides, the photochemical degradation of volatile organic compounds (VOC) leads to secondary gaseous and particulate products, such as ozone

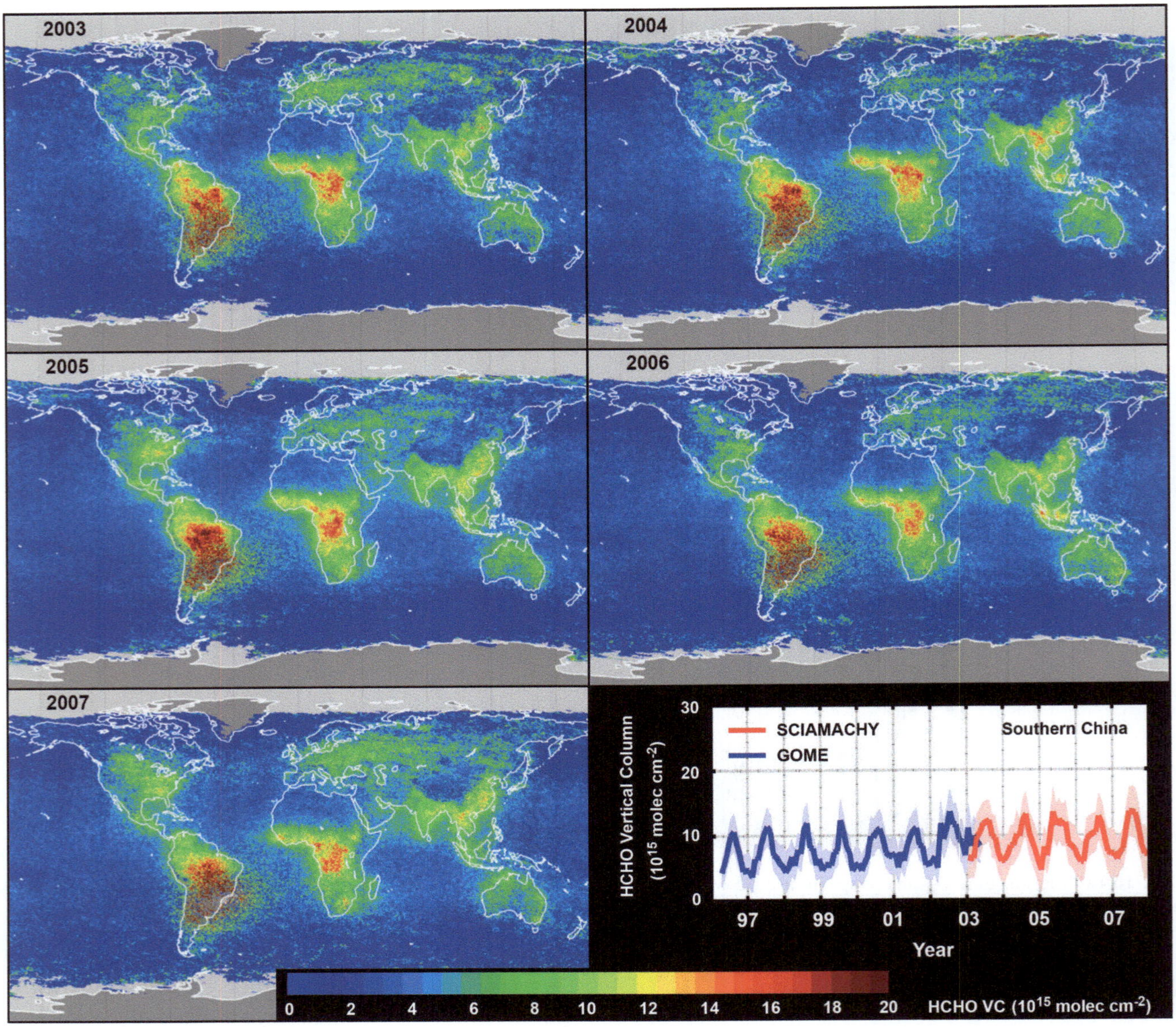

Fig. 10.13 Yearly averaged SCIAMACHY HCHO vertical columns from 2003–2007. In the bottom right panel the HCHO trend over China is indicated (Courtesy: adapted from De Smedt et al. 2008).

and secondary organic aerosols (SOA). Both are important contributors to air pollution with severe impacts on human health, ecosystems, and regional climate. Two important intermediate products in the oxidation of volatile organic compounds are HCHO and CHOCHO. Since both trace gases have short lifetimes their distribution mainly resembles the source areas and can be used as indicators of tropospheric VOC emissions. SCIAMACHY's capability to measure HCHO and CHOCHO on a global scale from space was demonstrated by Wittrock et al. (2006). Since then, several studies have been carried out using these data to compare directly with the output from chemical transport models (CTM) to improve substantially the models' accuracy and reliability. In addition, VOC emission strengths have been derived for biospheric, pyrogenic and anthropogenic sources, respectively.

A consistent dataset of global tropospheric HCHO (Fig. 10.13) has been created by De Smedt et al. (2008) using GOME and SCIAMACHY data covering more than a decade. This dataset has been utilised by Stavrakou et al. (2009) to evaluate the performance of pyrogenic and biogenic emission inventories and to investigate trends in HCHO over Asia and large cities worldwide (De Smedt et al. 2010). More regional aspects on the different emissions strengths of VOC, based on HCHO data, were investigated for Europe (Dufour et al. 2009) and Amazonia (Barkley et al. 2009).

In general, for regions with high biogenic emissions like tropical rainforests a reasonable agreement was found between modelled and measured HCHO columns. Other areas having more variable or less emissions of VOC reveal higher discrepancies. Recently Marbach et al. (2009) even succeeded in determining HCHO emissions from ships in the Indian Ocean.

Myriokefalitakis et al. (2008) adapted a global chemistry-transport model to simulate the temporal and spatial distribution of CHOCHO columns in the global troposphere focussing on the anthropogenic contribution. They found indication for a missing CHOCHO source of about 20 Tg/year or an overestimate of its sinks by the model. In addition Fu et al. (2008) examined the potential of CHOCHO as a source of secondary organic aerosol. It appears that irreversible uptake of CHOCHO by aqueous aerosols and clouds could make a significant contribution to the global SOA production. The long SCIAMACHY time series also reveal seasonal and year-to-year variability above several photochemical hot spots. This has been studied by Vrekoussis et al. (2009) for CHOCHO (Fig. 10.14). For the period 2002–2007 a significant annual increase in CHOCHO in addition to a seasonal cycle is reported over northeastern Asia. In general, the regions influenced by anthropogenic pollution encounter enhanced amounts of CHOCHO.

Carbon Monoxide – CO

CO plays a central role in tropospheric chemistry as CO is the leading sink of the hydroxyl radical, which itself largely determines the oxidising capacity of the troposphere. Therefore it is of prime importance for the troposphere's self-cleansing efficiency and the concentration of greenhouse gases such as CH_4. CO also has a large air quality impact as it is toxic and is a precursor of tropospheric ozone, a secondary pollutant which is associated with respiratory problems and decreased crop yields.

CO detection with SCIAMACHY is an ambitious task. Carbon monoxide vertical columns can be retrieved from a number of CO absorption lines located around 2.3 μm in the SWIR range. The retrieval is not straightforward because these lines are relatively weak, much weaker than the absorption structures of the overlapping absorbers water vapour and CH_4. In addition a number of calibration issues, mainly related to large variable dark signals and changing instrument characteristics as a result of the growth of an ice layer on the channel 8 detector, hamper the CO retrieval (e.g. Gloudemans et al. 2005). Nevertheless, a first survey of global CO over land had been performed which shows elevated CO in the case of fires due to biomass burning. These results are to a large degree consistent with MOPITT and model predictions (de Laat et al. 2006, Buchwitz et al.

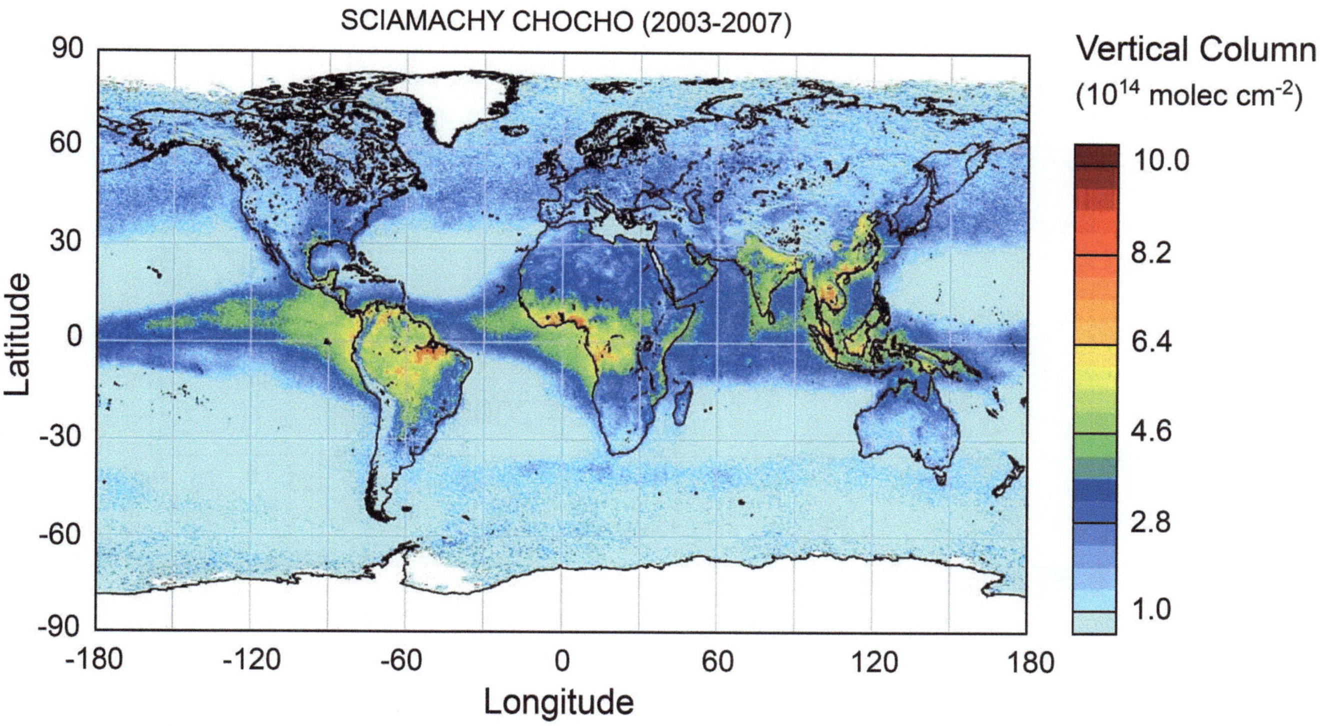

Fig. 10.14 Multiannual (2003–2007) SCIAMACHY CHOCHO vertical columns. The largest amounts are found over the tropics and sub-tropics where vegetation and biomass burning is abundant (Courtesy: Vrekoussis et al. 2009).

2007b). Gloudemans et al. (2009) succeeded in deriving CO also over the ocean, when using scenes with low maritime clouds. This yielded a 5-year CO dataset over both land and clouded ocean scenes between 2003 and 2007 (Fig. 10.15). The CO measurements over clouded ocean scenes have been compared with co-located modelled CO columns over the same clouds and agree well. Using clouded ocean scenes quadruples the number of useful CO measurements compared to land-only measurements. The 5-year dataset shows significant inter-annual variability over land and over clouded ocean

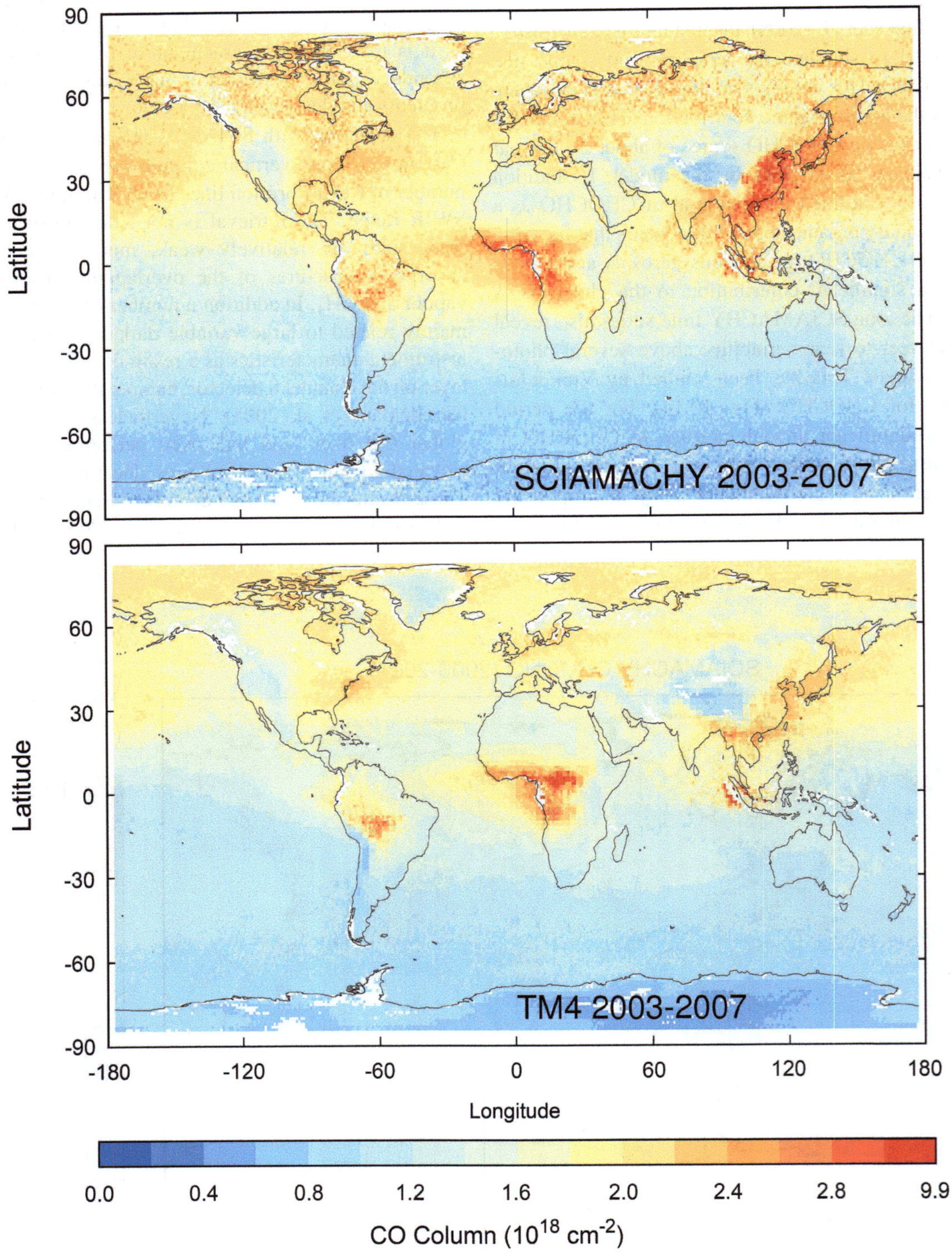

Fig. 10.15 Five year (2003–2007) average CO total columns on a 1° × 1° grid as retrieved by SCIAMACHY (*top*) and the TM4 chemistry-transport model (*bottom*). The SCIAMACHY CO columns above low clouds over sea are filled up with TM4 CO densities below the cloud to obtain total columns (Courtesy: Gloudemans et al. 2009).

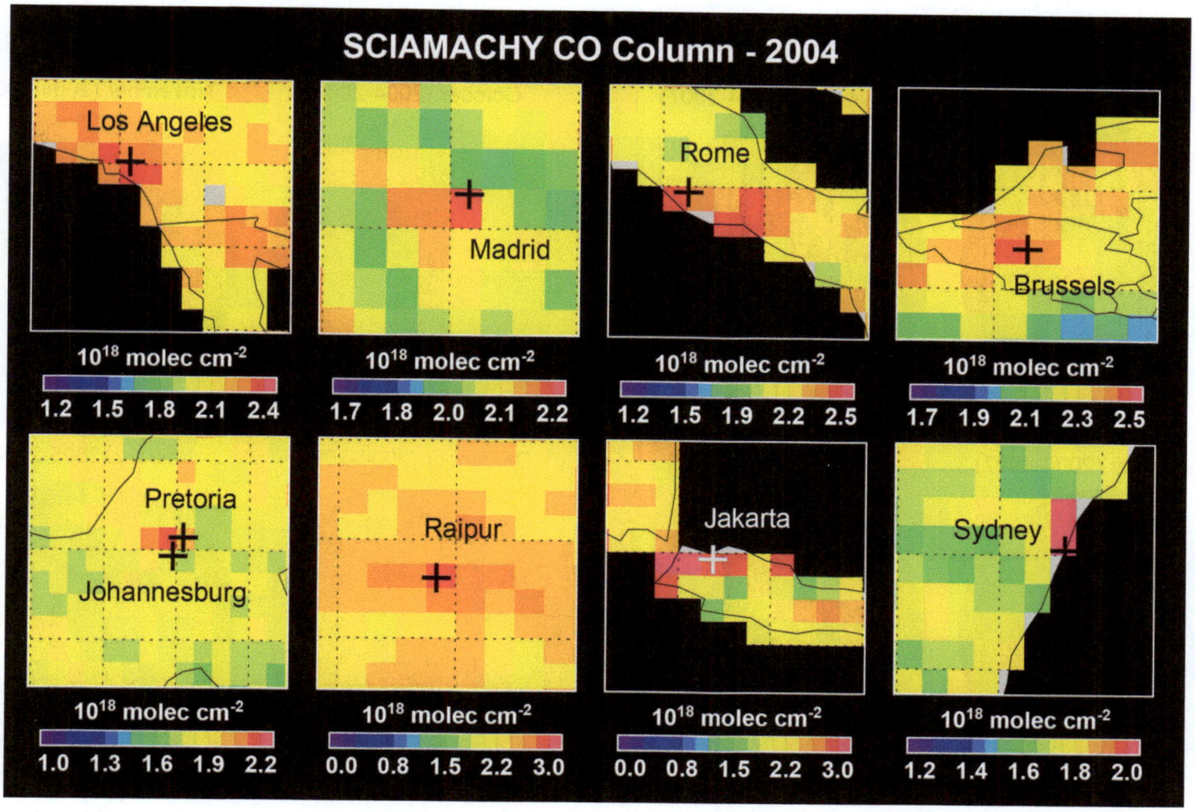

Fig. 10.16 A variety of urban areas as detected in SCIAMACHY CO data in 2004 (Courtesy: M. Buchwitz, IUP-IFE, University of Bremen).

areas, like Asian outflow of pollution over the northern Pacific, biomass-burning outflow over the Indian Ocean originating from Indonesia, and biomass burning in Brazil. In general there is good agreement between observed and modelled seasonal cycles and inter-annual variability.

Beside the global CO view, SCIAMACHY also allows to investigate CO over urban areas (Buchwitz et al. 2007b) as illustrated in Fig. 10.16. This information helps to constrain local to regional emission estimates and air quality studies. Assimilation of SCIAMACHY CO data was also used to improve CO emission estimates in the Middle East (Tangborn et al. 2009). One remarkable result is the finding that CO emissions over Dubai in 2004 have more than doubled compared to those in the available emissions inventory, based on data from 1998.

Tropospheric Halogen Oxides – BrO and IO

BrO and IO are reactive halogen radicals which have gained growing interest in recent years (Simpson et al. 2007). Both compounds impact on tropospheric chemistry. They react with ozone and change the oxidation pathways of several atmospheric species. Bromine compounds have been identified as initiator of strong boundary layer ozone depletion events in polar spring (Barrie et al. 1988). The IO molecule constitutes the starting point for iodine nucleation and the formation of fine aerosol particles, which affect the atmospheric radiation balance and which may potentially grow to cloud condensation nuclei (O'Dowd et al. 2002). Widespread plumes of enhanced BrO are regularly observed in the Arctic as well as in the Antarctic shortly after polar sunrise where they persist for several months. The spatial distributions and locations of these plumes move rapidly on a daily basis (Begoin et al. 2010) with the BrO probably not only being situated in the boundary layer, but also at higher altitudes. A strong link between the BrO patterns and sea ice cover exists. Sources of BrO are most likely of inorganic nature. Current discussions consider young sea ice, frost flowers, aerosols and brine (Kaleschke et al. 2004; Piot and von Glasow 2008).

Recent analyses of SCIAMACHY nadir observations using spectral data around 420 nm have enabled the detection of tropospheric IO columns (Saiz-Lopez et al. 2007; Schönhardt et al. 2008). IO amounts are small and close to the instrument's detection limit, but through efficient reaction cycles, even these small amounts still have a considerable impact on the polar tropospheric chemistry. Over the Antarctic polar region, a variety of details in the temporal

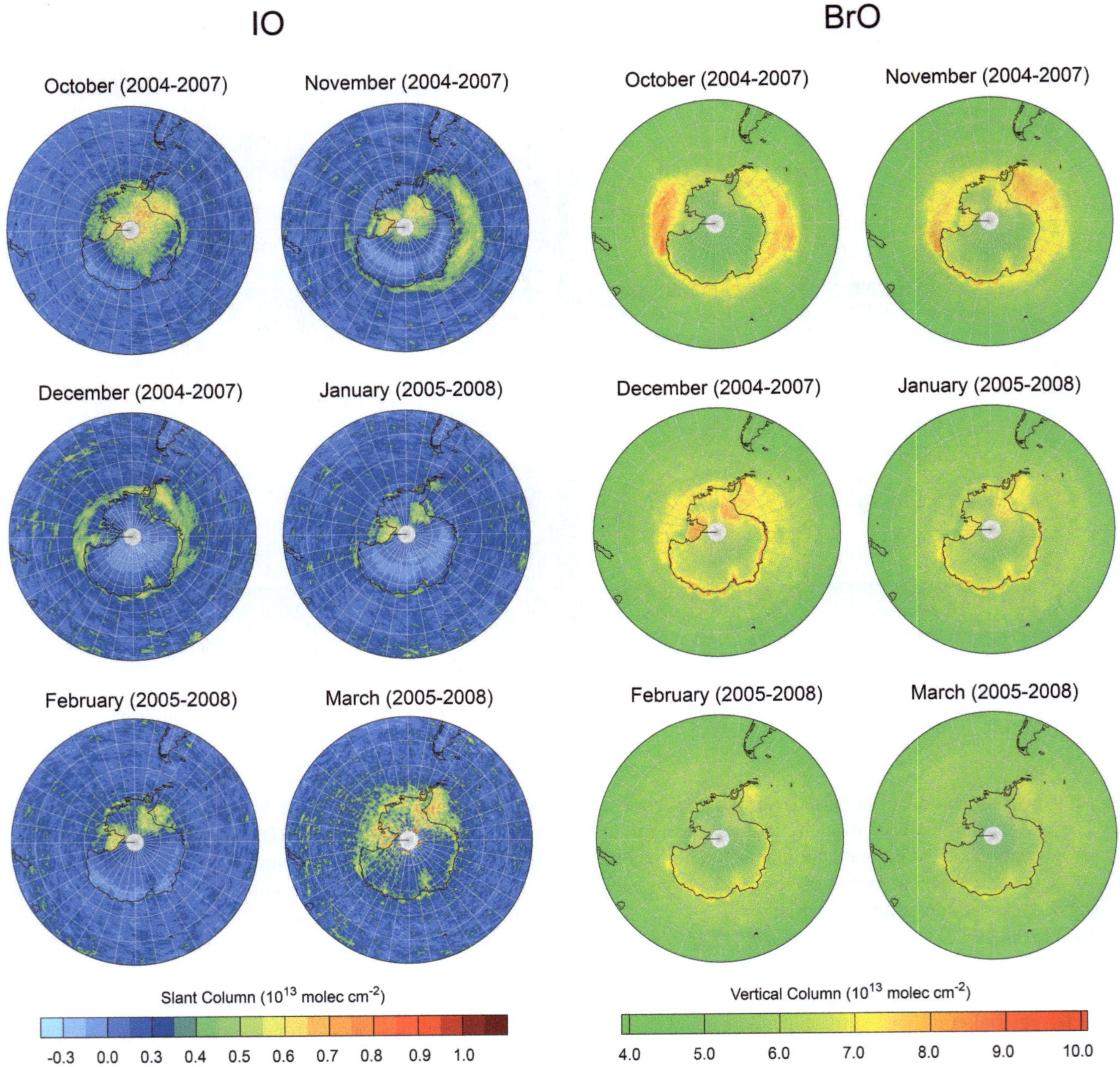

Fig. 10.17 Monthly maps of SCIAMACHY observations of IO (*left*) and BrO (*right*), averaged over four subsequent years from 2004–2008. A stratospheric air mass factor (AMF) is applied to the BrO columns only, leaving the patterns of IO and BrO still comparable (Courtesy: A. Schönhardt, IUP-IFE, University of Bremen).

and spatial distribution of both IO and BrO is revealed in Fig. 10.17. Monthly means from October through austral summer until March are shown, in each case data is averaged over four subsequent years. As for BrO, largest amounts of IO appear in Antarctic spring time. Besides that general similarity, spatial distributions are quite different. BrO is observed predominantly above sea ice regions during spring, and furthermore along coast lines and on shelf ice regions in summer. Abundances vanish towards autumn. IO however shows larger variability throughout the time series. Regions with enhanced IO include the sea ice, ice shelves, coast lines, but also the continent (in October). Enhanced IO above the sea ice in the characteristic ring-like pattern only occurs much later in spring (November) in contrast to BrO, where enhancements can already be observed in August well before the time period illustrated in Fig. 10.17. When southern autumn approaches, IO concentrations begin to increase again. While BrO behaves similarly on both hemispheres, no widespread enhanced IO abundances are observed in the Arctic spring time.

The details of the spatial and temporal patterns of IO in comparison to BrO are not well understood yet. However the observed differences in the distributions suggest that the two halogen oxides are released by different processes. While the BrO is produced by inorganic emissions and the bromine explosion cycle, it is an open question if the majority of IO is of biological origin. The cold Antarctic waters show high biological activity, and cold water diatoms may produce organic iodine species. Considerable differences between the South and North Polar regions might be linked to the fact that the biospheres are distinct and are emitting iodine compounds in different amounts and speciation.

10.3 The Stratospheric Ozone Layer

Already in the second half of the twentieth century the stratosphere was seen as fragile to human perturbation. Public interest grew even more with the detection of the Antarctic ozone hole in the mid-1980s. Until the mid-1990s a steady decrease has been observed in the ozone abundance over the South Pole, North Pole and the mid latitudes of up to 3–6% per decade. The most striking feature is the massive loss of stratospheric ozone over Antarctica each southern spring. This ozone loss is so large due to the fact that very low stratospheric temperatures over Antarctica in wintertime foster the underlying depletion processes. The polar vortex isolates the air during the polar night and allows that the cold conditions remain stable and Polar Stratospheric Clouds (PSC) grow (Fig. 10.18). In this environment the chemistry of ozone depletion begins with the conversion of reservoir species to chemically active molecules on the surface of the PSC. These react with ozone in catalytical reaction cycles resulting in the effective destruction of the O_3 molecules. In addition, ozone loss was also observed in the tropics and mid-latitudes. Today there is broad agreement that a continuous monitoring of stratospheric ozone, the ozone hole and of those species impacting the ozone chemistry is required in order to detect possible signs of recovery, and to answer the question in how far the cooling of the stratosphere and the strengthening of the Brewer-Dobson circulation as a consequence of climate change will delay or accelerate the recovery of the ozone layer (Rex et al. 2006, Newman et al. 2007).

Fig. 10.18 Polar Stratospheric Clouds over Southern Norway in January 2003, taken from the NASA DC-8 aircraft at an altitude of 11 km. The PSC hover well above the tropospheric cloud cover (Photo: P. Newman, NASA/GSFC).

SCIAMACHY allows exploitation of new opportunities using the limb backscatter as well as solar or lunar occultation measurement modes to determine PSC and vertically resolved concentration profiles of trace gases in the stratosphere, in addition to the established column measurements from the nadir mode.

Ozone – O_3

The important role of ozone in the Earth's atmosphere can be attributed to the fact that it absorbs solar UV radiation which would otherwise reach the surface where it can cause damage to the biosphere. In the wavelength range below 290 nm, UV photons are almost completely blocked. Radiation from 290 to 320 nm is strongly attenuated so that dose levels on ground become no longer harmful. Ozone does not only impact conditions at the bottom of the troposphere but also in the upper atmosphere through the effects of absorption of UV to IR radiation and subsequent heating. The heating produces a temperature profile which makes the stratosphere vertically stable. Even transport mechanisms in the layers above – the mesosphere and thermosphere – were found to be influenced by the energy content of the stratosphere.

In the year of ENVISAT's launch the ozone hole over Antarctica differed significantly from what had been observed before and after. Its extent was reduced in 2002 by 40% as compared to previous years (Fig. 10.19). However, this did not indicate a recovery of the ozone layer but originated from peculiar meteorological conditions where an unprecedented major stratospheric warming led to a split-up of the polar vortex which interrupted the heterogeneous processes leading usually to the massive ozone destruction. A more detailed view of this September 2002 event was obtained by retrieving stratospheric profiles over Antarctica (von Savigny et al. 2005a). The vertically resolved SCIAMACHY limb measurements showed that the ozone hole split did not occur throughout the entire stratosphere, but only above about 24 km. At 20 km there was still a single elongated area with low O_3 values (Fig. 10.20). In normal cold Antarctic winters however, the O_3 profiles display a more regularly shaped ozone hole throughout the altitude range, as depicted for 2005 in Fig. 10.20. The anomalous ozone hole in 2002 also developed quickly in time. Ozone losses computed from assimilated concentrations indicate that within a few days the situation changed drastically (Fig. 10.21). This ozone hole split-up was already predicted in the 9-day ozone forecast at KNMI (see Eskes et al. 2005). The following years displayed again an ozone

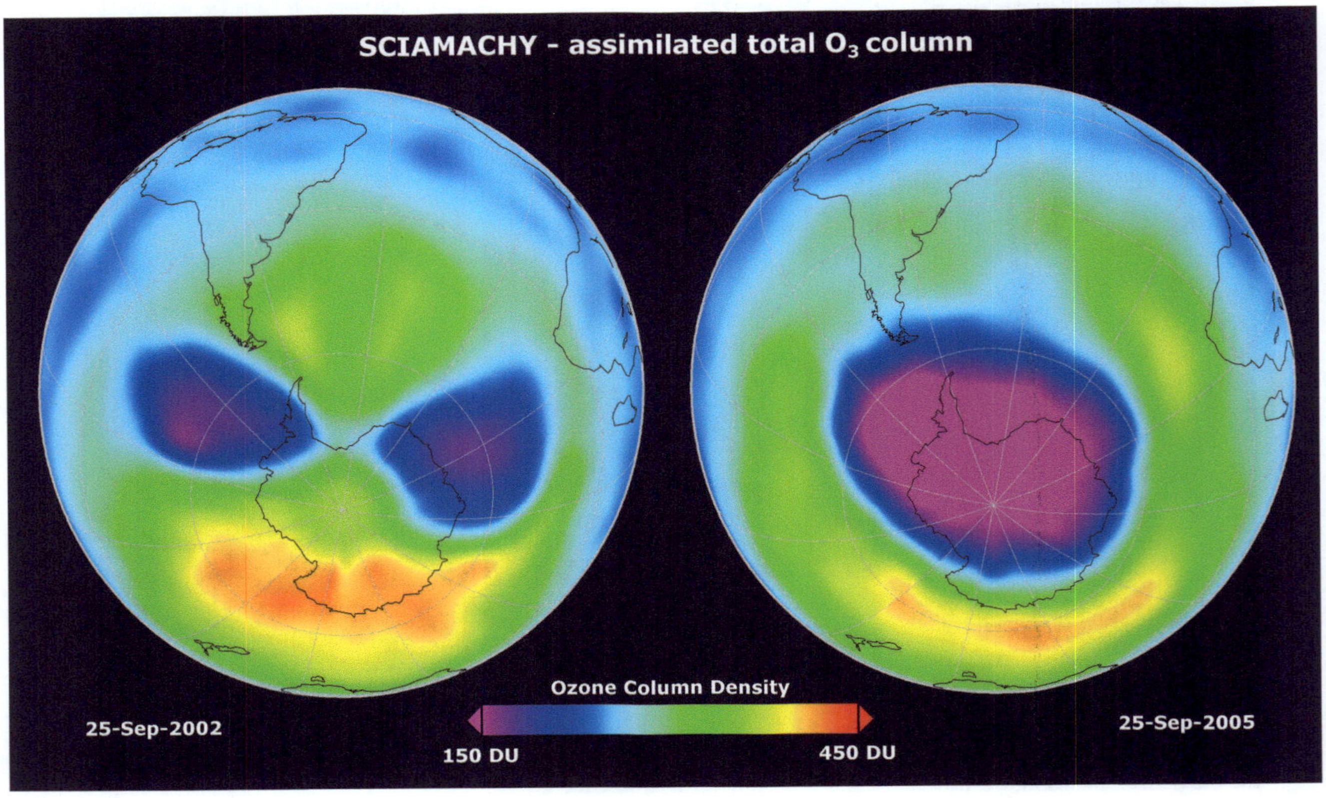

Fig. 10.19 The Antarctic ozone hole in 2002 (*left*) and 2005 (*right*). SCIAMACHY total column ozone measurements from 25 September have been analysed with the ROSE assimilation model to generate this view. Due to meteorological conditions the hole was split and reduced in size in 2002 but appeared 'as normal' in 2005 (Courtesy: T. Erbertseder, DLR-DFD).

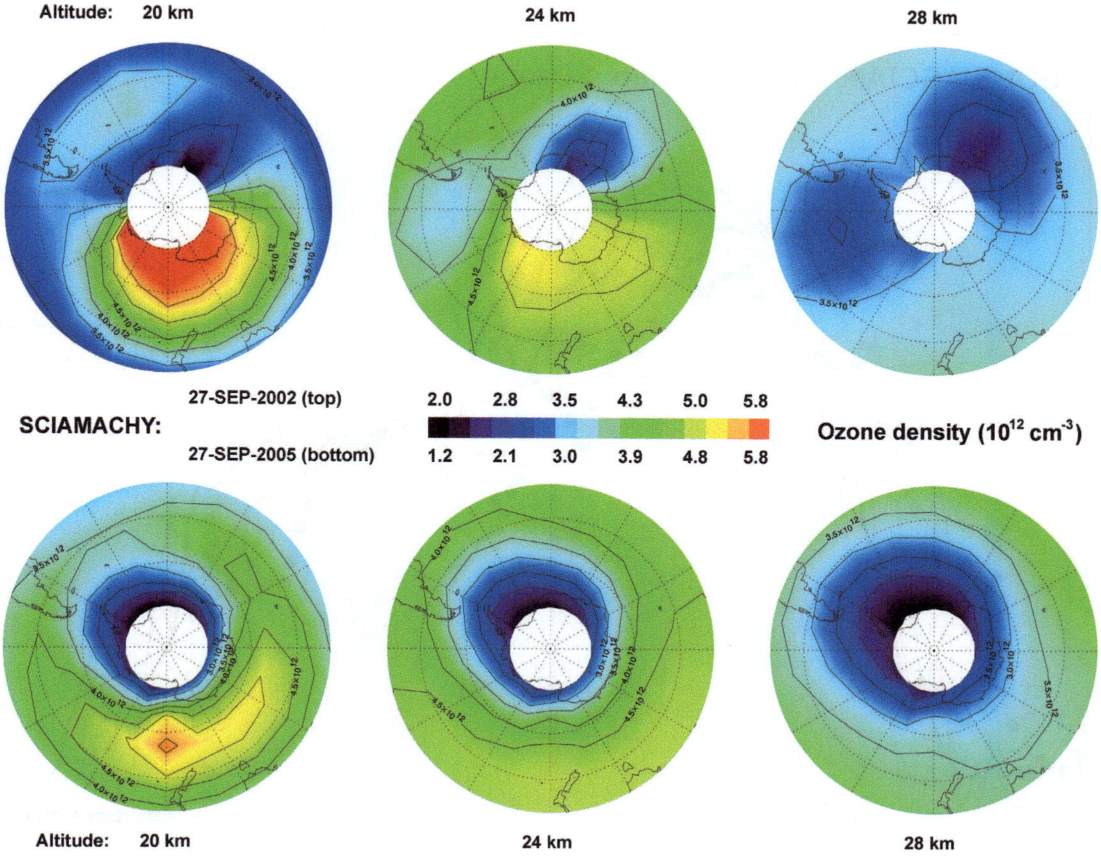

Fig. 10.20 Slices of the polar southern hemisphere ozone field at altitudes of approximately 20, 24 and 28 km on 27 September 2002 and 27 September 2005 as measured by SCIAMACHY. The observed split of the ozone hole in 2002 is not so obvious in the lower stratosphere around 20 km, but clearly visible at 24 and 28 km. In 2005 an ozone hole of 'normal shape' existed at all altitudes (Courtesy: C. von Savigny, IUP-IFE, University of Bremen).

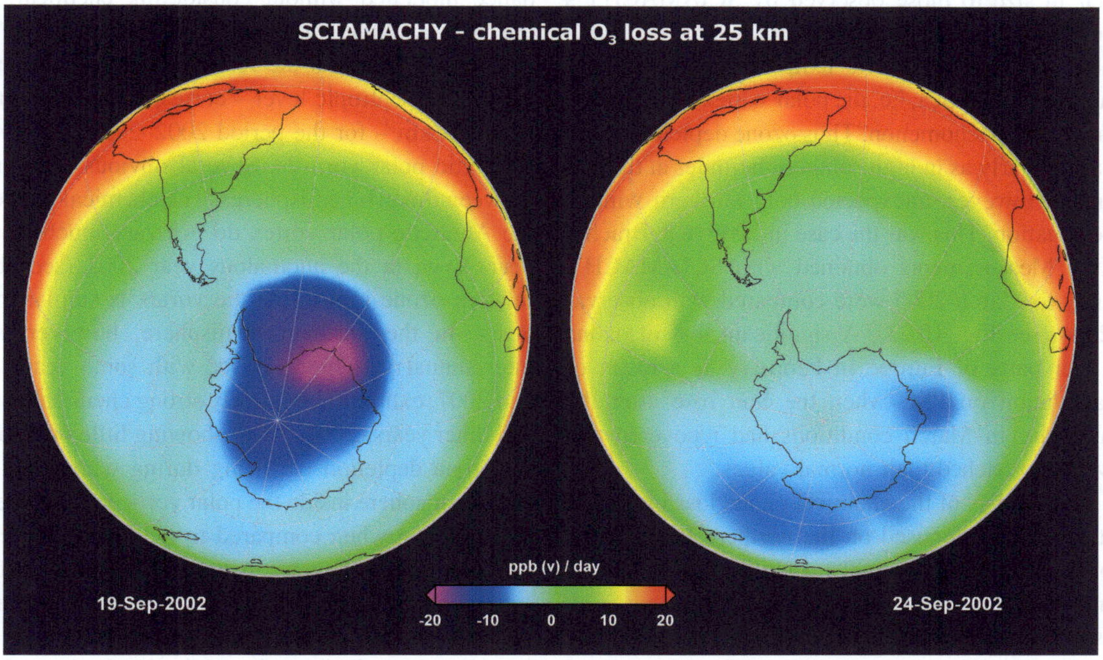

Fig. 10.21 Chemical ozone loss rates on two days in September 2002 in the mid-stratosphere at about 25 km, relative to the previous day. While on 19 September the ozone hole was still developing its usual shape, 5 days later the split-up of the vortex has reduced ozone loss rates (Courtesy: T. Erbertseder, DLR-DFD).

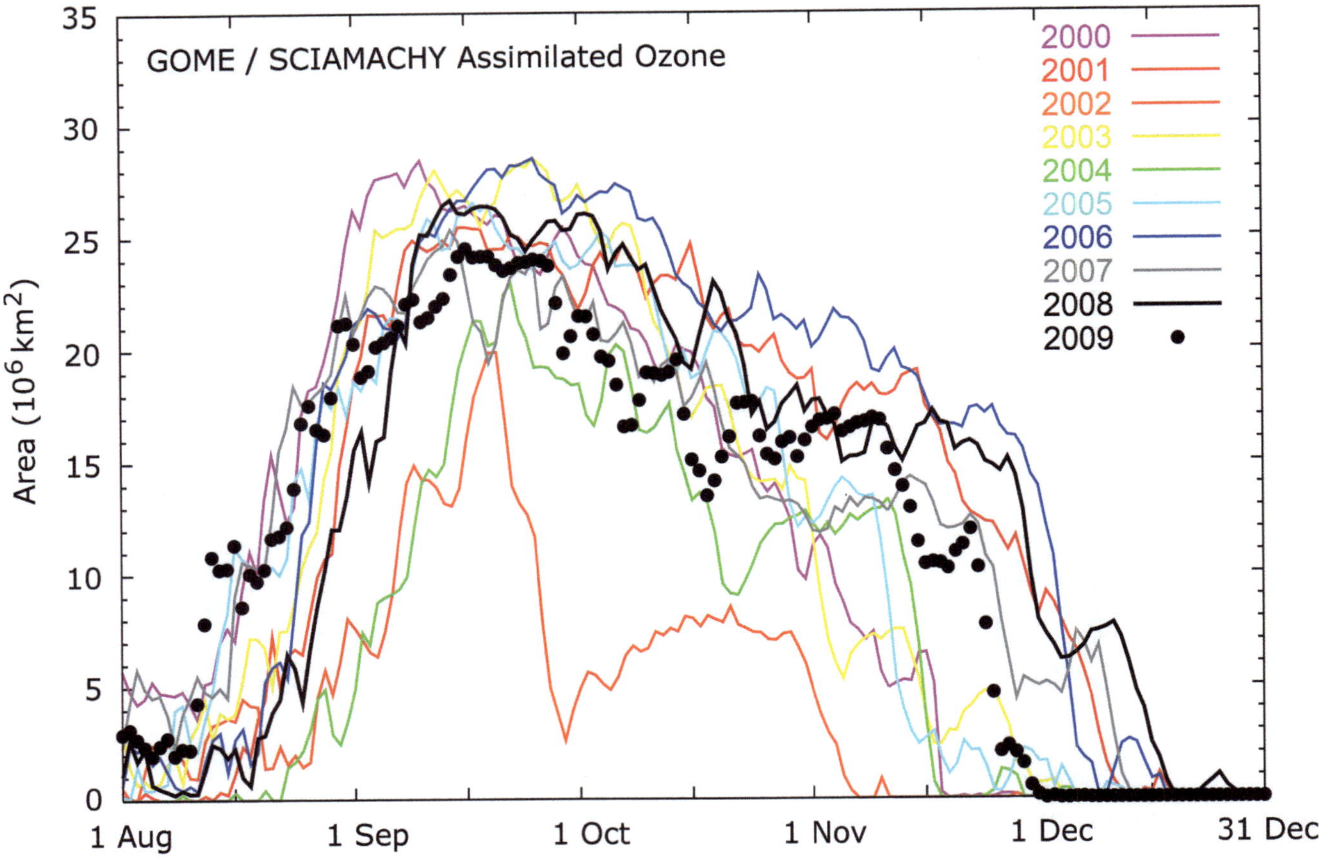

Fig. 10.22 Time series of the size of the Antarctic ozone hole from 2000–2009 based on observations of GOME and SCIAMACHY. The area includes ozone column values below 30°S lower than 220 Dobson Units (Courtesy: TEMIS KNMI/ESA).

hole similar in size to those observed by SCIAMACHY's predecessor GOME (see Fig. 10.22).

Over the Arctic, stratospheric temperatures are usually higher than over Antarctica, i.e. the polar vortex is less stable and PSC are a rare phenomenon. Thus ozone depletion is not observed as regularly as in high southern latitudes at the end of the winter. However, unexpected cold northern winters change the situation. This was the case in 2005 (e.g. Bracher et al. 2005). Measurements obtained with SCIAMACHY from January to March 2005 were compared with observations of 2004 (see Fig. 10.23). A significant loss of ozone occurred over parts of Europe in February as a result of chlorine activation on PSC when the Sun rose after the long polar night. In March conditions had recovered and the O_3 levels approached their normal values.

While most studies of the chemical ozone loss inside the polar vortices mainly focused on the northern hemisphere, because of the strong inter-annual variability in the stability of the vortex and the following ozone loss, Sonkaew et al. (2010) also analysed ozone depletion inside the Antarctic polar vortex by using limb ozone profiles. They determined the chemical ozone loss via the difference between the observed vortex-average ozone abundance and the abundance modelled without considering chemical processes, but with including dynamically induced ozone changes. Figure 10.24 depicts the relative chemical ozone losses at the 475 K isentropic level – corresponding to an altitude of about 18 km – for the period 2002–2009 in the Arctic and Antarctic polar vortices. Several obvious differences exist between the two hemispheres. The chemical ozone losses in the Antarctic polar vortex do not vary much from year to year. Even in the anomalous year 2002 (see above), the relative ozone loss inside the vortex is similar to all other years. In the northern hemisphere, however, significant inter-annual variability exists, with some years, e.g. 2005 and 2007, exhibiting relatively strong chemical ozone losses and other years (2004, 2006) showing little or no ozone loss.

Ozone depletion occurring during winter and spring in each hemisphere inside the polar vortices is a more localised phenomenon when compared to the global and continuous effects of anthropogenic halogen emissions on the stratospheric ozone layer. Again, O_3 limb profiles have proven a valuable tool for investigating long-term trends in stratospheric ozone. Steinbrecht et al. (2009) determined upper stratospheric ozone trends for several latitude bands from 1979 to 2008 using ground-based LIDAR and microwave as

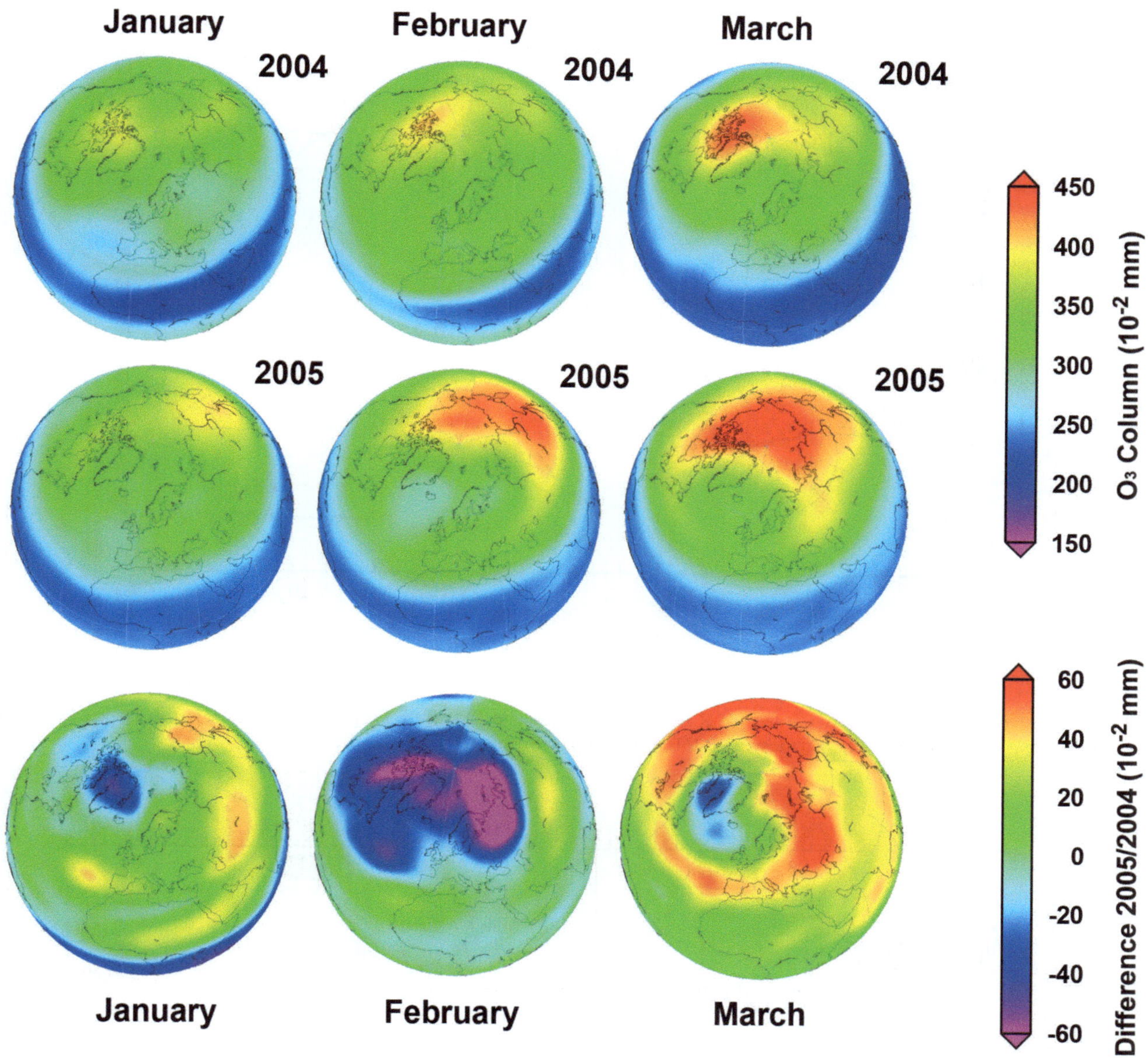

Fig. 10.23 Ozone columns over the northern hemisphere for late winter/early spring 2004 (*top row*), 2005 (*mid row*) and the difference between both years (*bottom row*). Caused by unusual low temperatures in the stratosphere over the Arctic, significant ozone loss occurred in February 2005 with a recovery of the ozone layer in March (Courtesy: J. Meyer-Arnek, DLR-DFD).

well as satellite observations with SAGE II, HALOE, SBUV, GOMOS and SCIAMACHY. Figure 10.25 shows ozone anomalies, i.e. monthly averaged ozone measurements minus average seasonal component, at five different latitudes. SCIAMACHY observations are indicated in yellow. The time series at all five locations display a decline until about the mid-1990s, followed by a stagnation of the negative trend or even a slight positive trend, in very good qualitative agreement with the time dependence of the effective stratospheric chlorine load. Since the quasi-biennial oscillation (QBO) variation has not been removed it is clearly visible in most time series.

As demonstrated by Steinbrecht et al. (2009) 'witnessing' the recovery of stratospheric ozone requires long-term datasets. These are usually not provided by a single instrument but by a series of – preferentially – similar sensors. Since SCIAMACHY is GOME heritage, combining GOME, GOME-2 and SCIAMACHY total columns from nadir measurements generates a unique repository. Loyola et al. (2009) formed a homogeneous dataset by merging O_3 columns from June 1995 to August 2009. Measurements from over 70 globally distributed Dobson and Brewer ground stations served as validation reference. Since the GOME data record is very stable it was used as a transfer standard and SCIAMACHY

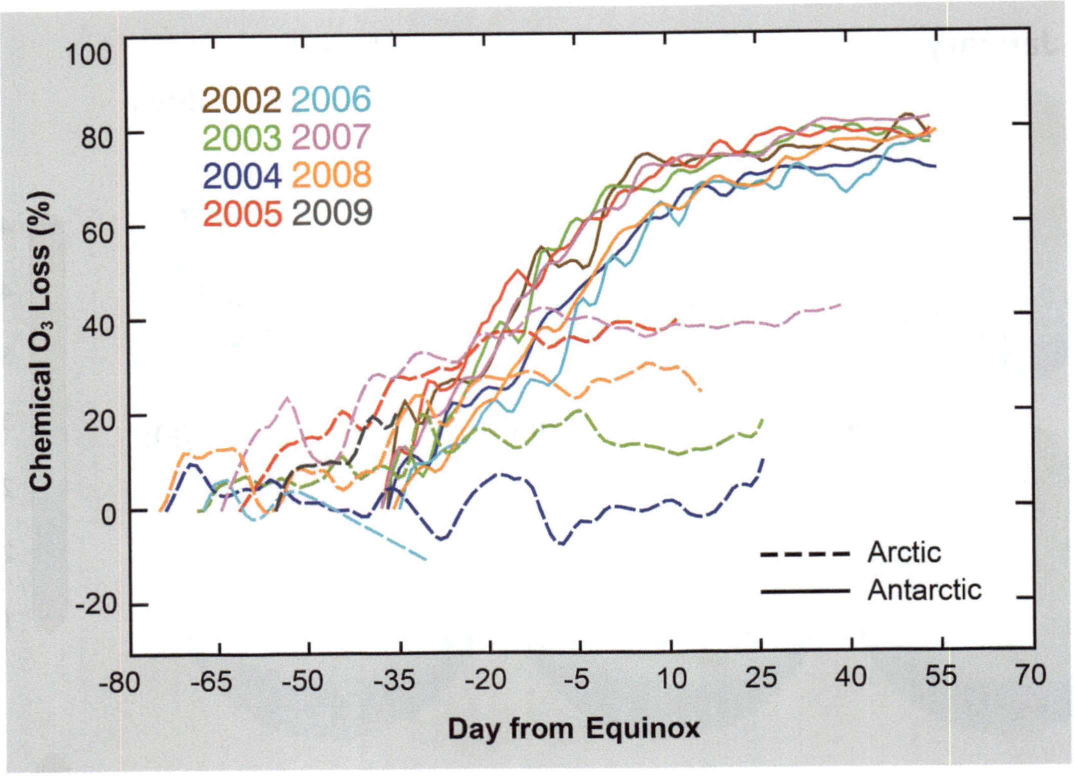

Fig. 10.24 Relative chemical ozone losses at the 475 K isentropic level (around 18 km) for the period 2002–2009 in the Arctic (*dotted lines*) and Antarctic (*solid lines*) polar vortices (Courtesy: IUP-IFE, University of Bremen).

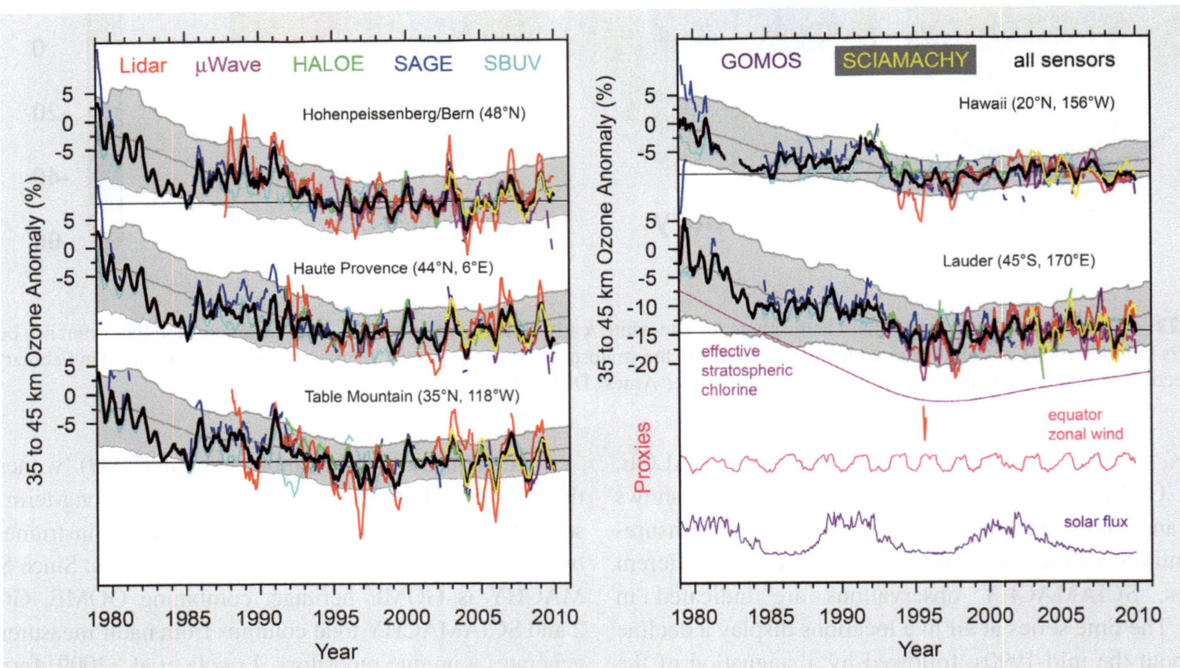

Fig. 10.25 Ozone anomalies from 1979 to 2009 derived from different datasets at five NDACC stations. The anomalies are averaged over the 35–45 km range. The all-instrument average is shown in black. All time series were smoothed by a 5-month running mean. At the bottom of the right hand panel three proxies are given: The negative 10 hPa zonal wind at the equator for the QBO, the 10.7 cm solar flux for the 11-year solar cycle and the inverted effective stratospheric chlorine for ozone destruction by chlorine. The grey line and the grey shaded area are showing model simulations and their 2-sigma standard deviation. (Courtesy: adapted from Steinbrecht et al. 2009. Reproduced/modified by permission of American Geophysical Union).

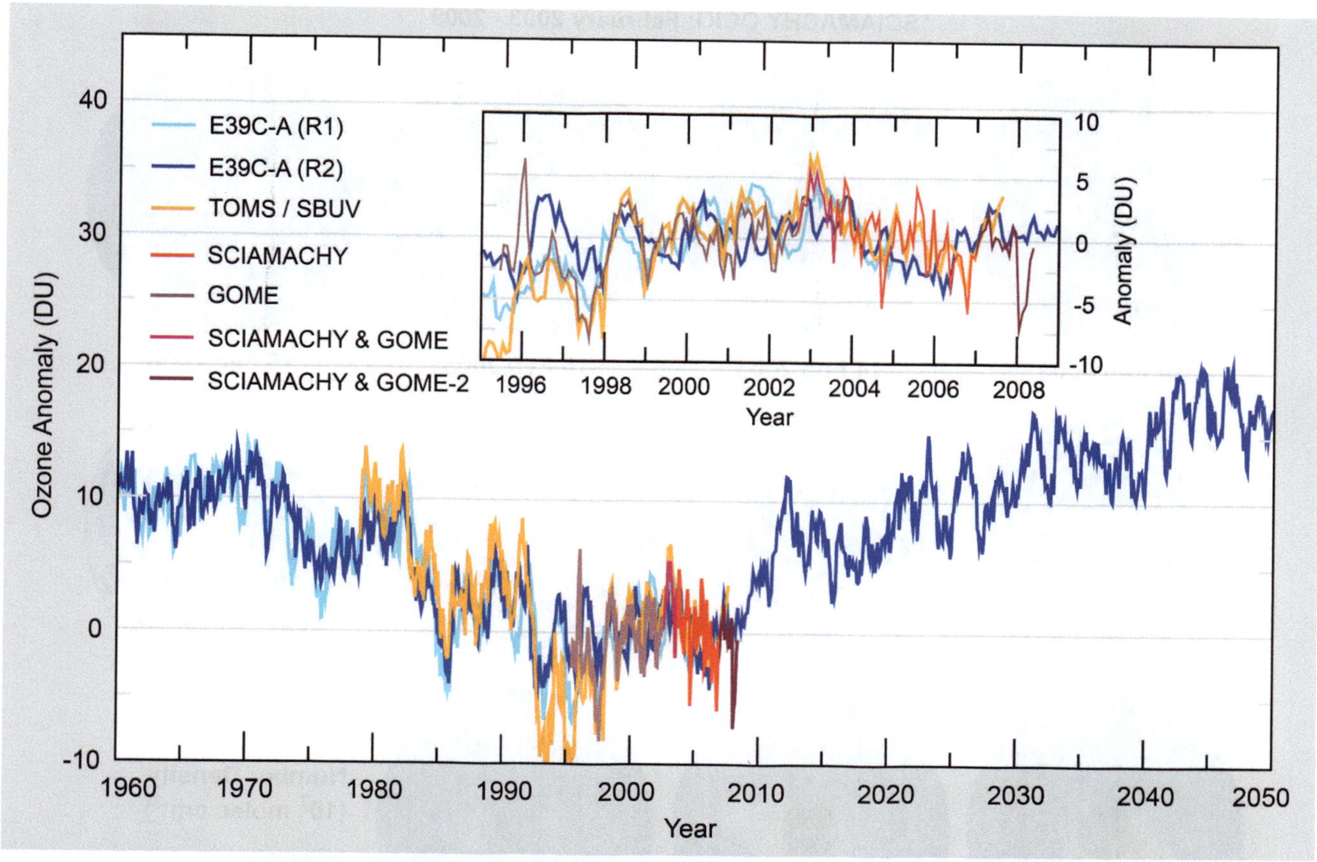

Fig. 10.26 Total ozone anomaly from 60°N to 60°S from the merged GOME/SCIAMACHY/GOME-2 dataset. For comparison the results from the merged TOMS/SBUV/OMI are given together with predictions from climate-chemistry model runs (Courtesy: Loyola et al. 2009. Reproduced/modified by permission of American Geophysical Union).

and GOME-2 data in periods of instrument overlap were adjusted accordingly. Global ozone trends were then derived by applying statistical methods, including the entire 60°N–60°S average serving as a near global mean. Figure 10.26 illustrates how well the merged GOME/GOME-2/SCIAMACHY dataset of total ozone columns compares with results from the chemistry-climate model (CCM) E39C-A. This figure again displays the so-called 'O_3 anomaly' which is the residual when subtracting the mean annual cycle from the satellite measurements. Apparently the phase of minimum stratospheric ozone is just occurring and a recovery can be expected in the next decades. The current good match between the CCM and the observations in Fig. 10.26 is a hint that the models' predications are trustworthy.

Chlorine Dioxide – OClO

One key question related to the expected recovery of stratospheric ozone is the degree of chlorine activation observed in polar winter and spring in both hemispheres. This effect depends not only on the total available inorganic chlorine amount but also on the presence of PSC for the activation of the chlorine reservoirs. The latter is a function of temperature and polar vortex stability and therefore is impacted by changes of the stratospheric dynamics and temperatures in response to increased concentrations of greenhouse gases. One good indicator for chlorine activation is the presence of OClO which is formed by reaction of BrO and ClO.

While SCIAMACHY observations in nadir (Kühl et al. 2006) continue the global measurements of total columns of OClO started with GOME (Wagner et al. 2001; Kühl et al. 2004; Richter et al. 2005b), also vertical profiles of OClO can be derived from the limb observations (Kühl et al. 2008). By applying a tomographic 2D approach (Puķīte et al. 2008) the retrieval can take into account horizontal gradients in the distribution of OClO, which is important in particular at the edge of the polar vortex. Figure 10.27 shows the OClO number density at 19 km altitude above the northern hemisphere derived from the SCIAMACHY limb observations (top) and the corresponding total column derived from nadir view (bottom) for mid of February in the Arctic winters 2002/2003–2008/2009. The data displays strong stratospheric chlorine activation only inside the polar vortex for cold

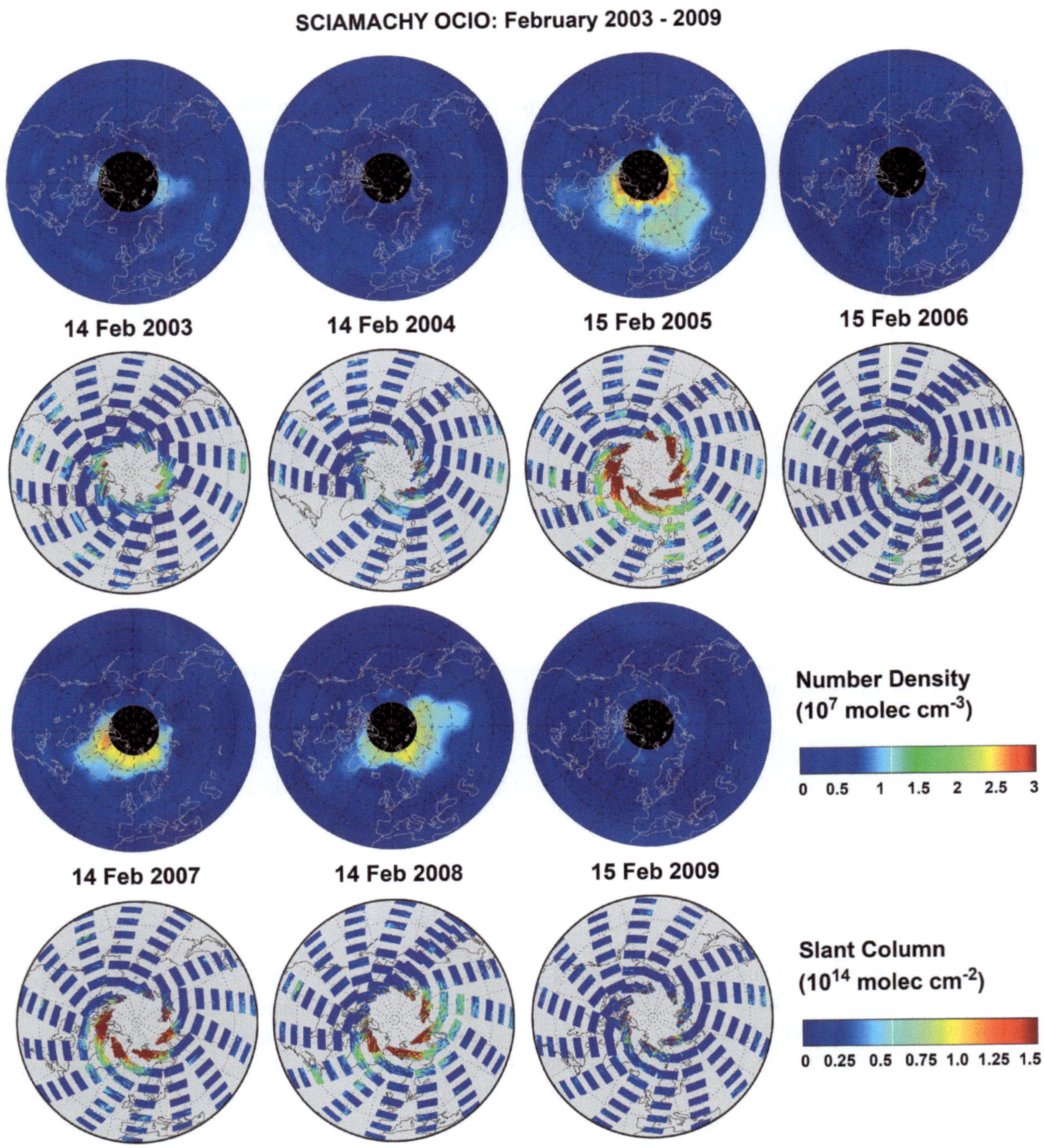

Fig. 10.27 OClO number density at 19 km altitude above the northern hemisphere derived from SCIAMACHY limb observations (*rows 1* and *3*) and the corresponding total OClO slant column derived from nadir views (*rows 2* and *4*) for mid of February in the Arctic winters 2002/2003–2008/2009 (Courtesy: S. Kühl; MPI for Chemistry, Mainz).

winters, reflecting the strong dependence of the degree of chlorine activation on the meteorological conditions of the respective winter, such as temperature and potential vorticity. While almost no OClO is found for the warm winters 2003, 2004, 2006 and 2009, much stronger activation of chlorine is observed for the cold winters 2005, 2007 and 2008. These findings are in good agreement with observations of the main active ozone depleting chlorine species ClO, the determining meteorological parameters, results from atmospheric chemistry models, and the related chemical ozone loss (compare Figs. 10.23 and 10.24 in the section on ozone). The strong inter-annual variability in the degree of chlorine activation reveals that despite the decrease of the stratospheric inorganic chlorine loading, large enhancements of OClO are still observed in the Arctic stratosphere for cold winters. Therefore, monitoring of stratospheric chlorine activation and detailed investigation of the relation to meteorological parameters is necessary to broaden the understanding of the related processes.

Bromine Oxide – BrO

Bromine compounds play an important role in the catalytic destruction of stratospheric ozone. Despite their importance, however, there are only few measurements of bromine compounds in the stratosphere. For the first time, SCIAMACHY provided global observations of stratospheric BrO profiles down to the lower stratosphere (Rozanov et al. 2005; Sioris et al. 2006; Kühl et al. 2008). The long-term changes of BrO as observed from SCIAMACHY agree well with ground-based observations at mid-latitudes (Hendrick et al. 2009). A comparison with ground-based UV-VIS zenith-sky observations by BIRA-IASB at Harestua, Norway (60°N) has already been presented in Fig. 9.18. The seasonal cycle, including the activation during winter and long-term changes, are consistently seen by both instruments. When comparing zonal mean BrO at selected stratospheric altitudes obtained from limb observations and from model simulations, an additional source of stratospheric bromine from very short-lived substances (VSLS) is required (Fig. 10.28) to explain the difference. This component amounts to about three to six parts per trillion by volume (pptv), or about 20% (WMO 2007, Sinnhuber et al. 2005). In an opposite analysis approach Theys et al. (2009) used the SCIAMACHY limb stratospheric BrO observations to demonstrate the validity of a new stratospheric BrO profile climatology.

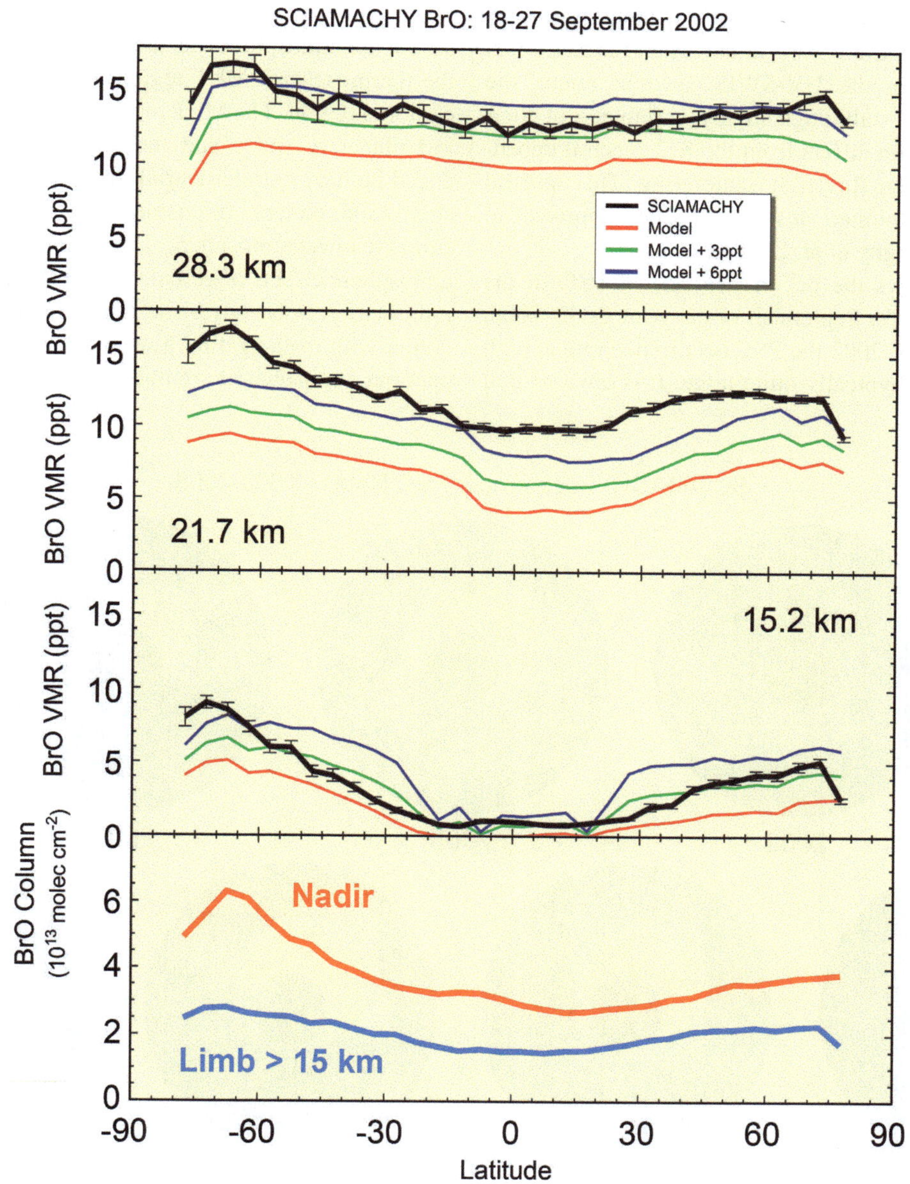

Fig. 10.28 Zonal mean BrO at selected altitudes obtained from SCIAMACHY limb observations and from model calculations. Additional bromine of 0, 3, and 6 pptv, respectively, has been added to the model calculations to account for the contribution from very short-lived substances (VSLS) (Courtesy: adapted from WMO 2007 – based on Sinnhuber et al. 2005).

Polar Stratospheric Clouds – PSC

Polar Stratospheric Clouds play a key role in the chemical processes which lead to severe ozone depletion in the polar stratosphere. These clouds are necessary to transfer inactive chlorine compound reservoirs such as HCl and $ClONO_2$ to active Cl that participates in different catalytic O_3 destruction cycles. PSC form at altitudes of about 15–25 km and exist as different types. Type Ia consists of crystalline NAT (nitric acid tri-hydrate) particles, the liquid type Ib PSC consist of ternary solutions of nitric acid, sulphuric acid and water. Type II PSC are made of water ice. A common feature of all types is that they only form at very low temperatures of less than about −78°C (195 K). PSC scatter solar radiation and thus affect the measured limb radiance spectra. Since PSC are rather Mie-scatterers than Rayleigh-scatterers in the UV-SWIR spectral range, the spectral dependence – although highly variable – of their scattering cross section differs from the λ^{-4} spectral dependence of the molecular Rayleigh scatterering. This spectral difference can be exploited in a colour-index approach to detect PSC (von Savigny et al. 2005b).

Figure 10.29 shows the occurrence rate of PSC in the southern hemisphere for September of the years 2002–2008. With the exception of 2002 the PSC occurrence rate at high southern latitudes is typically quite large. It is obvious that PSC are not symmetrically distributed around the South Pole, but the distribution is characterised by a wave-1 structure with a maximum in the South Atlantic sector and a minimum in the Australian sector. The low PSC occurrence rates in September 2002 are due to the anomalous mid-winter major stratospheric warming after 22 September, which caused PSC to disappear. In most cases there is a good correspondence between the detected PSC and their formation threshold of −78° C.

10.4 The Upper Atmosphere and Solar Activity

The upper atmosphere, i.e. the mesosphere and lower part of the thermosphere (MLT region), is still a relatively poorly explored region. The MLT region forms a transition between interplanetary space and the terrestrial atmosphere influenced both by extraterrestrial impacts – e.g. solar radiation, solar wind, meteors or cosmic dust – as well as by impacts from the lower atmosphere. Thus signs exist that in the upper atmosphere effects of global climate change can be detected rather early. Solar-terrestrial interactions can be studied together with investigating how anthropogenic emissions impact our atmosphere at remote altitudes.

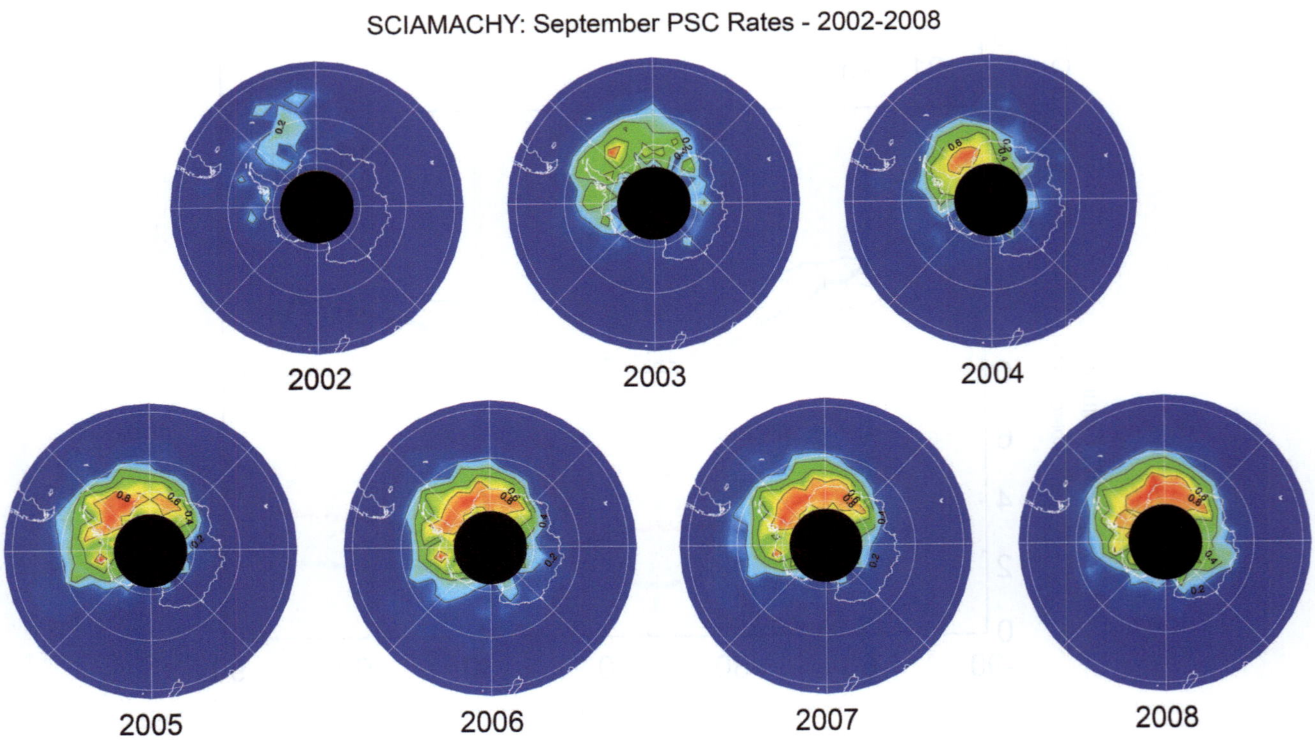

Fig. 10.29 Maps of PSC occurrence rates for September of the years 2002–2008. Contours levels correspond to 0.2, 0.4, 0.6, and 0.8. Red areas indicate occurrence rates exceeding 0.8 (Courtesy: C. von Savigny, IUP-IFE, University of Bremen).

Noctilucent Clouds – NLC

Noctilucent Clouds, also referred to as *Polar Mesospheric Clouds*, are a high latitude summertime mesospheric phenomenon, even observable from ground. They occur at altitudes of about 83–85 km near the polar summer mesopause and consist of H_2O ice particles with radii ranging from a few nm up to about 80–100 nm. NLC received a significant amount of scientific interest in recent years, since they may be early indicators of global change. This is because they react very sensitively to small changes in ambient conditions, particularly to changes in temperature and H_2O abundance. The scattering properties of the NLC particles allow mapping of these high altitude clouds. Since they scatter solar radiation efficiently, they affect the measured limb radiance profiles significantly, especially in the northern hemisphere where scattering angles at polar latitudes are especially small for SCIAMACHY limb observations. The NLC component of the measured limb signal, the blue curve in Fig. 10.30, can be two orders of magnitude larger than the Rayleigh signal (red curve in Fig. 10.30) at the NLC altitude, allowing for a simple detection of NLC.

SCIAMACHY observations of NLC have contributed in different ways to the current research on the polar summer mesopause. The main focus of these investigations was to improve the scientific understanding of the natural variability in NLC in order to better understand the role of NLC as indicators of global change. The natural variability is partially driven by the solar input and by dynamical processes such as planetary waves.

SCIAMACHY NLC observations were used to observe for the first time a depletion of NLC during a solar proton event (SPE, von Savigny et al. 2007a). Associated with such an event in January 2005, highly energetic solar protons precipitated into the Earth's polar cap areas. The January 2005 SPE is included in the period covered by Fig. 10.31. This graph shows SCIAMACHY NLC occurrence rates at different latitudes in the southern hemisphere during the 2004/2005 NLC season together with the ionisation caused by the precipitating solar protons. At the time of the SPE, the NLC occurrence rate in both southern latitude bands decreased rapidly. A mechanism for SPE induced NLC depletion is proposed by Becker and von Savigny (2010) using model simulations with a General Circulation Model (GCM). It suggests that a polar mesopause warming is caused by the SPE-driven catalytic ozone loss in the middle mesosphere followed by different stages of dynamic processes finally leading to a reduced upwelling above the pole, i.e. reduced adiabatic cooling. The January 2005 event is not the only event where NLC depletion has been detected. Rahpoe et al. (2010) demonstrated that a depletion of NLC did also occur during some of the other strong SPEs in the last three decades.

Solar proton events are rather intermittent and irregular events mainly occurring during solar maximum. They are not the only cause for NLC variability linked to solar impacts seen in SCIAMACHY observations. Robert et al. (2010) identified for the first time a 27-day solar cycle signature in NLC. Maxima in solar activity associated with the 27-day solar cycle – quantified for example using the MgII index described below (Skupin et al. 2004) – coincide with minima in the NLC occurrence frequency. Using MLS (Microwave Limb Sounder) observations of middle atmospheric temperatures, a 27-day solar cycle signature in the polar mesopause temperature was identified as the immediate cause of the apparent 27-day signature in NLC.

Another important driver for variability in NLC are planetary wave signatures. The most important of these signatures are the quasi-2-day-wave and the quasi-5-day-wave. Both of them are caused by instabilities of the summer mesosphere jet

Fig. 10.30 NLC as seen from the International Space Station on 22 July 2008 (*left panel*) together with the NLC signature in a SCIAMACHY UV limb radiance profile from 3 July 2002 (*right panel*). The scattering by ice particles in the NLC leads to a significant increase of the observed limb radiance profile peaking at the characteristic NLC altitude of about 83–85 km (*blue line*). Shown in red is for comparison a limb radiance measurement without NLC being present (Courtesy: C. von Savigny, IUP-IFE, University of Bremen; ISS photo: NASA).

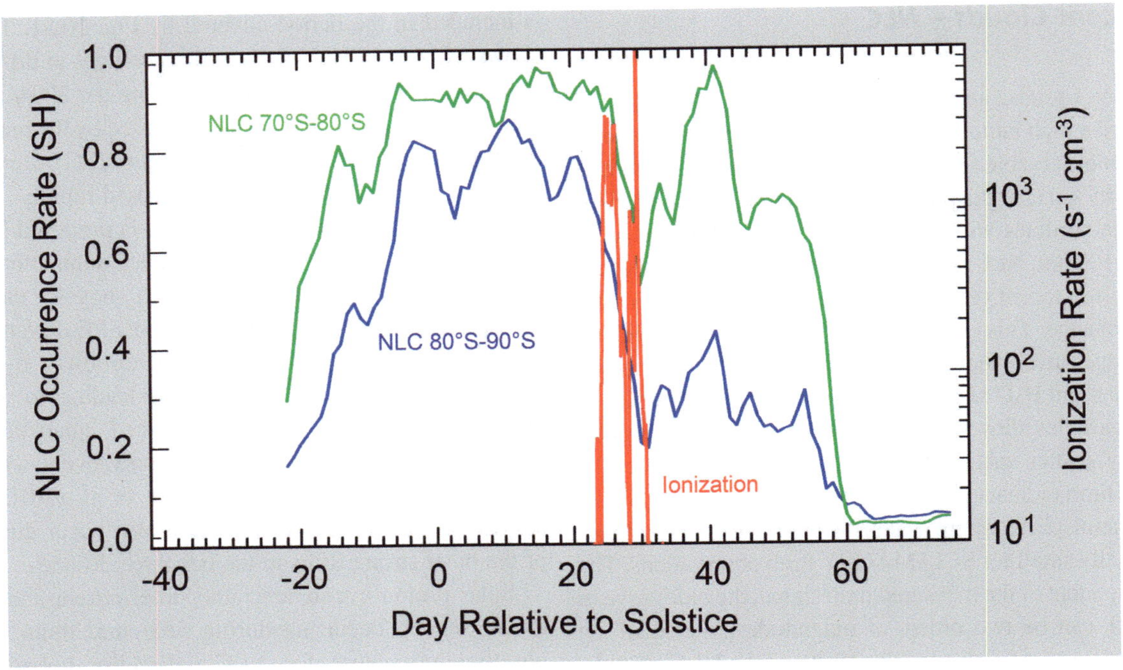

Fig. 10.31 Smoothed zonally averaged NLC occurrence rates in the southern hemisphere NLC season 2004/2005 (*left abscissa*) and ionisation rates at 82 km (*right abscissa*). At the time of the solar proton event the ionisation rate increases and causes a drop in the NLC rate (Courtesy: C. von Savigny, IUP-IFE, University of Bremen).

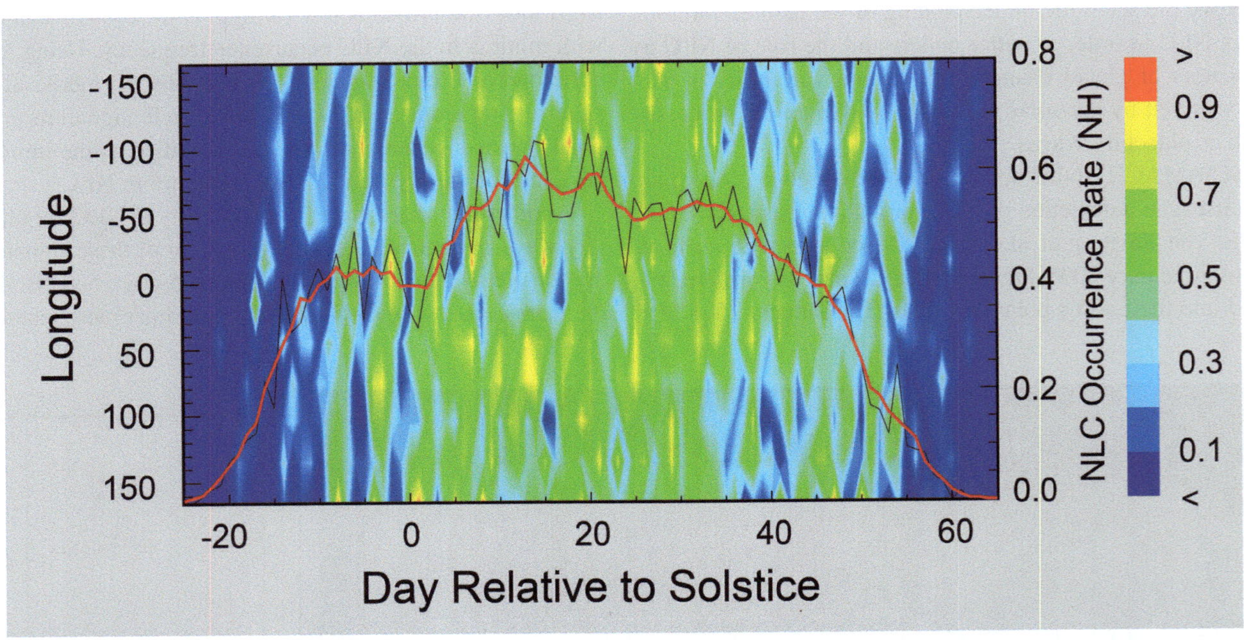

Fig. 10.32 Contour plot of the longitude and time dependence of the NLC occurrence rate for 2005 in the northern hemisphere with the zonally averaged NLC occurrence rate overlaid (red curve, right abscissa). The 5-day wave pattern is clearly visible at the beginning of the NLC season as a westward propagating signature with wavenumber 1 and a period of 5 days (Courtesy: C. von Savigny, IUP-IFE, University of Bremen).

and occur intermittently during the NLC seasons in both hemispheres. SCIAMACHY NLC observations, again in combination with MLS temperature observations, showed for the first time, that the quasi 5-day-wave signatures in NLC are caused by similar wave signatures in mesopause temperatures (von Savigny et al. 2007b). The 5-day-wave signatures appear as fairly regular westward propagating perturbations on the NLC field (Fig. 10.32) with a period of 5-days, and are most pronounced during the beginning and the end of the NLC seasons for the case shown.

Apart from the detection and mapping of NLC, SCIAMACHY observations also permit the estimation of the NLC particle size. For wavelengths below about 310 nm, the multiple scattering and surface reflection components to the limb signal are negligible. In single scattering approximation, the spectral exponent of the NLC scattering spectrum can be related to the NLC particle size assuming for example Mie theory and the refractive index of H_2O ice (von Savigny et al. 2004a). NLC particle sizes of 40–50 nm were determined from a distance of about 3300 km. The SCIAMACHY NLC size dataset currently presents the most comprehensive satellite dataset of NLC particle sizes. The derived particle sizes are in good agreement with independent observations (von Savigny and Burrows 2007; von Savigny et al. 2009).

Mesospheric Ozone and the October/ November 2003 Solar Storm

Highly energetic protons ejected from the Sun during phases of high coronal activity – solar flares or solar coronal mass ejections – reach the Earth with the solar wind, ionise the atmosphere and lead to the formation of HO_x and NO_x in the mesosphere and upper stratosphere. Both families participate in catalytic O_3 destruction cycles, with HO_x being more efficient above about 50 km and NO_x below about 50 km. Consequently enhanced O_3 destruction is expected after strong solar proton events. A good opportunity to study the impact of solar activity on mesospheric ozone fields was a period in October/November 2003 – also known as the 'Halloween Storm' – when the Sun exhibited extremely large coronal SPE. Figure 10.33 presents an analysis of the impact of the solar proton event at the end of October 2003 on the upper atmospheric O_3 (Rohen et al. 2005). A strong ozone depletion of more than 50% even deep in the stratosphere is observed at high geomagnetic latitudes in the northern hemisphere, whereas the observed ozone depletion in the more sunlit southern hemisphere is much weaker. SCIAMACHY measurements of the O_3 loss due to SPE agree well with model simulations, indicating that the main processes leading to the O_3 loss are fairly well understood.

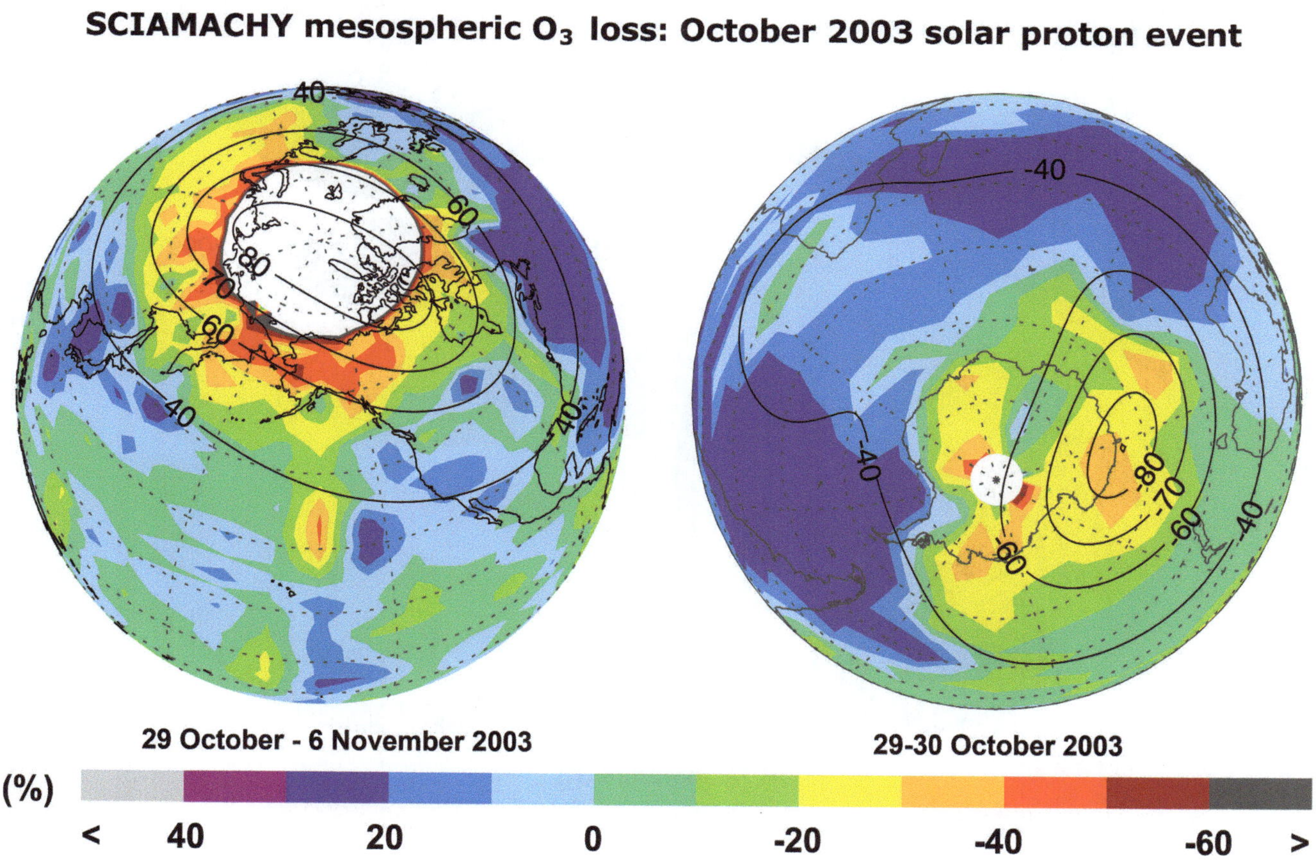

Fig. 10.33 Measured change of ozone concentration at 49 km altitude due to the strong solar proton events end of October 2003 in the northern and southern hemisphere relative to the reference period of 20–24 October 2003. White areas depict regions with no observations. The black solid lines are the Earth's magnetic latitudes at 60 km altitude for 2003 (Courtesy: Rohen et al. 2005).

Mesopause Temperatures, Thermospheric NO and Metal Layers

A number of atomic and molecular emission signals from the mesosphere and lower thermosphere (MLT) can be detected throughout the SCIAMACHY spectral range, e.g. neutral and ionised magnesium (Mg) lines as well as NO gamma bands in the UV, OH Meinel band emissions in the SWIR, several transitions from excited-state O_2 or sodium lines and atomic oxygen in the VIS range. These emission signals can be used to characterise the distribution of the atoms, ions and molecules in the upper atmosphere.

SCIAMACHY measurements of molecular emissions in the upper atmosphere provided for the first time the retrieval of OH^* rotational temperatures at the mesopause from satellite measurements during night. OH^* is vibrationally excited at the mesopause through the reaction of H and O_3. This produces an OH^* emission layer centred at about 87 km with a width of 8–10 km. Several of the OH^* Meinel emission bands are observable in the SCIAMACHY spectral range, e.g. the OH^* (3–1) band at around 1500 nm. This emission band is used for the retrieval of OH^* rotational temperatures because it is one of the most intense emission bands. From the relative intensity of two or more rotational lines the effective temperature of the emitting layer can be retrieved

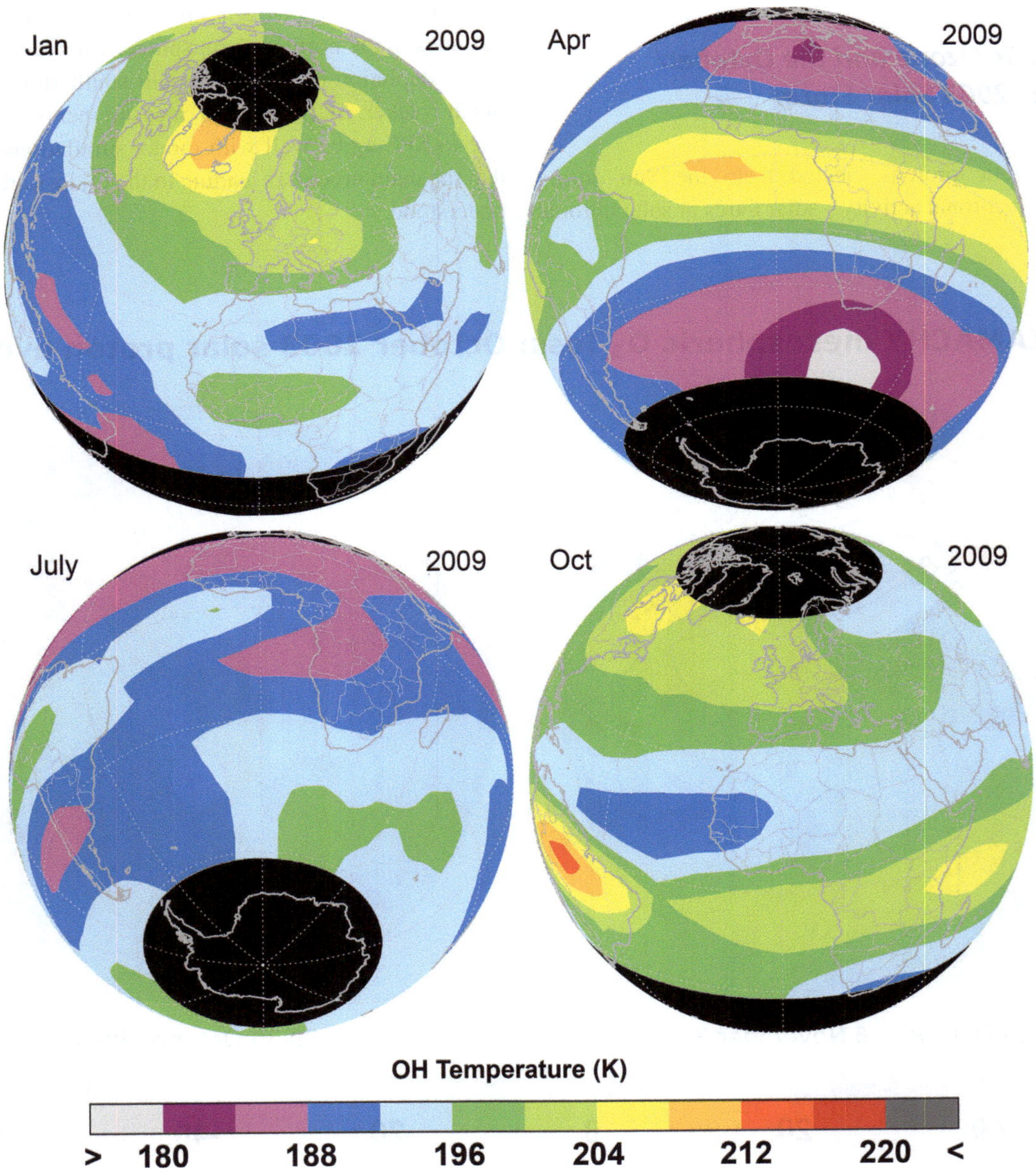

Fig. 10.34 Monthly averaged SCIAMACHY retrievals of OH rotational temperatures at about 87 km for January, April, July and October 2009 (Courtesy: K.-U. Eichmann, IUP-IFE, University of Bremen).

from SCIAMACHY data. This method has been applied for several decades to retrieve mesopause OH^* rotational temperatures from ground measurements (e.g. Bittner et al. 2002). Therefore coincident ground-based OH^* rotational temperature measurements over Germany in Wuppertal and at Hohenpeissenberg, as well as Hawaii, had originally been used to validate the novel SCIAMACHY space-based approach (von Savigny et al. 2004b). A comparison of both methods yielded very good agreement. Figure 10.34 shows as an example the monthly averaged mesopause temperatures retrieved from SCIAMACHY OH^* (3–1) emission measurements for January, April, July and October 2009. As the observations used for the retrievals are made on the Earth's nightside, the OH rotational temperatures are only available at higher latitudes during wintertime.

Retrieval of OH^* rotational temperatures is not the only scientific application of SCIAMACHY's OH Meinel-band emission observations. The data can also be used to study atmospheric wave signatures (Ern et al. 2009) or to determine chemical heating rates associated with the exothermic chemical reaction $H + O_3 \rightarrow OH^* + O_2$, which forms the vibrationally excited OH molecules as mentioned above (Kaufmann et al. 2007).

Especially interesting are also UV wavelengths. Here, emissions from the NO gamma-bands and atomic oxygen are observed as well as a number of metallic species and their ions, e.g. MgI/MgII and FeI/FeII (Fig. 10.35) which have the bulk of their abundances in the lower thermosphere around 100–200 km. In order to fully exploit SCIAMACHY's capabilities in this exotic atmospheric region, limb measurements are specifically adjusted since summer 2008 for about 60 orbits per month to scan the region between 60 km up to 160 km.

From the NO gamma-band emissions, slant column densities of NO along the line-of-sight can be derived. NO is produced in the thermosphere and upper mesosphere by ionisation processes related to auroral electron precipitation in polar regions and to photoionisation in low latitudes. As the lifetime of NO in the MLT region is restricted by photolysis, NO abundances are largest in high-latitude winter. However short increases at high latitudes can occur due to energetic particle precipitation events related to geomagnetic disturbances, leading to the well-known correlation between thermospheric NO and the geomagnetic activity index (Fig. 10.36).

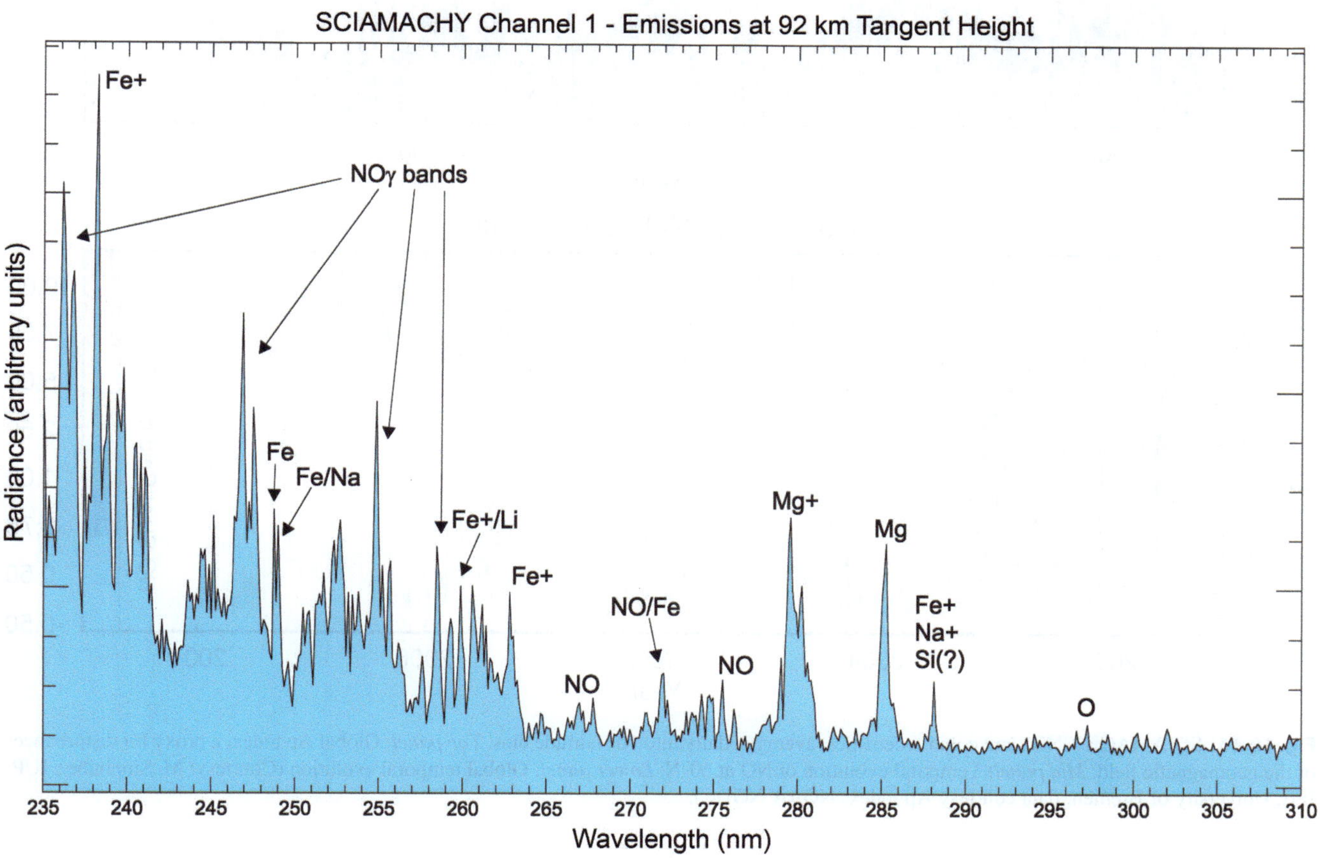

Fig. 10.35 Emission lines in SCIAMACHY limb radiance in channel 1 at the highest limb tangent altitude, normalised to the solar irradiance, in arbitrary units. A number of emission signals are detected. The most dominant features are the NO gamma-bands and MgII, but signals from neutral Mg, atomic oxygen, Si, Fe and Fe+ are observed as well (Courtesy: M. Sinnhuber, IUP-IFE, University of Bremen).

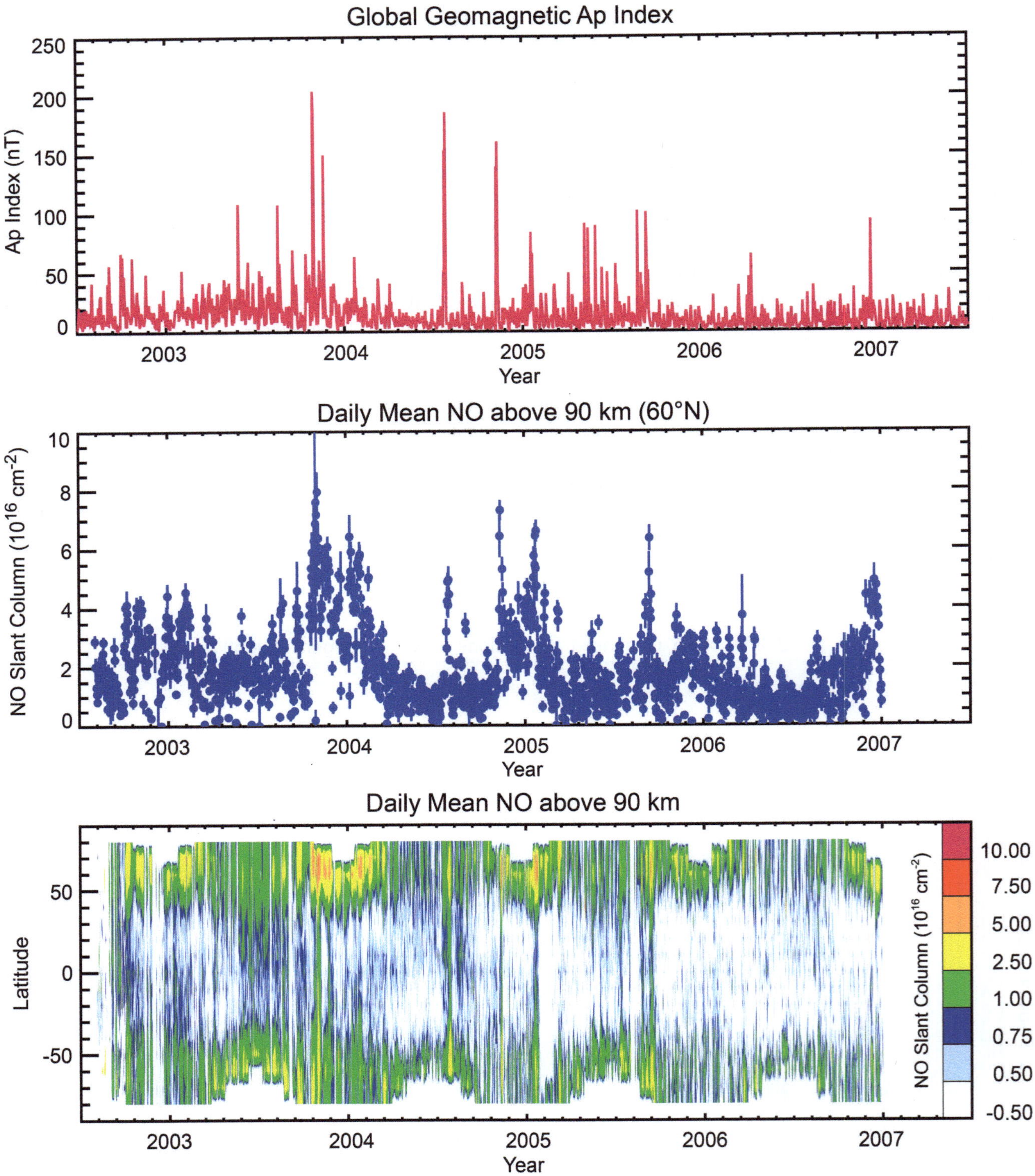

Fig. 10.36 SCIAMACHY NO slant column densities, averaged daily into 10° latitude bins. *Top panel*: Global Ap index, a proxy for disturbances of the geomagnetic field. *Mid panel*: Temporal evolution of NO at 60°N. *Lower panel*: Global temporal evolution (Courtesy: M. Sinnhuber, IUP-IFE, University of Bremen; data courtesy Ap index: NOAA NGDC).

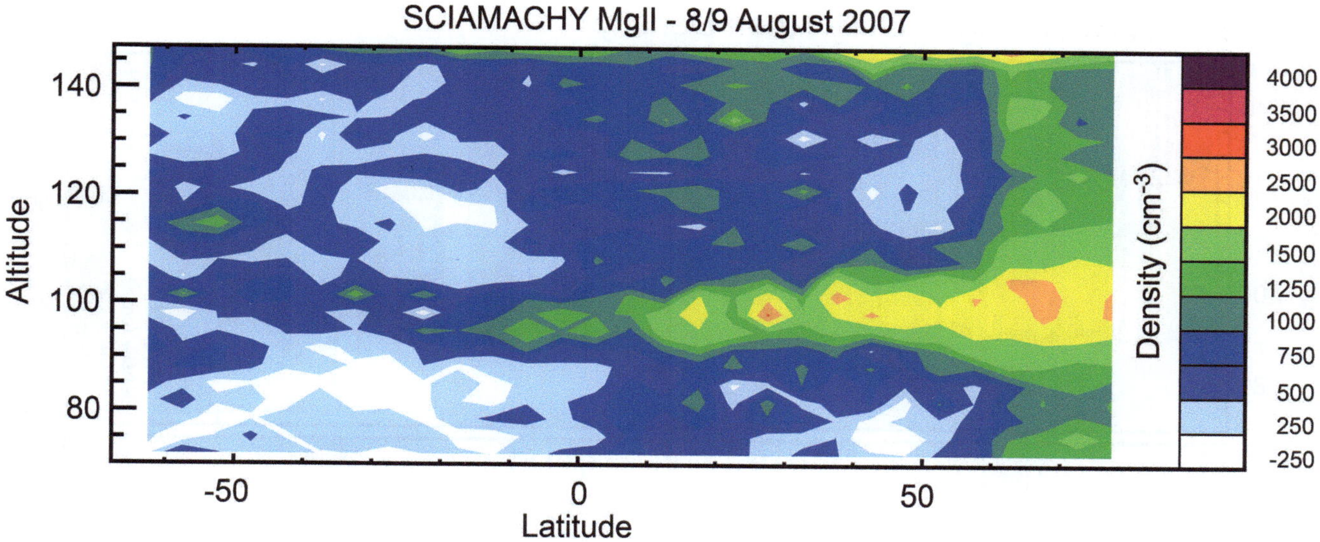

Fig. 10.37 Atmospheric abundance of MgII derived from 20 orbits scanning in the mesosphere-thermosphere limb mode from 60 to 160 km in August 2007. The most striking feature is the strong MgII layer around 100 km extending from high northern latitudes to the equator (Courtesy: M. Sinnhuber, IUP-IFE, University of Bremen).

Neutral and singly ionised magnesium can be derived from their emissions at 280 and 285 nm, respectively (see Fig. 10.35). Vertical profiles of number density of atmospheric Mg species were derived from a satellite sensor for the first time using SCIAMACHY measurements (Scharringhausen et al. 2008). While atmospheric abundances of many species – including NO and most metals and metal ions – are largest in the lower thermosphere, the topmost tangent altitude of SCIAMACHY in the nominal limb mode remains below this region. To overcome that limitation a 'tomographic' two-dimensional retrieval was developed using simultaneous information from all limb- and nadir measurements of one orbit to derive number densities up to altitudes of 400 km, albeit with a limited vertical resolution above 93 km (Scharringhausen 2007). The vertical structure of MgII derived from this tomographic retrieval is very comparable to that obtained from the mesosphere-thermosphere mode (Fig. 10.37), with a pronounced maximum around ~100–110 km, but generally showing lower peak values due to the strong smoothing of the profile above 90 km for the nominal limb mode. In the tropics, MgII is quite stable, with a sharp increase above 90 km and no distinguishable annual cycle. In mid-latitudes, the MgII maximum shows a seasonal cycle with maxima closely following solar illumination, as shown, e.g. for total column values using GOME data by Correira et al. (2008). This seasonal cycle apparently extends down to altitudes of ~85 km, where enhanced values are observed around mid-summer. At high latitudes, the maximum is shifted to autumn/winter. Around the winter solstice values in the upper mesosphere are again increased (Fig. 10.38). No correlation of MgII or Mg with meteor activity is observed, emphasising the dependency of the ionised metal on local chemistry rather than on meteoric influx.

Observing the Active Sun – The Mg II Index

SCIAMACHY's scientific objective to explore atmospheric trace constituents is achieved by analysing solar radiation, both in terms of scattered and reflected sunlight but also in direct viewing for calibration purposes. Due to the high sensitivity and spectral stability it is, in addition, feasible to retrieve information about those aspects of solar activity which manifest themselves in the emitted radiation. Therefore solar observations are analysed on a regular time grid offering the possibility to monitor solar variations and their impact on the atmosphere. The solar activity shows some well known periodic variations such as the 27-day cycle caused by solar rotation. Another is the 11-year solar cycle, coupled with the 22-year magnetic cycle which correlates with changes in sunspots and Fraunhofer lines. During phases of high solar activity an increase in the number of sunspots in the photosphere and large chromospheric plage areas are observed. The plage areas are hotter than the surrounding areas and cause the enhancement of the emission core within the absorption features of many solar Fraunhofer lines. Thus solar proxy indicators can be given by the core-to-wing ratio of selected Fraunhofer lines. The Mg II index is defined as the core-to-wing ratio of the Mg II Fraunhofer line centered

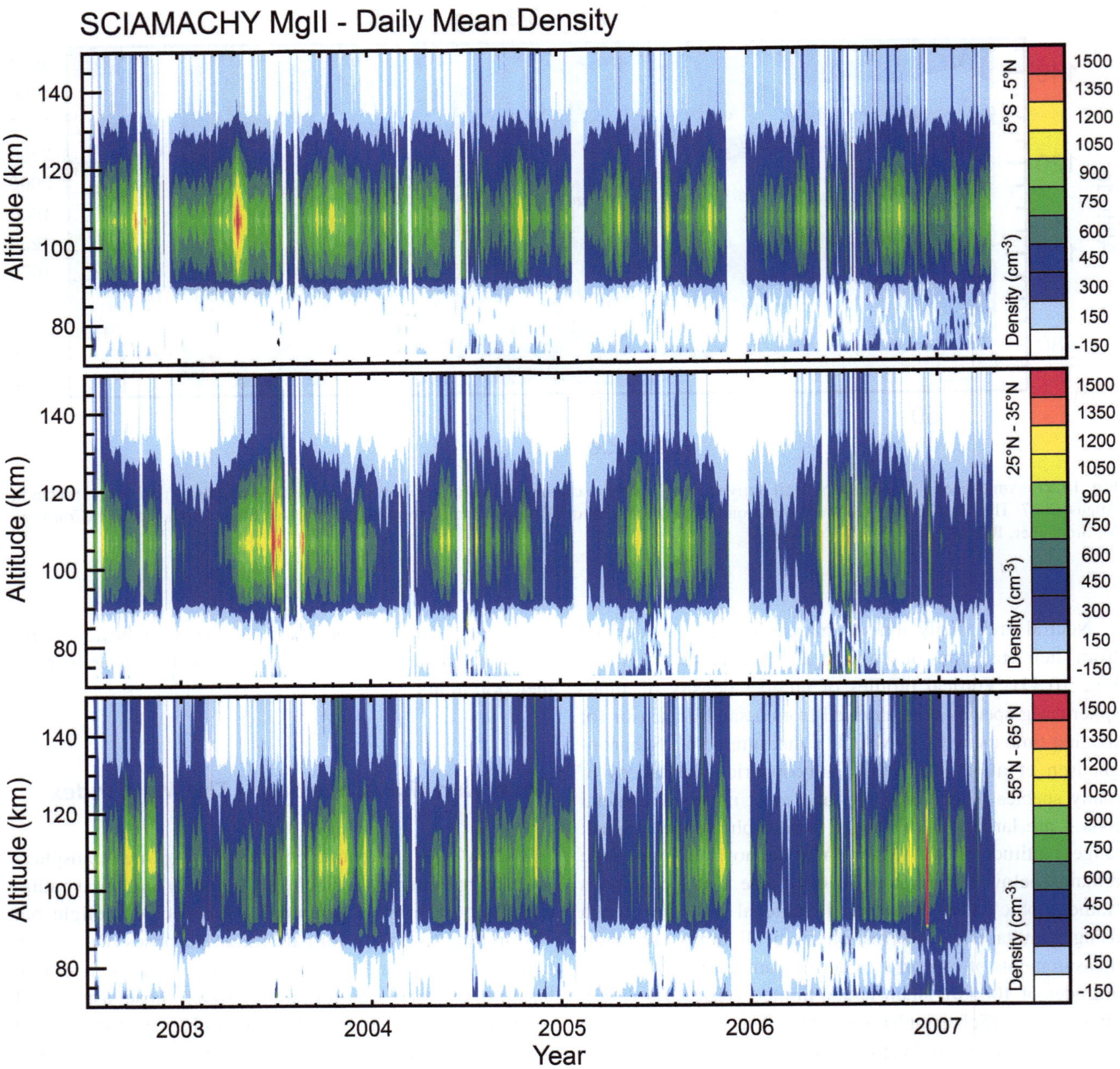

Fig. 10.38 Atmospheric abundances of MgII derived from the nominal limb scanning from the surface up to 93 km. The strong maximum around 100 km is reproduced very well, although with lower values due to the restricted vertical resolution above 93 km. The three panels show the MgII densities at different latitude bands: 5°S–5°N (*top*), 25°N–35°N (*mid*) and 55°N–65°N (*bottom*) (Courtesy: M. Scharringhausen/M. Sinnhuber, IUP-IFE, University of Bremen).

at 279.9 nm. It can be used as a proxy for spectral variations in the solar extreme UV (EUV, Viereck et al. 2001) and correlates with atmospheric ozone variations and other relevant atmospheric quantities.

For the understanding of the solar-terrestrial climate interaction the establishment of long time series covering several solar cycles is important. Due to the limited lifetime of spaceborne missions this has to be constructed from different satellite experiments. Figure 10.39 combines the Mg-II indices for the solar cycles 21–23, using data from NOAA missions (Viereck et al. 2004), GOME (Weber 1999), and SCIAMACHY (Skupin et al. 2004). Both the 27-day periodicity and the declining phase of solar cycle 23 are clearly visible. Differences between SCIAMACHY and GOME are mostly below ±0.5%, between SCIAMACHY and NOAA below ±0.25%. The derived MgII index was used to identify a correlation of stratospheric ozone with the 27-day solar cycle (Dikty et al. 2010). SCIAMACHY is

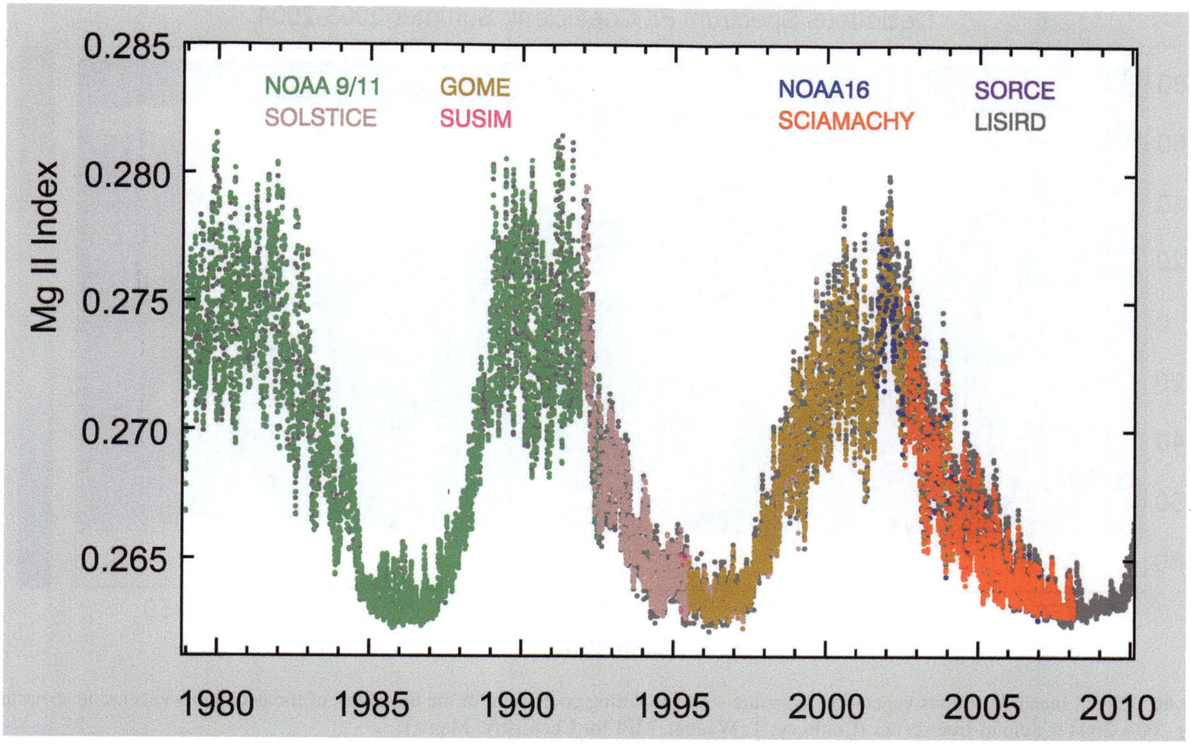

Fig. 10.39 Solar activity measured via the Mg II index by several satellite instruments including GOME and SCIAMACHY. By combining various satellite instruments, the composite Mg II index covers more than three complete 11-year solar cycles (Courtesy: M. Weber, IUP-IFE, University of Bremen).

the first spaceborne instrument observing daily solar spectral irradiance (SSI) continuously between 230 and 1750 nm. In order to address how much the irradiance changes in the UV-VIS-NIR and SWIR range on 27-day and 11-year timescales, short-term SSI variations were parameterised in terms of the proxies faculae brightening, i.e. MgII index, and sunspot darkening, i.e. photometric sunspot index (Pagaran et al. (2009).

10.5 The Earth Surface and Beneath

SCIAMACHY's realm is the Earth's atmosphere. However, the measured earthshine spectra are also affected by surface reflection and absorption, i.e. by the broadband ground albedo, and by narrowband spectral structures of different origin. These features went for a long time unnoticed in trace gas retrieval algorithms. The initial approach is certainly to take them into account for improving the tropospheric trace gas information (see Chapter 7). However, these features can also be used to derive various surface parameters over land or even phytoplankton properties of the oceans.

Land Vegetation Characteristics

Vegetation is a unique property of the Earth. Understanding how it changes contributes to many applications, ranging from global climate change to predicting crop yield. Usually vegetation indices exploit spectrally broadband differences in reflectivity between the red and NIR wavelength range. Atmospheric sensors like SCIAMACHY provide a higher spectral resolution, i.e. such differences can be analysed on a much finer spectral scale. While optimising DOAS trace gas retrievals Wagner et al. (2007) noticed that the resulting residuals displayed distinct structures, particularly over vegetated land. These structures could be reduced in amplitude when vegetation data such as spectral reflectances of conifers, deciduous trees or grass were included in the spectral fitting. Figure 10.40 shows the fit results for deciduous vegetation for summer 2003–2004. A correspondence to deciduous vegetation is obvious, but interference with other vegetation types and coastal waters is still present. Since retrieving vegetation information from remotely sensed atmospheric data is a rather novel approach, current results are still preliminary and require further investigations. There is a clear need to collect more spectral reference data at spectral resolutions similar to SCIAMACHY.

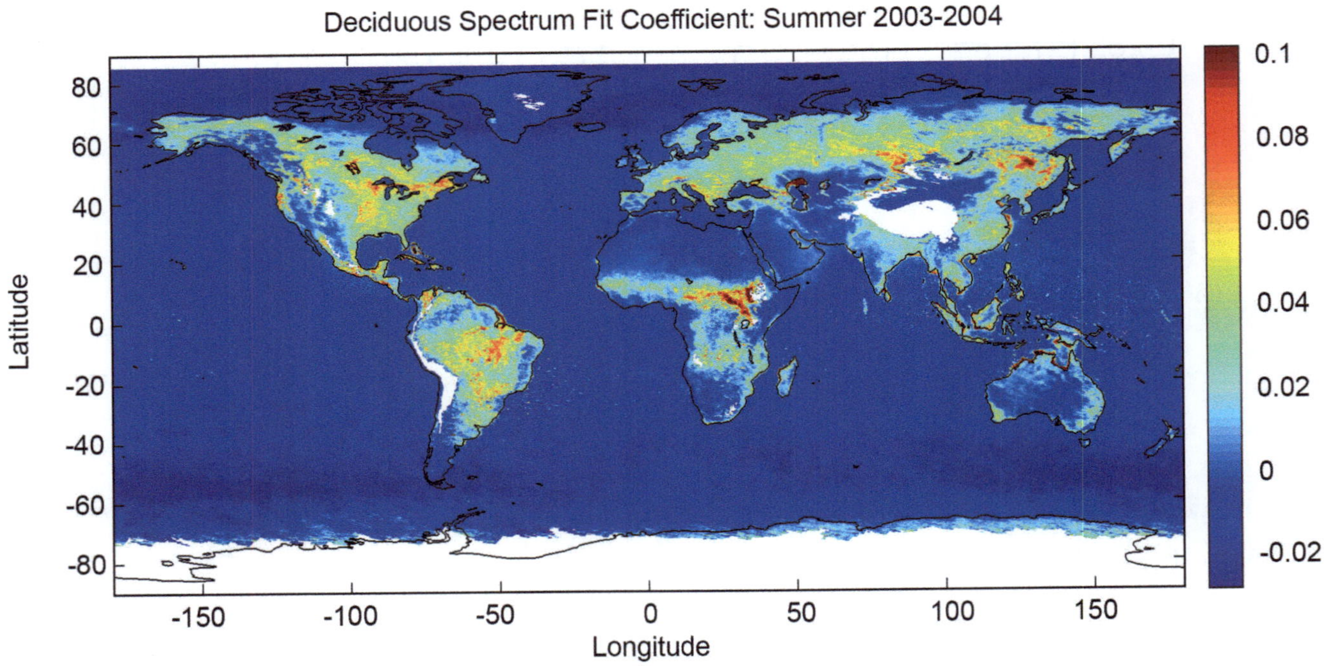

Fig. 10.40 Global mean deciduous vegetation signature – DOAS fitting coefficient of the logarithm of the deciduous vegetation spectrum – for summer 2003–2004 for cloud-free scenes (Courtesy: T. Wagner, MPI for Chemistry, Mainz).

Oligotrophic Oceanic Regions

Certain regions of the world's oceans are oligotrophic, i.e. they contain very low levels of nutrients and are thus almost free of biogenic activity. As a consequence, solar irradiation partly penetrates into the uppermost water layers. This has two effects on the spectra of the backscattered light: first, the ocean water causes rather broadband absorptions according to its cross section. In addition, vibrational Raman scattering occurs causing high-frequent spectral structures (Vasilkov et al. 2002; Vountas et al. 2003) similar to the atmospheric *Ring* effect (see Chapter 7). Global maps of fit results for vibrational Raman scattering show consistent global patterns with high values for the oligotrophic oceanic regions and have been obtained from SCIAMACHY data by applying differential optical absorption spectroscopy in the UV-A region (Vountas et al. 2007).

Oceanic Phytoplankton Characteristics

In open water, marine phytoplankton is the basis of the marine food web. They are contributing 50% to the global primary production via photosynthesis. Microscopic algae also play an important role in the global carbon cycle. For photosynthesis sunlight is absorbed by certain pigments, such as chlorophyll. The absorption spectrum is typical for a particular groups of algae due to their charcteristic pigment composition. The absorption signatures can be identified in SCIAMACHY data, allowing the quantitative evaluation of the global distribution of phytoplankton. Including a phytoplankton absorption spectrum in the DOAS retrieval, Vountas et al. (2007) successfully retrieved global maps of marine chlorophyll concentrations from SCIAMACHY data. Furthermore, Bracher et al. (2009) could even distinguish different types of phytoplankton, namely diatoms from cyanobacteria (Fig. 10.41). The resulting global maps are in good agreement with biochemical models and with independent in situ measurements obtained during cruises with the German research vessels *Polarstern*, *Maria S. Merian* and *Sonne* in the Atlantic and Pacific Oceans. These comparisons again proved the validity of the phytoplankton concentrations retrieved from SCIAMACHY measurements.

10.6 SCIAMACHY and the Goddess of Love

Although the Earth's atmosphere is unique, most of the other planets of the solar system possess a gaseous 'envelope' as well. Interplanetary spacecraft, equipped with spectrometers to study such extraterrestrial atmospheres, sometimes steer their instruments towards our home planet. The acquired Earth spectra usually serve calibration purposes. In an opposite approach SCIAMACHY was using Venus as a point source to better characterise the slit width of the spectrometer. During these observations Venusian spectra

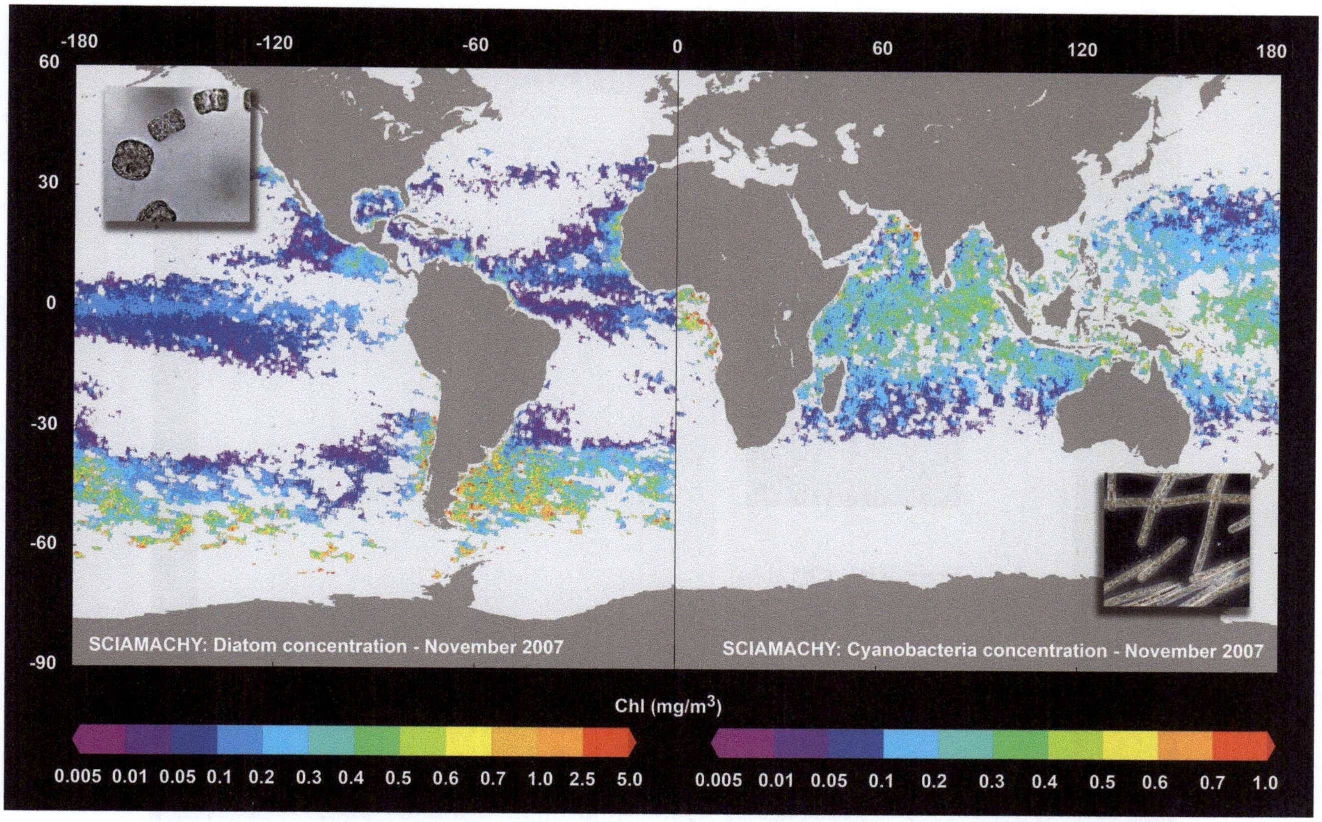

Fig. 10.41 Global biomass distributions of diatoms (*left hemisphere*) and cyanobacteria (*right hemisphere*) in November 2007 as derived from SCIAMACHY data using the PhytoDOAS method. The insets show members of the two algae groups (Courtesy: A. Bracher IUP-IFE, University of Bremen and Alfred Wegener Institute for Polar and Marine Research, adapted from Bracher et al. 2009; photo diatoms: E. Allhusen, cyanobacteria: S. Kranz; both Alfred Wegener Institute for Polar and Marine Research).

were obtained which proved very valuable in terms of planetary science (Gottwald et al. 2009).

With a diameter of 12100 km and being an inner planet at distances from Earth ranging from 42 to 256 million km Venus exhibits continuously changing viewing conditions during a full solar revolution. On 20 March 2009 when the first measurement was scheduled Venus was close to inferior conjunction at its minimum distance with the planet passing between the Earth and the Sun. SCIAMACHY mainly looked at the night side of Venus and only a very small part of the planet was lit by sunlight. Three months later on 25 June 2009, Venus was much farther away at 127 million km, with Sun-Venus-Earth forming a right-angled triangle. Thus, about half of Venus' sunlit side was in view. In March the observation geometry resembled a limb configuration while in June it was more of nadir-type. Both occasions yielded spectra with detailed absorption features of carbon dioxide (CO_2), the gas that forms Venus' atmosphere and sustains its extreme greenhouse effect. As expected the features are much stronger in June, because the line of sight was penetrating much deeper into the atmosphere (Fig. 10.42).

SCIAMACHY's Venus observations serve two purposes. They supplement close-up measurements by the instruments SPICAV and VIRTIS onboard ESA's Venus Express (VEX) mission orbiting the Earth's neighbour since 2006. Because SCIAMACHY viewing geometries are different from those of VEX, new information about the Venusian atmosphere can be inferred. The main benefit is, however, that the SCIAMACHY measurements provide another example of how the spectral signatures of a terrestrial planet might appear when viewed from far away. Since in 1995 the first planet around a solar-type star was discovered, finding an Earth-like planet around another star – a *Second Earth* or *Exo-Earth* – has become one of the big challenges in astronomy. Most of the currently known exoplanets are giant planets, like Jupiter. In the coming years, improved and dedicated instrumentation will bring smaller, terrestrial planets into reach. Because of the exoplanets' enormous distance, they will always appear point-like. Only measuring and analysing their spectra permits finding out whether they might be suitable for harbouring life. This is why spectroscopy of solar system bodies is a greatly appreciated test case. By fully exploiting SCIAMACHY's spectral and operational capabilities, even the other bright solar system bodies – Mars, Jupiter and Saturn – come into reach.

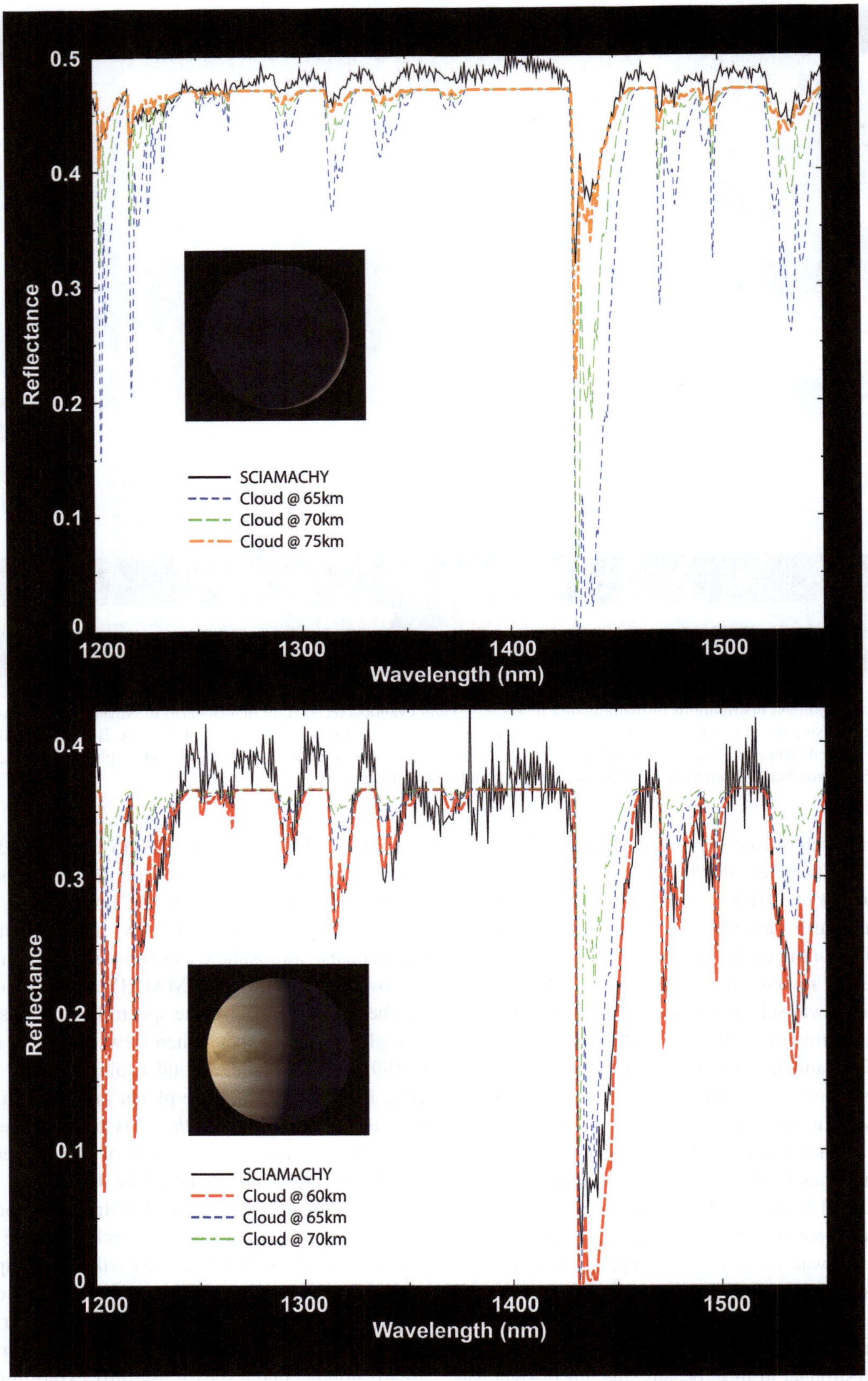

Fig. 10.42 Venus reflectance from 1200 to 1550 nm as measured in March (*top*) and June (*bottom*) 2009. Most absorption features are due to CO_2. The spectra are modelled by using CO_2 as absorber and H_2SO_4 clouds at different altitudes as scatterer. The illumination conditions in the insets have been derived from the NASA/JPL Solar System Simulator (Courtesy: M. Vasquez, DLR-IMF).

References

Barkley MP, Monks PS, Hewitt AJ, Machida T, Desai A, Vinnichenko N, Nakazawa T, Yu Arshinov M, Fedoseev N, Watai T (2007) Assessing the near surface sensitivity of SCIAMACHY atmospheric CO_2 retrieved using (FSI) WFM-DOAS. Atmos Chem Phys 7:3597–3619

Barkley MP, Palmer PI, De Smedt I, Karl T, Guenther A, Van Roozendael M (2009) Regulated large-scale annual shutdown of Amazonian isoprene emissions? Geophys Res Lett 36:L04803. doi:10.1029/2008GL036843

Barrie LA, Bottenheim JW, Schnell RC, Crutzen PJ, Rasmussen RA (1988) Ozone destruction and photochemical reactions at polar sunrise in the lower Arctic atmosphere. Nature 334:138–141

Becker E, von Savigny C (2010) Dynamical heating of the polar summer mesopause induced by solar proton events. J Geophys Res 115: D00I18, doi:10.1029/2009JD012561

Begoin M, Richter A, Kaleschke L, Tian-Kunze X, Stohl A, Burrows JP (2010) Satellite observations of long range transport of a large BrO cloud in the Arctic. Atmos Chem Phys 10:6515–6526

Beirle S, Platt U, Wagner T (2004) Monitoring nitrogen oxides with satellite instruments: High resolution maps from GOME narrow swath mode and SCIAMACHY. Proceedings of ENVISAT & ERS symposium 2004, Salzburg, Austria, ESA SP-572

Bergamaschi P, Frankenberg C, Meirink JF, Krol M, Villani MG, Houweling S, Dentener F, Dlugokencky EJ, Miller KB, Gatti LV, Engel A, Levin I (2009) Inverse modeling of global and regional CH_4 emissions using SCIAMACHY satellite retrievals. J Geophys Res 114:D22301. doi:10.1029/2009JD012287

Bittner M, Offermann D, Graef HH, Donner M, Hamilton K (2002) An 18 year time series of OH rotational temperatures and middle atmosphere decadal variations. J Atmos Sol Terr Phys 64:1147–1166

Bösch H, Toon GC, Sen B, Washenfelder RA, Wennberg PO, Buchwitz M, de Beek R, Burrows JP, Crisp D, Christi M, Connor BJ, Natraj V, Yung YL (2006) Space-based near-infrared CO_2 measurements: Testing the Orbiting Carbon Observatory retrieval algorithm and validation concept using SCIAMACHY observations over Park Falls, Wisconsin. J Geophys Res 111:D23302. doi:10.1029/2006JD007080

Boersma KF, Jacob DJ, Eskes HJ, Pinder RW, Wang J, van der A RJ (2008) Intercomparison of SCIAMACHY and OMI tropospheric NO_2 columns: Observing the diurnal evolution of chemistry and emissions from space. J Geophys Res 113:D16S26. doi: doi:10.1029/2007JD008816

Bracher A, Eichmann K-U, von Savigny C, Sinnhuber B-M, Weber M, Bramstedt K, Burrows JP (2005) Polar ozone distributions in the arctic winter/spring 2004/05 as measured by the Envisat instruments GOMOS and SCIAMACHY. Geophys Res Abstr 7:08693

Bracher A, Vountas M, Dinter T, Burrows JP, Röttgers R, Peeken I (2009) Quantitative Observation of cyanobacteria and diatoms from space using PhytoDOAS on SCIAMACHY data. Biogeosciences 6:751–764

Buchwitz M, de Beek R, Noël S, Burrows JP, Bovensmann H, Bremer H, Bergamaschi P, Körner S, Heimann M (2005a) Carbon monoxide, methane and carbon dioxide columns retrieved from SCIAMACHY by WFM-DOAS: Year 2003 initial data set. Atmos Chem Phys 5:3313–3329

Buchwitz M, de Beek R, Burrows JP, Bovensmann H, Warneke T, Notholt J, Meirink JF, Goede APH, Bergamaschi P, Körner S, Heimann M, Schulz A (2005b) Atmospheric methane and carbon dioxide from SCIAMACHY satellite data: Initial comparison with chemistry and transport models. Atmos Chem Phys 5:941–962

Buchwitz M, Schneising O, Burrows JP, Bovensmann H, Reuter M, Notholt J (2007a) Corrigendum to 'First direct observation of the atmospheric CO_2 year-to-year increase from space.' published in Atmos Chem Phys 7:4249–4256. Atmos Chem Phys 7:5341–5342

Buchwitz M, Khlystova I, Bovensmann H, Burrows JP (2007b) Three years of global carbon monoxide from SCIAMACHY: Comparison with MOPITT and first results related to the detection of enhanced CO over cities. Atmos Chem Phys 7:2399–2411

Correira J, Aikin AC, Grebowsky JM, Pesnell WD, Burrows JP (2008) Seasonal variation of magnesium atoms in the mesosphere-thermosphere. Geophys Res Lett 35. doi:10.1029/2007GL033047

De Graaf M, Tilstra LG, Aben I, Stammes P (2010) Satellite observations of the seasonal cycle of absorbing aerosols in Africa related to the monsoon rainfall. Atmos Env 44:1274–1283. doi:10.1016/j.atmosenv.2009.12.038

de Laat ATJ, Gloudemans AMS, Schrijver H, van den Broek MMP, Meirink JF, Aben I, Krol M (2006) Quantitative analysis of SCIAMACHY total carbon monoxide column measurements. Geophys Res Lett 33:L07807. doi:10.1029/2005GL025530

De Smedt I, Müller J-F, Stavrakou T, van der A R, Eskes H, Van Roozendael M (2008) Twelve years of global observations of formaldehyde in the troposphere using GOME and SCIAMACHY sensors. Atmos Chem Phys 8:4947–4963

De Smedt I, Stavrakou T, Müller J-F, van der A R J, and Van Roozendael M (2010) Trend detection in satellite observations of formaldehyde tropospheric columns. Geophys Res Lett 37: L18808, doi:10.1029/2010GL044245

Dikty S, Weber M, von Savigny C, Sonkaew T, Rozanov A, Burrows JP (2010) Modulations of the 27-day solar cycle signal in stratospheric ozone from SCIAMACHY. J Geophys Res 115:D00I15. doi: doi:10.1029/2009JD012379

Dufour G, Wittrock F, Camredon M, Beekmann M, Richter A, Burrows JP (2009) SCIAMACHY formaldehyde observations: Constraint for isoprene emissions over Europe? Atmos Chem Phys 9:1647–1664

Ern M, Lehmann C, Kaufmann M, Riese M (2009) Spectral wave analysis at the mesopause from SCIAMACHY airglow data compared to SABER temperature spectra. Annal Geophysic 27(1):407–416

Eskes H, Segers A, van Velthoven PFJ (2005) Ozone forecasts of the stratospheric polar vortex-splitting event in September 2002. J Atmos Sci 62:812–821

Franke K, Richter A, Bovensmann H, Eyring V, Jöckel P, Burrows JP (2009) Ship emitted NO_2 in the Indian Ocean: Comparison of model results with satellite data. Atmos Chem Phys 9:7289–7301

Frankenberg C, Meirink JF, van Weele M, Platt U, Wagner T (2005) Assessing methane emissions from global space-borne observations. Science 308:1010–1014

Frankenberg C, Bergamaschi P, Butz A, Houweling S, Meirink JF, Notholt J, Petersen AK, Schrijver H, Warneke T, Aben I (2008) Tropical methane emissions: A revised view from SCIAMACHY onboard ENVISAT. Geophys Res Lett 35:L15811. doi:10.1029/2008GL034300

Frankenberg C, Yoshimura K, Warneke T, Aben I, Butz A, Deutscher N, Griffith D, Hase F, Notholt J, Schneider M, Schrijver H, Röckmann T (2009) Dynamic processes governing lower-tropospheric HDO/H2O ratios as observed from space and ground. Science 325:1374. doi:10.1126/science.1173791

Fu T-M, Jacob DJ, Wittrock F, Burrows JP, Vrekoussis M, Henze DK (2008) Global budgets of atmospheric glyoxal and methylglyoxal, and implications for formation of secondary organic aerosols. J Geophys Res 113:D15303. doi:10.1029/2007JD009505

Gloudemans AMS, Schrijver H, Kleipool Q, van den Broek MMP, Straume AG, Lichtenberg G, van Hees RM, Aben I, Meirink JF (2005) The impact of SCIAMACHY instrument calibration on CH_4 and CO total columns. Atmos Chem Phys 5:2369–2383

Gloudemans AMS, de Laat ATJ, Schrijver H, Aben I, Meirink JF, van der Werf GR (2009) SCIAMACHY CO over land and oceans: 2003–2007 interannual variability. Atmos Chem Phys 9:3799–3813

Gottwald M, Slijkhuis S, Krieg E, Schreier F, Lichtenberg G, Vasquez M, Snel R, Stam D, de Kok R (2009) Venus observations with SCIAMACHY. Procedings of Atmospheric science conference, Barcelona, Spain, ESA SP-676

Hendrick F, Rozanov A, Johnston PV, Bovensmann H, De Maziere M, Fayt C, Hermans C, Kreher K, Lotz W, Sinnhuber B-M, Theys N, Thomas A, Burrows JP, Van Roozendael M (2009) Multi-year comparison of stratospheric BrO vertical profiles retrieved from SCIAMACHY limb and ground-based UV-visible measurements. Atmos Meas Tech 2:273–285

Kaleschke L, Richter A, Burrows JP, Afe O, Heygster G, Notholt J, Rankin AM, Roscoe HK, Hollwedel J, Wagner T, Jacobi H-W (2004) Frost flowers on sea ice as a source of sea salt and their influence on tropospheric halogen chemistry. Geophys Res Lett 31. doi:10.1029/2004GL020655

Kaufmann M, Lehmann C, Hoffmann L, Riese M, Lopez-Puertas M, Funke B, von Savigny C (2007) Chemical heating rates derived from SCIAMACHY vibrationally excited OH limb emission spectra. Adv Space Res 41(11):1914–1920

Kim S-W, Heckel A, McKeen SA, Frost GJ, Hsie E-Y, Trainer MK, Richter A, Burrows JP, Peckham SE, Grell GA (2006) Satellite observed U.S. power plant NOx emission reductions and their impact on air quality. Geophys Res Lett 33:L22812. doi: doi:10.1029/2006GL027749

Kim S-W, Heckel A, Frost GJ, Richter A, Gleason J, Burrows JP, McKeen S, Hsie E-Y, Granier C, Trainer M (2009) NO_2 columns in the western United States observed from space and simulated by a regional chemistry model and their implications for NOx emissions. J Geophys Res 114:D11301. doi:10.1029/2008JD011343

Konovalov IB, Beekmann M, Burrows JP, Richter A (2008) Satellite measurement based estimates of decadal changes in European nitrogen oxides emissions. Atmos Chem Phys 8:2623–2641

Konovalov IB, Beekmann M, Richter A, Burrows JP, Hilboll A (2010) Multi-annual changes of NOx emissions in megacity regions: Nonlinear trend analysis of satellite measurement based estimates. Atmos Chem Phys 10:8481-8498

Kühl S, Dörnbrack A, Wilms-Grabe W, Sinnhuber B-M, Platt U, Wagner T (2004) Observational evidence of rapid chlorine activation by mountain waves above northern Scandinavia. J Geophys Res 109:D22309. doi:10.1029/2004JD004797

Kühl S, Wilms-Grabe W, Frankenberg C, Grzegorski M, Platt U, Wagner T (2006) Comparison of OClO Nadir Measurements from SCIAMACHY and GOME. Adv Space Res 37:2247–2253

Kühl S, Pukite J, Deutschmann T, Platt U, Wagner T (2008) SCIAMACHY limb measurements of NO_2, BrO and OClO. Retrieval of vertical profiles: Algorithm, first results, sensitivity and comparison studies. Adv Space Res 42:1747–1764

Loyola D, Coldewey-Egbers M, Dameris M, Garny H, Stenke A, Van Roozendael M, Lerot C, Balis D, Koukouli M (2009) Global long-term monitoring of the ozone layer – a prerequisite for predictions. Int J Rem Sens 30(15):4295–4318

Marbach T, Beirle S, Platt U, Hoor P, Wittrock F, Richter A, Vrekoussis M, Grzegorski M, Burrows JP, Wagner T (2009) Satellite measurements of formaldehyde from shipping emissions. Atmos Chem Phys 9:8223–8234

Meirink JF, Bergamaschi P, Frankenberg C, d'Amelio MTS, Dlugokencky EJ, Gatti LV, Houweling S, Miller JB, Röckmann T, Villani MG, Krol M (2008) Four-dimensional variational data assimilation for inverse modelling of atmospheric methane emissions: Analysis of SCIAMACHY observations. J Geophys Res 113:D17301. doi:10.1029/2007JD009740

Mieruch S, Noël S, Bovensmann H, Burrows JP (2008) Analysis of global water vapour trends from satellite measurements in the visible spectral range. Atmos Chem Phys 8:491–504

Myriokefalitakis S, Vrekoussis M, Tsigaridis K, Wittrock F, Richter A, Brühl C, Volkamer R, Burrows JP, Kanakidou M (2008) The influence of natural and anthropogenic secondary sources on the glyoxal global distribution. Atmos Chem Phys 8:4965–4981

Newman PA, Daniel JS, Waugh DW, Nash ER (2007) A new formulation of equivalent effective stratospheric chlorine (EESC). Atmos Chem Phys 7:4537–4552

Noël S, Buchwitz M, Bovensmann H, Burrows JP (2004) First retrieval of global water vapour column amounts from SCIAMACHY measurements. Atmos Chem Phys 4:111–125

O'Dowd CD, Jimenez JL, Bahreini R, Flagan RC, Seinfeld JH, Hameri K, Pirjola L, Kulmala M, Jennings SG, Hoffmann T (2002) Marine aerosol formation from biogenic iodine emissions. Nature 417:632–636

Pagaran J, Weber M, Burrows JP (2009) Solar variability from 240 to 1750 nm in terms of faculae brightening and sunspot darkening from SCIAMACHY. Astrophys J 700:1884–1895

Piot M, von Glasow R (2008) The potential importance of frost flowers, recycling on snow, and open leads for ozone depletion events. Atmos Chem Phys 8:2437–2467

Puķīte J, Kühl S, Deutschmann T, Platt U, Wagner T (2008) Accounting for the effect of horizontal gradients in limb measurements of scattered sunlight. Atmos Chem Phys 8:3045–3060

Rahpoe N, von Savigny C, Robert CE, DeLand MT, Burrows JP (2010) Impact of solar proton events on noctilucent clouds. J Atmosph Sol Terr Phys doi:10.1016/j.jastp.2010.07.017 (in press)

Reuter M, Buchwitz M, Schneising O, Heymann J, Bovensmann H, Burrows JP (2010) A method for improved SCIAMACHY CO_2 retrieval in the presence of optically thin clouds. Atmos Meas Tech 3:209–232

Rex M, Salawitch RJ, Deckelmann H, von der Gathen P, Harris NRP, Chipperfield MP, Naujokat B, Reimer E, Allaart M, Andersen SB, Bevilacqua R, Braathen GO, Claude H, Davies J, De Backer H, Dier H, Dorokhov V, Fast H, Gerding M, Godin-Beekmann S, Hoppel K, Johnson B, Kyrö E, Litynska Z, Moore D, Nakane H, Parrondo MC, Risley AD, Skrivankova P, Stübi R, Viatte P, Yushkov V, Zerefos C (2006) Arctic winter 2005: Implications for stratospheric ozone loss and climate change. Geophys Res Lett 33:L23808. doi:10.1029/2006GL026731

Richter A, Eyring V, Burrows JP, Bovensmann H, Lauer A, Sierk B, Crutzen PJ (2004) Satellite measurements of NO_2 from international shipping emissions. Geophys Res Lett 31:L23110. doi:10.1029/2004GL020822

Richter A, Burrows JP, Nüß H, Granier C, Niemeier U (2005a) Increase in nitrogen dioxide over China observed from space. Nature 437:129–132

Richter A, Wittrock F, Weber M, Beirle S, Kühl S, Platt U, Wagner T, Wilms-Grabe W, Burrows JP (2005b) GOME observations of stratospheric trace gas distributions during the splitting vortex event in the Antarctic winter 2002 Part I: Measurements. J Atm Sci 62:778–785

Robert CE, von Savigny C, Rahpoe N, Burrows JP, DeLand MT, Schwartz MJ (2010) First evidence of a 27-day signature in noctilucent cloud occurrence frequency. J Geophys Res 115, D00I12. doi:10.1029/2009JD012359

Rohen GJ, von Savigny C, Sinnhuber M, Eichmann K-U, Kaiser JW, Llewellyn EJ, Rozanov A, Bovensmann H, Burrows JP (2005) Impact of the October/November 2003 Solar proton events on mesospheric ozone: SCIAMACHY measurement and model results. J Geophys Res 110, A09S39. doi:10.1029/2004JA010984

Rozanov A, Bovensmann H, Bracher A, Hrechanyy S, Rozanov V, Sinnhuber M, Stroh F, Burrows JP (2005) NO_2 and BrO vertical profiles retrieval from SCIAMACHY limb measurements: Sensitivity studies. Adv Space Res 36:846–854. doi:10.1016/j.asr.2005.03.013

Saiz-Lopez A, Chance K, Liu X, Kurosu TP, Sander SP (2007) First observations of iodine oxide from space. Geophys Res Lett 34: L12812. doi:10.1029/2007GL030111

Scharringhausen M (2007) Investigation of mesospheric and thermospheric magnesium species from space. Ph.D. thesis, University of Bremen

Scharringhausen M, Aikin AC, Burrows JP, Sinnhuber M (2008) Space-borne measurements of mesospheric magnesium species – a retrieval algorithm and preliminary profiles. Atmos Chem Phys 8:1963–1983

Schneising O, Buchwitz M, Burrows JP, Bovensmann H, Reuter M, Notholt J, Macatangay R, Warneke T (2008) Three years of greenhouse gas column-averaged dry air mole fractions retrieved from satellite – Part 1: Carbon dioxide. Atmos Chem Phys 8:3827–3853

Schneising O, Buchwitz M, Burrows JP, Bovensmann H, Bergamaschi P, Peters W (2009) Three years of greenhouse gas column-averaged dry air mole fractions retrieved from satellite – Part 2: Methane. Atmos Chem Phys 9:443–465

Schönhardt A, Richter A, Wittrock F, Kirk H, Oetjen H, Roscoe HK, Burrows JP (2008) Observations of iodine monoxide columns from satellite. Atmos Chem Phys 8:637–653

Schrijver H, Gloudemans AMS, Frankenberg C, Aben I (2009) Water vapour total columns from SCIAMACHY spectra in the 2.36 μm window. Atmos Meas Tech 2:561–571

Simpson WR, von Glasow R, Riedel K, Anderson P, Ariya P, Bottenheim J, Burrows J, Carpenter LJ, Frieß U, Goodsite ME, Heard D, Hutterli M, Jacobi H-W, Kaleschke L, Neff B, Plane J, Platt U, Richter A, Roscoe H, Sander R, Shepson P, Sodeau J, Steffen A, Wagner T, Wolff E (2007) Halogens and their role in polar boundary-layer ozone depletion. Atmos Chem Phys 7:4375–4418

Sinnhuber B-M, Rozanov A, Sheode N, Afe OT, Richter A, Sinnhuber M, Wittrock F, Burrows JP, Stiller GP, von Clarmann T, Linden A (2005) Global observations of stratospheric bromine monoxide from SCIAMACHY. Geophys Res Lett 32:L20810. doi:10.1029/2005GL023839

Sioris CE, Kovalenko LK, McLinden CA, Salawitch RJ, Van Roozendael M, Goutail F, Dorf M, Pfeilsticker K, Chance K, von Savigny C, Liu X, Kurosu TP, Pommereau JP, Bösch H, Frerick J (2006) Latitudinal and vertical distribution of bromine monoxide in the lower stratosphere from scanning imaging absorption spectrometer for atmospheric chartography limb scattering measurements. J Geophys Res 111:D14301. doi:10.1029/2005JD006479

Skupin J, Weber M, Bovensmann H, Burrows JP (2004) The Mg II solar activity proxy indicator derived from GOME and SCIAMACHY. Proceedings of ENVISAT and ERS Symposium, Salzburg, Austria, ESA SP-572

Sonkaew T, von Savigny C, Eichmann K-U, Rozanov A, Weber M, Bovensmann H, Burrows JP (2010) Chemical ozone loss in Arctic and Antarctic polar winter/spring derived from SCIAMACHY limb measurements 2002–2009 (in prep)

Stavrakou T, Müller J-F, Boersma KF, De Smedt I, van der A RJ (2008) Assessing the distribution and growth rates of NOx emission sources by inverting a 10-year record of NO_2 satellite columns. Geophys Res Lett 35, L10801. doi:10.1029/ 2008GL033521

Stavrakou T, Müller J-F, De Smedt I, Van Roozendael M, van der Werf GR, Giglio L, Guenther A (2009) Evaluating the performance of pyrogenic and biogenic emission inventories against one decade of space-based formaldehyde columns. Atmos Chem Phys 9:1037–1060

Steinbrecht W, Claude H, Schönenborn F, McDermid IS, Leblanc T, Godin S, Keckhut P, van Gijsel A, Swart DPJ, Bodeker G, Parrish A, Boyd I, Kämpfer N, Hocke C, Stolarski RS, Frith SM, Thomason LW, Remsberg EE, von Savigny C, Rozanov A, Burrows JP (2009) Ozone and temperature trends in the upper stratosphere at five stations of the network for the detection of atmospheric composition change. Int J Rem Sens 30(15–16):3875–3886

Tangborn A, Stajner I, Buchwitz M, Khlystova I, Pawson S, Burrows J, Hudman R, Nedelec P (2009) Assimilation of SCIAMACHY total column CO observations: Global and regional analysis of data impact. J Geophys Res 114:D07307. doi10.1029/2008JD010781

Theys N, Van Roozendael M, Errera Q, Hendrick F, Daerden F, Chabrillat S, Dorf M, Pfeilsticker K, Rozanov A, Lotz W, Burrows JP, Lambert J-C, Goutail F, Roscoe HK, De Mazière M (2009) A global stratospheric bromine monoxide climatology based on the BASCOE chemical transport model. Atmos Chem Phys 9:831–848

van der A RJ, Eskes HJ, Boersma KF, van Noije TPC, Van Roozendael M, De Smedt I, Peters DHMU, Meijer EW (2008) Trends, seasonal variability and dominant NOx source derived from a ten year record of NO_2 measured from space. J Geophys Res 113:D04302. doi: doi:10.1029/2007JD009021

Vasilkov AP, Joiner J, Gleason J, Bhartia PK (2002) Ocean Raman scattering in satellite backscatter UV measurements. Geophys Res Lett 29(17):1837. doi:10.1029/2002GL014955

Viereck RA, Puga LC, McMullin D, Judge D, Weber M, Tobiska WK (2001) The Mg II index: A proxy for solar EUV. Geophys Res Lett 28:1343–1346

Viereck RA, Floyd LE, Crane PC, Woods TN, Knapp BG, Rottman G, Weber M, Puga LC, DeLand MT (2004) A composite Mg II index spanning from 1978 to 2003. Space Weather 2:S10005. doi:10.1029/2004SW000084

von Savigny C, Kokhanovsky A, Bovensmann H, Eichmann K-U, Kaiser JW, Noël S, Rozanov AV, Skupin J, Burrows JP (2004a) NLC detection and particle size determination: First results from SCIAMACHY on ENVISAT. Adv Space Res 34:851–856

von Savigny C, Eichmann K-U, Llewellyn EJ, Bovensmann H, Burrows JP, Bittner M, Höppner K, Offermann D, Steinbrecht W, Winkler P, Taylor MJ, Cheng Y (2004b) First near-global retrieval of OH rotational temperatures from satellite-based Meinel band emission measurements. Geophys Res Lett 31:L15111. doi:10.1029/2004GL020410

von Savigny C, Rozanov A, Bovensmann H, Eichmann K-U, Noël S, Rozanov VV, Sinnhuber B-M, Weber M, Burrows JP (2005a) The ozone hole break-up in September 2002 as seen by SCIAMACHY on ENVISAT. J Atmosph Sci 62:721–734

von Savigny C, Ulasi EP, Eichmann K-U, Bovensmann H, Burrows JP (2005b) Detection and mapping of polar stratospheric clouds using limb scattering observations. Atmos Chem Phys 5:3071–3079

von Savigny C, Sinnhuber M, Bovensmann H, Burrows JP, Kallenrode M-B, Schwartz MJ (2007a) On the disappearance of Noctilucent Clouds during the January 2005 solar proton events. Geophys Res Lett 34:L02805. doi:10.1029/2006GL128106

von Savigny C, Robert C, Bovensmann H, Burrows JP, Schwartz M (2007b) Satellite observations of the quasi 5-day wave in noctilucent clouds and mesopause temperatures. Geophys Res Lett 34: L24808. doi:10.1029/2007GL030987

von Savigny C, Burrows JP (2007) Latitudinal variation of NLC particle radii derived from northern hemisphere SCIAMACHY/Envisat limb measurements. Adv Space Res 40(6):765–771. doi:10.1016/j.asr.2007.12.032

von Savigny C, Robert CE, Baumgarten G, Bovensmann H, Burrows JP (2009) Comparison of NLC particle sizes derived from SCIAMACHY/Envisat observations with ground-based LIDAR measurements at ALOMAR (69°N). Atmos Meas Tech 2:523–531

Vountas M, Richter A, Wittrock F, Burrows JP (2003) Inelastic scattering in ocean water and its impact on trace gas retrievals from satellite data. Atmos Chem Phys 3:1365–1375

Vountas M, Dinter T, Bracher A, Sierk B, Burrows JP (2007) Spectral studies of ocean water with space-borne sensor SCIAMACHY using Differential Optical Absorption Spectroscopy (DOAS). Ocean Sci 3:429–440

Vrekoussis M, Wittrock F, Richter A, Burrows JP (2009) Temporal and spatial variability of glyoxal as observed from space. Atmos Chem Phys 9:4485–4504

Wagner T, Leue C, Pfeilsticker K, Platt U (2001) Monitoring of the stratospheric chlorine activation by Global ozone monitoring experiment (GOME) OClO measurements in the austral and boreal winters 1995 through 1999. J Geophys Res 106:4971–4986

Wagner T, Beirle S, Deutschmann T, Grzegorski M, Platt U (2007) Satellite monitoring of different vegetation types by differential optical absorption spectroscopy (DOAS) in the red spectral range. Atmos Chem Phys 7:69–79

Weber M (1999) Solar activity during solar cycle 23 monitored by GOME. Proceedings of European symposium atmospheric measurements from space (ESAMS'99)

Wittrock F, Richter A, Oetjen H, Burrows JP, Kanakidou M, Myriokefalitakis S, Volkamer R, Beirle S, Platt U, Wagner T (2006) Simultaneous global observations of glyoxal and formaldehyde from space. Geophys Res Lett 33:L16804. doi:10.1029/2006GL026310

WMO (World Meteorological Organization) (2007) Scientific assessment of ozone depletion: 2006. Global Ozone Research and Monitoring Project – Report No. 50, p 572, Geneva, Switzerland

Appendix

Atmospheric Gases

BrO	Bromine oxide
CFC	Chlorofluorocarbon
CFC11	Trichlorofluoromethane
CHOCHO	Glyoxal
CH_4	Methane
CH_3Cl	Chloromethane
CH_3CN	Acetonitrile
C_2H_6	Ethane
ClO	Chlorine monoxide
$ClONO_2$	Chlorine nitrate
CO	Carbon monoxide
CO_2	Carbon dioxide
HCHO	Formaldehyde
HCl	Hydrogen chloride
HCN	Hydrogen cyanide
HDO	Deuterium protium oxide (heavy water)
HF	Hydrofluor
HFC	Hydrofluorocarbon
HNO_3	Nitric acid
H_2O	Water vapour
IO	Iodine oxide
NAT	Nitric acid trihydrate
NO	Nitric oxide
NO_2	Nitrogen dioxide
NO_3	Nitrate radical
N_2O	Nitrous oxide
N_2O_5	Dinitrogen pentoxide
OClO	Chlorine dioxide
OH	Hydroxyl radical
O_3	Ozone
O_2	Oxygen (molecular)
$O_2(^1\Delta)$	Oxygen (singlet delta)
O_4	Oxygen (dimer)
PAN	Peroxyacetyl nitrate
SO_2	Sulphur dioxide

Abbreviations and Acronyms

AAI	Absorbing Aerosol Index
AATSR	Advanced ATSR
ACE	Atmosphere Climate Experiment
ACVE	Atmospheric Chemistry Validation of ENVISAT
ACVT	Atmospheric Chemistry Validation Team
ADC	Analogue-to-Digital Converter
ADEOS	Advanced Earth Observing System
ADF	Auxiliary Data File
ADS	Annotation Data Set
AGU	American Geophysical Union
AIRS	Atmospheric Infrared Sounder
ALC	Absolute Limb Correction
AMAX-DOAS	Airborne MAX-DOAS
AMC-DOAS	Air Mass Corrected DOAS
AMF	Airmass Factor
AMON	Absorption par les Minoritaires Ozone et NOx
ANX	Ascending Node Crossing
AO	Announcement of Opportunity
AO	Analogue Offset
AOCS	Attitude and Orbit Control System
AOI	Announcement of Opportunity Instrument
AOP	AO Instrument Provider
AOT	Aerosol Optical Thickness
APSM	Aperture Stop Mechanism
ARCF	Absolute Radiometric Calibration Facility
ASAR	Advanced Synthetic Aperture Radar
ASI	Agenzia Spaziale Italiana
ASM	Azimuth Scan Mechanism
ASUR	Airborne Submillimeter Radiometer
ATBD	Algorithm Theoretical Basis Document
ATC	Active Thermal Control
ATSR	Along-Track Scanning Radiometer
AZACM	Azimuth Aperture Cover Mechanism
BAER	Bremen Aerosol Retrieval
BCPS	Broadcast Pulse
BDPM	Bad and Dead Pixel Mask

M. Gottwald and H. Bovensmann (eds.), *SCIAMACHY - Exploring the Changing Earth's Atmosphere*,
DOI 10.1007/978-90-481-9896-2, © Springer Science+Business Media B.V. 2011

BIRA-IASB	Belgisch Instituut voor Ruimte-Aëronomie /Institut d'Aéronomie Spatiale de Belgique
BIRRA	Beer InfraRed Retrieval Algorithm
BMFT	Bundesministerium für Forschung und Technologie
BOL	Begin-of-Life
BONBON	Cryogenic whole air sampler for the collection of airsamples in the stratosphere
BU	Binary Unit
BUV	Backscattered Ultraviolet
CA	Corrective Action
CARIBIC	Civil Aircraft for the Regular Investigation of the Atmosphere Based on an Instrument Container
C&C	Command & Control
CCA	Communication Area
CCM	Chemistry-Climate Model
CF	Cloud Fraction
CGT	Cloud Geometrical Thickness
CINDI	Cabauw Intercomparison Campaign of Nitrogen Dioxide Measuring Instruments
CLRTAP	Convention on Long-Range Transboundary Air Pollution
CNES	Centre Nationale d'Etudes Spatiales
COT	Cloud Optical Thickness
CPI	Cloud Phase Index
CSA	Canadian Space Agency
CTH	Cloud Top Height
CTI	Configurable Transfer Item
CTM	Chemical Transport Model
CTP	Cloud Top Pressure
C1	Category 1
C2	Category 2
DAK	Double-Adding code KNMI
DANDELIONS	Dutch Aerosol and Nitrogen Dioxide Experiments for Validation of OMI and SCIAMACHY
DARA	Deutsche Agentur für Raumfahrtangelegenheiten
DBU	Digital Bus Unit
DDS	Data Dissemination System
DFD	Deutsches Fernerkundungs-Datenzentrum
DHCM	Decontamination Heater Control Module
DIAL	Differential Absorption Lidar
DLR	Deutsches Zentrum für Luft- und Raumfahrt
DME	Detector Module Electronics
DNPM	Deutsch-Niederländisches Projektmanagement
DNX	Descending Node Crossing
DOAS	Differential Optical Absorption Spectroscopy
DORIS	Doppler Orbitography and Radiopositioning by Satellite
D-PAC	German PAC
DU	Dobson Unit
EA	Electronic Assembly
EADS	European Aeronautic Defense and Space Company
ECMWF	European Centre for Medium-Range Weather Forecasts
ECPL	Environmental Chemical Processes Laboratory
EDI	ESA Developed Instrument
EEPROM	Electrical Erasable Programmable Read Only Memory
ELACM	Elevation Aperture Cover Mechanism
ELHYSA	Etude l'Hygromérie Stratosphérique
EMC	Electromagnetic Compatibility
ENVISAT	Environmental Satellite
ENVISOLAR	Environmental Information Services for Solar Energy Industries
EO	Earth Observation
EOL	End-of-Life
EOS	Earth Observing System
E-PAC	Spanish PAC
EPS	EUMETSAT Polar System
ERBS	Earth Radiation Budget Satellite
ERS	European Remote Sensing Satellite
ESA	European Space Agency
ESABC	ENVISAT Stratospheric Aircraft and Balloon Campaign
ESM	Elevation Scan Mechanism
ESOC	European Space Operation Centre
ESRIN	European Space Research Institute
ESTEC	European Space Research and Technology Centre
EU	European Union
EUMETSAT	European Organisation for the Exploitation of Meteorological Satellites
EURAD	European Air Pollution Dispersion model
EURECA	European Retrievable Carrier
EUTELSAT	European Telecommunications Satellite
EUV	Extreme UV
EVDC	ENVISAT Validation Data Centre
FD	Fast Delivery
FMI	Finnish Meteorological Institute
Fin CoPAC	Finnish Co-PAC
FIRS	Far-Infrared Spectrometer
FISH	Fast In Situ Stratospheric Hygrometer
FOCC	Flight Operation Control Centre
FODP	Flight Operation and Data Plan
FOS	Flight Operation Segment
FoV	Field of View

F-PAC	French PAC
FRESCO	Fast Retrieval Scheme for Clouds from the Oxygen A-band
FTIR	Fourier Transform Infrared
FTS	Fourier Transform Spectrometer
GAW	Global Atmospheric Watch
GBMCD	Ground-Based Measurement and Campaign Database
GCM	General Circulation Model
GDF	General Distribution Function
GDP	GOME Data Processor
GEO	Global Earth Observation
GeoFIS	Geostationary Fourier Transform Interferometer
GEOSS	Global Earth Observation System of Systems
GeoSCIA	Geostationary Scanning Imaging Absorption Spectrometer
GeoTROPE	Geostationary Tropospheric Explorer
GMES	Global Monitoring for Environment and Security
GOME	Global Ozone Monitoring Experiment
GOMETRAN	radiative transfer model for GOME
GOMOS	Global Ozone Monitoring by Occultation of Stars
GOSAT	Greenhouse Gas Observing Satellite
GSFC	Goddard Space Flight Center
HALOE	Halogen Occultation Experiment
HALOX	Halogen Oxides balloon experiment
HALOX-B	HALOX for Balloons
HICRU	Heidelberg Iterative Cloud Retrieval Utilities
HK	Housekeeping
HSM	High Speed Multiplexer
IABG	Industrieanlagen-Betriebsgesellschaft
IASI	Infrared Atmospheric Sounding Interferometer
ICARTT	International Consortium of Atmospheric Research on Transport and Transformation
ICU	Instrument Control Unit
ID	Identifier
IECF	Instrument Engineering and Calibration Facility
IFoV	Instantaneous Field of View
IGACO	Integrated Global Atmospheric Chemistry Observations
ILAS	Improved Limb Atmospheric Spectrometer
ILoS	Instantaneous Line of Sight
IMAP	Iterative Maximum a Posteriori
IMAU	Institute of Marine and Atmospheric Research Utrecht
IMF	Institut für Methodik der Fernerkundung
IMG	Interferometric Monitor for Greenhouse Gases
IMIA	Instrument Mission Implementation Agreement
IMLM	Iterative Maximum Likelihood Method
INCA	Interaction of Chemistry and Aerosol
INTA	Instituto Nacional de Técnica Aerospacial
INTEX	Intercontinental Chemical Transport Experiment
INTEX-NA	Intercontinental Chemical Transport Experiment – North America
IOM	Instrument Operation Manual
I-PAC	Italian PAC
IPCC	Intergovernmental Panel of Climate Change
IR	Infrared
IRS	Infrared Sounder
ISCCP	International Satellite Cloud Climatology Project
IST	Integrated System Team
IT	Integration Time
ITCT	Intercontinental Transport and Chemical Transformation
ITOP	Intercontinental Transport of Ozone and Precursors
IUP	Institut für Umweltphysik (Heidelberg)
IUP-IFE	Institut für Umweltphysik/Institut für Fernerkundung (Bremen)
JAXA	Japan Aerospace Exploration Agency
JPL	Jet Propulsion Laboratory
KNMI	Koninklijk Nederlands Meteorologisch Instituut
LC	Leakage Current
LEOP	Launch and Early Operation Phase
LIDORT	Linearized Discrete Ordinate Radiative Transfer
LLI	Life Limited Item
LoS	Line-of-Sight
LPMA	Laboratoire de Physique Moleculaire et Applications
LPMA-DOAS	Limb Profile Monitoring of the Atmosphere DOAS
LRAC	Low Rate Reference Archive Centre
LRR	Laser Retro-Reflector
LSCE	Laboratoire des Sciences du Climat et de l'Environnement
MACC	Monitoring Atmospheric Composition and Climate
MANTRA	Middle Atmosphere Nitrogen Trend Assessment balloon experiment
MAP	Measurement of Atmospheric Pollution
MASI	Models and Data Assimilation, Satellite Intercomparison

MAX-DOAS	Multi-Axis DOAS
McArtim	Monte Carlo Atmospheric Radiative Transfer Inversion Model
MCFA	MERIS Cloud Fraction Algorithm
MCMD	Macrocommand
MDS	Measurement Data Set
MERIS	Medium Resolution Imaging Spectrometer
METEOSAT	Meteorological Satellite
METOP	Meteorological Operational Satellite
MIPAS	Michelson Interferometer for Passive Atmospheric Sounding
MIPAS-B	MIPAS for Balloons
MIR	Montgolfier Infra-Red
MLI	Multilayer Insulation
MLS	Microwave Limb Sounder
MLST	Mean Local Solar Time
MLT	Mesosphere Lower Thermosphere
MO&C	Moon Occultation & Calibration
MODIS	Moderate Resolution Imaging Spectroradiometer
MOPITT	Measurements of Pollution in the Troposphere
MOZAIC	Measurement of Ozone and Water Vapour by Airbus In-service Aircraft
MPH	Main Product Header
MPS	Mission Planning System
MSG	METEOSAT Second Generation
MTG	METEOSAT Third Generation
MWR	Microwave Radiometer
NADIR	NILU Atmospheric Database for Interactive Retrieval
NASA	National Aeronautics and Space Administration
NASDA	National Space Development Agency of Japan
NCW	Nadir Calibration Window
NCWM	Nadir Calibration Window Mechanism
NDACC	Network for the Detection of Atmospheric Composition Change
NDF	Neutral Density Filter
NDFM	Neutral Density Filter Mechanism
NDSC	Network for the Detection of Stratospheric Change
NEAQS	New England Air Quality Study
NGDC	National Geophysical Data Center
NH	Northern Hemisphere
NILU	Norsk Institutt for Luftforskning
NIR	Near Infrared
NIS	New Independent States
NIVR	Nederlands Instituut voor Vliegtuigontwikkeling en Ruimtevaart
NLC	Noctilucent Cloud
NNDEC	Non-nominal Decontamination
NOAA	National Oceanic and Atmospheric Administration
NOXAR	Measurements of Nitrogen Oxides and Ozone along Air Routes
NPP	NPOESS Preparatory Project
NPOESS	National Polar-orbiting Operational Environmental Satellite System
NRL	Naval Research Laboratory
NRSC	National Remote Sensing Centre Ltd
NRT	Near-realtime
NSO	Netherlands Space Office
OA	Optical Assembly
OBM	Optical Bench Module
OBT	On-Board Time
OCM	Orbit Control Manoeuvre
OCO	Orbiting Carbon Observatory
OCR	Operation Change Request
OCRA	Optical Cloud Recognition Algorithm
OE	Optimal Estimation
OIP	Optique et Instruments de Précision
OL	Offline
OLEX	Ozone Lidar Experiment
OMI	Ozone Monitoring Instrument
OMPS	Ozone Monitoring and Profiling Suite
OPTEC	Optical Test Facility
OSIRIS	Optical Spectrograph and Infrared Imaging System
OU	Optical Unit
PAC	Processing and Archiving Facility
PDAS	Payload Data Acquisition Station
PDCC	Payload Data Control Center
PDHS	Payload Data Handling Station
PDHS-E	Payload Data Handling Station – ESRIN
PDHS-K	Payload Data Handling Station – Kiruna
PDS	Payload Data Segment
PET	Pixel Exposure Time
PFM	Proto Flight Model
PI	Principle Investigator
PMC	Polar Mesospheric Cloud
PMC	Payload Management Computer
PMD	Polarisation Measurement Device
PMTC	Power Mechanism and Thermal Control Unit
POAM	Polar Ozone and Aerosol Measurement
POEM	Polar Orbit Earth Observation Mission
POLARCAT	Polar Study using Aircraft, Remote Sensing, Surface Measurements and Modelling of Climate, Chemistry, Aerosols and Transport
POLDER	Polarization and Directionality of the Earth's Radiance

PPG	Pixel-to-Pixel Gain
PROMOTE	Protocol Monitoring for the GMES Service Element
PSC	Polar Stratospheric Cloud
QBO	Quasi-biennial Oscillation
QWG	Quality Working Group
RA-2	Radar Altimeter 2
RAD A	Radiator A
RADIBAL	Radiomètre Balloon
RAM	Random Access Memory
RASA	Russian Aviation and Space Agency
RE	Radiated Emission
RGB	Red, Green, Blue
RH	Relative Humidity
RLC	Relative Limb Correction
RMS	Root Mean Square
ROSE	Research on Ozone in the Stratosphere and its Evolution
RR	Reduced Resolution
RRU	Radiant Reflector Unit
RS	Radiated Susceptibility
RSM	Reference Sector Method
RT	Random Telegraph
RTCS	Relative Time Command Sequence
RTM	Radiative Transfer Model
SAA	South Atlantic Anomaly
SABER	Sounding of the Atmosphere using Broadband Emission Radiometry
SACURA	Semi-Analytical Cloud Retrieval Algorithm
SAGE	Stratospheric Aerosol and Gas Experiment
SALOMON	UV-VIS Spectrometer
SAM	Stratospheric Aerosol Measurement
SAOZ	Système d'Analyse par Observations Zénithale
SAUNA	Sodankylä Total Column Ozone Intercomparison
SBUV	Solar Backscatter Ultraviolet instrument
SCCVT	SCIAMACHY Calibration and Verification Team
SCD	Slant Column Density
SCIA-BAER	SCIAMACHY Bremen Aerosol Retrieval
SCIAMACHY	Scanning Imaging Absorption Spectrometer for Atmospheric Chartography
SCIARAYS	toolbox for radiative transfer modeling and atmospheric parameter retrieval in the UV-VIS
SCIATRAN	radiative transfer model for SCIAMACHY
SCIAVALIG	SCIAMACHY Validation and Interpretation Group
SCIA-VALUE	SCIAMACHY Validation and Utilization Experiment
SCOOP	SCIAMACHY On-board Operation Plan
SDLA-LAMA	Spectromètre à Diode Laser Accordable Laser pour l'Analyse du Méthane Atmosphérique
SDOAS	SCIAMACHY DOAS
SDPU	Science Data Processing Unit
SEM	Simplified Engineering Model
SEU	Single Event Upset
SF	Sun Follower
SFCM	Stellar Fine Control Manoeuvre
SFM	Spectrophotometer
SGP	SCIAMACHY Ground Processor
SH	Southern Hemisphere
SIM	Spectral Irradiance Monitor
SIRD	SCIAMACHY Instrument Requirements Document
SJT	SCIAMACHY Joint Team
SLS	Spectral Line Source
SME	Solar Mesospheric Explorer
SMR	Sun Mean Reference
SMR	Submillimeter Radiometer
SNR	Signal-to-Noise Ratio
SNSB	Swedish National Space Board
SOA	Secondary Organic Aerosol
SO&C	Sun Occultation & Calibration
SODAP	Switch-on and Data Acquisition Phase
SOLSPEC	Solar Spectrum Measurement
SOLSTICE	Solar Stellar Irradiance Comparison Experiment
SOR	SCIAMACHY Operations Request
SOST	SCIAMACHY Operations Support Team
S-PAC	Swedish PAC
SPCA	SCIAMACHY PMD Cloud Algorithm
SPE	Solar Proton Event
SPH	Special Product Header
SPICAV	Spectroscopy for Investigation of Characteristics of the Atmosphere of Venus
SPICI	SCIAMACHY PMD Identification of Clouds and Ice/snow
SPICS	SCIAMACHY PMD based Identification and Classification of Clouds and Surfaces
SPIRALE	Spectroscopie Infra-Rouge par Absorption de Laser Embarqué
SQWG	SCIAMACHY QWG
SRC	SCIAMACHY Radiant Cooler
SRON	SRON Netherlands Institute for Space Research
SSAG	SCIAMACHY Science Advisory Group
SSC	Swedish Space Corporation
SSI	Solar Spectral Radiance
SSM/I	Special Sensor Microwave Imager
STM	Structural Model

SUSIM	Solar Ultraviolet Spectral Irradiance Monitor
SWIR	Short-Wave Infrared
SYNEAR	Synergetic Aerosol Retrieval
SYSM	Stellar Yaw Steering Mode
SZA	Solar Zenith Angle
TANSO	Thermal and Near Infrared Sensor for Carbon Observation
TB	Thermal Balance
TC	Thermal Control
TCFoV	Total Clear Field of View
TCCON	Total Carbon Column Observing Network
TEMIS	Tropospheric Emission Monitoring Internet Service
TES	Tropospheric Emission Spectrometer
TM	Transport Model
TOA	Top-of-Atmosphere
TOMS	Total Ozone Mapping Spectrometer
TOSOMI	Total Ozone retrieval scheme for SCIAMACHY based on the OMI DOAS algorithm
TNO-TPD	Technisch-Naturwetenschappelijk Onderzoek – Technisch Physische Dienst
TRACY	Trace Gas Radiative Transfer Monte Carlo Y(I)mplementation
TRIPLE	Multi-instrument balloon payload
TRUE	Tangent Height Retrieval by UV-B
TV	Thermal Vacuum
TWIN	Multi-instrument balloon payload
UARS	Upper Atmosphere Research Satellite
UK	United Kingdom
UK-PAC	United Kingdom PAC
US	United States
UTC	Universal Time Coordinated
UV	Ultraviolet
UVAI	UV Aerosol Index
UVN	UV-VIS-NIR
UVNS	UV-VIS-NIR-SWIR
UVSCI	UV Scattering Index
VA	Value-added
VCD	Vertical Column Density
VEX	Venus Express
VIRTIS	Visible and Infrared Thermal Imaging Spectrometer
VIS	Visible
VOC	Volatile Organic Compound
VSLS	Very Short-Lived Substances
WFM-DOAS	Weighting Function Modified DOAS
WGS84	World Geodetic System 1984
WLS	White Light Source
WMO	World Meteorological Organization

Index